D1442895

The Experimenters

Accademia del Cimento fiorita in Firen.ᵉ sotto la protezione della Real Casa dei Medici nel Secolo XVII. C. Vascellini inc.ᵉ scolpì

Imaginary group of the Accademia del Cimento, from an engraving by G. Vascellini in *Serie di ritratti d'uomini illustri toscani . . .* (Florence, 1773. The bust of Prince Leopold as cardinal shown in this drawing is of course an egregious anachronism, as he was made a cardinal only after the Academy had ceased to meet.)

The Experimenters

A STUDY OF THE ACCADEMIA DEL CIMENTO

W. E. KNOWLES MIDDLETON

THE JOHNS HOPKINS PRESS   BALTIMORE AND LONDON

CONTENTS

PLATES

FIGURES

PREFACE

In writing a history of the barometer nearly ten years ago, I was introduced to the magnificent assemblage of seventeenth-century scientific manuscripts that constitutes the so-called "Galileo" collection of the Biblioteca Nazionale Centrale in Florence. A large number of these deal with that short-lived but unique institution, the Accademia del Cimento, and after working in that excellent library on several later occasions while studying the history of the thermometer and other instruments, I began to be impressed with the paucity of published material about that Academy and to think about supplying that lack, which in English is almost total. My resolution was much strengthened when I found that even less has been published about the founder and protector of the Accademia del Cimento, Prince Leopoldo de' Medici, later cardinal, one of the most intelligent and attractive members of that distinguished family, then in its decline.

The extent to which this particular period has been neglected, even by Italian historians, is remarkable. The indexes of the excellent *Archivio Storico Italiano* bear this out, showing an almost total absence of published documents from the middle third of the *seicento*. The period may not have the fascination of the times of Dante or of Savonarola or of Garibaldi, but it is not devoid of interest. The advent of the journal *Studi Secenteschi* in the last dozen years raises the hope that this neglect of the seventeenth century may be decreasing.

I began to read background material for this present book in 1966, not long before the disastrous flood in November of that year which caused such ruin in many of the libraries and museums of Florence. By great good fortune, the "Galileo" collection at the National Library and the seventeenth-century documents in the Archivio di Stato were beyond the reach of the flood waters, and in a surprisingly short time these collections were again made available to scholars. In the summer of 1967, then, I began a serious study of the Accademia del Cimento.

The only publication of the Academy is a sumptuous book, the *Saggi di naturali esperienze*, which appeared in 1667. I soon came to the conclusion that a translation of the *Saggi* would have to be a central feature of a book about the Accademia del Cimento, for otherwise the reader would have to provide himself with a copy of that work. Although the translation into English by Richard Waller, published in 1684, has recently been reprinted (see Appendix A, item 14), it is not entirely satisfactory from the standpoint of the twentieth-century reader, and I have ventured to make a new translation into the English of our own day. I have provided this with copious notes that have two distinct purposes: the first is to give references to the manuscripts and to date the experiments where possible; the second is to relate the work of the Academy to other seventeenth-century science.

If it seems presumptuous for a non-Italian to write on such a subject and necessarily make judgments about it, I can only plead that I have tried to be entirely objective, having no motive whatsoever for being otherwise. I trust that the book is well enough documented to permit the discovery of any serious errors of interpretation that I may have committed.

In studying the Accademia del Cimento the temptation to rely on secondary sources was greatly reduced by their scarcity, and I have gone to the manuscripts whenever possible, making use of printed collections, especially of letters, though not without verifying references wherever I could do so. With a few exceptions, noted as they occur, all translations are my own. In Appendix D I have given the originals of all the translated manuscript passages that as far as I know have remained unpublished up to now.

In transcribing such passages I have modernized the use of the letters "u" and "v," and written out the ubiquitous contractions for *per*, *questo*, *quello*, etc. In translating titles I have generally ignored the obligatory *Serenissimo*, which falls strangely on the ears of Anglophones; thus *V.A.S.* becomes simply "Your Highness." Similarly, in private letters *V. S. Illustrissima* ("your very illustrious lordship") has been rendered as "you." I have given some thought to the translation or otherwise of Italian proper names. It has seemed natural to refer to the princes, whom we shall come to know very well, as Ferdinand and Leopold; on the other hand I have decided against Anglicizing such names as Vincenzio (Viviani) and Carlo (Dati). Names of cities I have Anglicized where they are so well known that it would be an affectation to use the foreign forms (Florence, Rome, Naples, Leyden).

I have to acknowledge help from many people and institutions. The research for this book would have been impossible without grants from the Canada Council in 1968 and 1969. I also had the great privilege of a term during the winter of 1968–69 at the Institute for Advanced Study in Princeton, New Jersey, and of the friendship and advice of Professors Marshall Clagett, Harold F. Cherniss, and Felix Gilbert of that Institute, as well as of Professors Charles Gillispie and Thomas Kuhn of Princeton University. I wish to thank Professor Stillman Drake of the University of Toronto for some stimulating discussions. The University of British Columbia has kindly provided me library facilities. In Italy, I have been much indebted to Professoressa Maria Luisa Righini Bonelli, director of the Museo di Storia della Scienza in Florence, and to her staff; also to Signor Giuseppe di Pietro of Florence and to Dr. Carlo Maccagni of the Domus Galilaeana in Pisa, as well as to numerous librarians in these cities and elsewhere in Italy. In London, Professors A. Rupert Hall and Marie Boas Hall have been most helpful. My old friend Stephen K. Marshall has once again acted as a touchstone for my uncertain Latinity.

Extracts from manuscripts in the Archivio di Stato, the Biblioteca Mediceo-Laurenziana, and the Biblioteca Nazionale Centrale in Florence, the Library of the Royal Society of London, that of Leyden University, and the Bibliothèque Nationale in Paris, have been printed with the kind permission of these institutions. The portrait of Prince Leopold was furnished through the kindness of the Director of the Uffizi Gallery.

I am grateful to the publishers of *Physis* and to the British Society for the History of Science for permission to reprint extracts from papers of mine published in their journals, and to Pantheon Books, a Division of Random House, Inc., for permission to quote from a translation of Dante by Lawrence Grant White.

Finally, I cannot find words to thank my wife, not only for her continual encouragement but for the strenuous task of typing the entire manuscript twice—some of it three times—and for help with proofs and index.

Vancouver, Canada   W.E.K.M.

ABBREVIATIONS

BM   London, Library of the British Museum.

BN   Paris, Bibliothèque Nationale.

DIS   *Le Opere dei discepoli di Galileo Galilei*, Edizione Nazionale, Vol. I: *L'Accademia del Cimento. Parte Prima* (Florence, S.A.G. Barbèra Editore, 1942).

FAS   Florence, Archivio di Stato.

FBNC   Florence, Biblioteca Nazionale Centrale.

FMSS   Florence, Museo di Storia della Scienza.

G.   (followed by a number) One of the 347 volumes of "Galileo" manuscripts in *FBNC*.

G.OP.   *Le Opere di Galileo Galilei*, Edizione Nazionale, 20 vols. (Florence, Barbèra, 1890–1909, and later reprints).

LIUI   A. Fabroni, ed., *Lettere inedite d'uomini illustri*, 2 vols. (Florence, 1773 and 1775).

Saggi   *Saggi di naturali esperienze fatte nell'Accademia del Cimento* . . . ed. princ. (Florence, 1667), unless otherwise stated.

TT   G. Targioni Tozzetti, *Notizie degli aggrandimenti delle scienze fisiche accaduti in Toscana nel corso di anni LX. del secolo XVII* . . . , 3 volumes in 4 (Florence, 1780). *TT* 1 refers to volume 1; *TT* 2 to volume 2 of this.

FLORENTINE WEIGHTS AND MEASURES

It is difficult to find a translation for some of the terms, now obsolete, used in the Florentine metrology of the period. Pounds, ounces, and grains are obvious, as is the mile, but *denaro* has two distinct meanings, as will be seen below, and with the exception of *braccio* the remaining terms had better be left alone, especially as the present-day meanings of some of them are entirely different. As to *braccio*, the Cambridge Italian Dictionary suggests "yard" or "ell"; I have also seen the translation "cubit." As the *braccio* is much shorter than our yard, I have selected the word "ell.'

WEIGHTS.
1 *libbra* (pound) = 12 *once* = 339.5 grams
1 *oncia* (ounce) = 24 *denari* = 28.29 grams
1 *denaro* = 24 *grani* = 1.179 grams
1 *grano* (grain) = 0.049 grams

LENGTHS. There were two kinds of ells in use at the time of the Accademia del Cimento: the "land ell" (*braccio a terra*) and the "cloth ell" (*braccio a panna*). The mile (*miglio*) was apparently defined as 3,000 *braccia a terra*, and the *braccio a panna*, the more usual one, was 18/17 as long as the other. *Braccio* by itself always means "cloth ell," and we have:

1 *miglio* (mile) = 2,833-1/3 *braccia* (ells) = 1,654 meters.
1 *braccio* (ell) = 20 *soldi* = 58.36 cm.
1 *soldo* = 12 *denari* [= 6 *piccioli* = 3 *quattrini*] = 2.918 cm.
1 *quattrino* = 4 *denari* = 0.9727 cm.
1 *denaro* = 12 *punti* = 0.2432 cm.
1 *punto* = 0.0203 cm.

The *palmo* (palm) also appears, and probably means the Roman palm, 22.34 cm according to the editors of *DIS* (see abbreviations, p. xii), p. 75, from which all these equivalents have been taken.

I have given the metric equivalents to four significant figures, although even a slight acquaintance with the history of metrology will suggest that standardization cannot possibly have reached that pitch in the seventeenth century.

TIME. The hour was divided into minutes (*minuti*) and seconds (*minuti secondi*) as at present. The day began one hour after sunset, the hours being numbered from 1 to 24.

A NOTE ON DATES

At the period dealt with in this work the same event might be dated in at least five ways in various regions of Europe and for different purposes:

The New Style or Gregorian calendar had been in general use in Catholic Europe since its introduction by Pope Gregory XIII in 1582. During our period this was also in use in the Netherlands. In the ordinary application of this calendar the year was considered to begin on January the first.

In Florence, and to some extent elsewhere, especially in ecclesiastical circles, the so-called *stile fiorentino* was employed. In this the dates were the same as in the New Style, but the year began on the *following* March 25 (the feast of the Annunciation). Sometimes dates on this system were indicated by the words "Ab Inc.[arnatione]," but not always. The effect was, for example, that January 28, 1662 *Ab Inc.* would be January 28, 1663 in the ordinary New Style, or *stile commune.*

According to C. R. Cheney, *Handbook of Dates for Students of English History* (London, 1961), pp. 4–5, a related calendar, but with the year beginning—more logically—on the *preceding* March 25, was called the *calculus pisanus* and survived to some extent at Pisa until 1750. I have not detected its use in any of the documents that I have examined.

In England the New Style was adopted only in 1752, so that in the seventeenth century the Old Style or Julian calendar was still in use, and was 10 days behind the New Style; for example, July 9, 1667, O.S. is July 19, 1667, N.S. Englishmen wishing to avoid ambiguity would generally write "July 9/19, 1667." At this period the year officially began on the *following* March 25 in England, so that an Old Style date between January 1 and March 25 would be dated as in the *stile fiorentino,* but to avoid trouble the form "February 25, 1665/6 was frequently employed, or sometimes even "February 25/March 7, 1665/6," to put dates beyond all doubt in international correspondence. If only one date is used in English letters it is reasonable to suppose, in the absence of contrary evidence, that it is Old Style.

Letters and works in Latin were often dated on the Roman system; for example, "XII Kal. Oct. MDCLVII" is Sept. 20, 1657. A table is given in the *Handbook of Dates* already cited, pp. 76–81, but it is often impossible to be sure that any particular document actually follows this table.

In order to help the nonspecialist reader through this jungle I have uniformly converted dates to the New Style with the year beginning on the first of January, except in Section 2 of chapter 6, where the dates of meetings of the Royal Society and of letters written in England are kept in the Old Style, and one or two difficulties explained.

1.

INTRODUCTION:

THE BACKGROUND AND THE SOURCES

1. THE SCOPE OF THIS BOOK

This book is intended as a study of the first organization founded for the sole purpose of making scientific experiments, the Accademia del Cimento, or Academy of Experiment, set up at Florence in the year 1657 and dissolved, or at least ceasing to exist, only ten years later.

After dealing with the *ambiance* in which the Academy came into being, and the character of its founder and protector Prince Leopold de' Medici and that of his brother the reigning Grand Duke Ferdinand II of Tuscany, the individual academicians—there were only ten who can be so identified—will have our attention. This will be followed by an interpretation, perhaps unorthodox, of the records that tell us how the Academy came into being and how it operated. Next its sole publication, the *Saggi di naturali esperienze,* will be presented in a new translation, with introduction and notes, and this will be succeeded by an account of the more important of the unpublished experiments and observations. A chapter will be devoted to the relations of the Academy with scientific workers on the other side of the Alps, and especially with the Royal Society of London and the Académie Montmor at Paris. What little is known about the dissolution of the Academy will be presented, and finally an attempt will be made in the last chapter to evaluate the Academy's achievements and its place in the history of science.

2. EXPERIMENTAL SCIENCE BEFORE 1657

As an introduction to the main subject of this book it will be well to consider briefly the state of experimental science in the period immediately before the foundation of the Accademia del Cimento in 1657. I say *experimental* science because, as we shall see, the corporate activities of the Academy were almost entirely confined to the making

1

of experiments and observations, with one exception that will be discussed in chapter 5. At least two of the members were highly competent mathematicians, but from the published and unpublished records of the Academy itself we should never discover this.

While physical experiments of great interest had been made from time to time in the later Middle Ages—for example the magnetic experiments made in about the year 1260 by Peter of Maricourt[1] and the experiments on refraction of the Polish scientist Witelo at about the same period[2]—it can fairly be said that serious experimentation was sporadic before about 1580, except in the hands of the alchemists, whose peculiar preoccupations gave their work such a flavor of sorcery that Lynn Thorndike was constrained to entitle his great work *A History of Magic and Experimental Science*.[3] The *Magia naturalis libri XX* of Giovanni Batista della Porta[4] advertises a similar bias in its title. I shall return to Della Porta later.

A different attitude to experimentation developed rapidly in the last quarter of the sixteenth century. This new trend started as the systematic exploration of a specific subject by means of purposeful experiments. One of the first to do this successfully was the English physician William Gilbert, who worked on the properties of the lodestone and of substances such as amber, publishing his results in the year 1600.[5] The title of his book reveals his original and very important hypothesis that the earth is a great magnet, a hypothesis that he purported to test by experimenting with a large spherical lodestone. He did not, however, attempt to apply mathematics to the subject, and in the *Dialogue concerning the two chief World Systems* we find Galileo, in the person of Salviati, regretting that Gilbert had not been a better mathematician.[6]

Gilbert died in 1603, and in the succeeding twenty years the most influential works of Francis Bacon appeared.[7] Twentieth-century esti-

---

[1] *Epistle to Sygerus of Foncaucourt, Soldier, concerning the Magnet*, trans. S. P. Thompson (London, 1902).

[2] Kepler's *Ad Vitellionem paralipomena* (Frankfurt, 1604) will be remembered.

[3] Eight vols. (New York, 1923–58).

[4] (Naples, 1589); trans. as *Natural Magic in twenty Books*, etc. (London, 1658).

[5] Gilbert, *De magnete, magneticisque corporibus, et de magno magnete tellure*, etc. (London, 1600).

[6] *G.OP.*, VII, 432.

[7] *The two Bookes of . . . the proficience and advancement of learning* (London, 1605); *Novum organum* (London, 1620); *De dignitate et augmentis scientiarum* (London, 1623).

mates of Bacon's position in the history of science are extraordinarily diverse, varying from Lynn Thorndike's dismissal of him as "a crooked chancellor in a moral sense and a crooked naturalist in an intellectual and scientific sense," who "did not think straight,"[8] to the equally emphatic adulation of Margery Purver, who will have him the founder of a new system of natural philosophy.[9] In my judgment this latter position is completely untenable, and it has been attacked by several reviewers of Dr. Purver's book. Bacon was not a scientific man but a philosopher, and he seems not to have been in touch with what was going on in science, even in experimental science, during his own times. What he did not understand was the process by which mathematical reasoning and experiment are made to complement one another, a process that was being exploited in Italy during his lifetime.

To what extent Bacon affected the outlook of the Accademia del Cimento is a point worthy of study, but in Italy, and especially in Tuscany, the fame of Bacon, and indeed of every other philosopher or scientist,[10] was completely eclipsed by that of the "immortal Galileo." Now every society since the dawn of history has surrounded its divinities with myths, and it may fairly be said that the ancient Greeks were almost silent about their highly entertaining gods and goddesses in comparison with the historians of science, especially those from Italy, who have dealt with this great man. Many of their writings can only be classified as hagiography.

The works and the surviving correspondence of Galileo have been published in twenty sumptuous volumes,[11] and it might be supposed that it would be less than a lifetime's work to study these and to find out what experiments Galileo made, what deductions he drew from them, and what attitude he took toward experimentation. Would it were so! Unfortunately it is not only possible but fatally easy to agree (in general) about what Galileo said and did, and at the same time to disagree violently and (which is worse) sarcastically about the significance of these words and deeds. One extreme view is taken by the followers of that distinguished historian Alexandre Koyré, who would persuade us that Galileo performed scarcely any experiments, and

[8] *A History of Magic*, etc., VII, 88. Thorndike employs a whole chapter (*ibid.*, pp. 63–88) to document this conviction.

[9] Margery Purver, *The Royal Society: Concept and Creation* (London, 1967), pp. 20–62.

[10] I am well aware that this is a nineteenth-century word and thus an anachronism; but the reader may pronounce it "natural philosopher" if he so desires.

[11] *Le Opere, edizione nazionale*, ed. A. Favaro, 20 vols. (Florence, 1890–1909); reprinted with additions, 1929–40. This will be referred to as *G.OP.*

these of very little value.[12] Others have thought that he was an excellent experimenter and observer.[13]

The reason for this situation is that each reader of the works of Galileo brings to the task his own philosophical predilections. Now Galileo was not primarily a philosopher (in the modern sense), and, therefore, as far as I can see, he never developed any consistent philosophical point of view. Hence, as Dr. Crombie writes: "Philosophers looking for historical precedent for some interpretation or reform of science which they themselves are advocating, have all, however much they have differed from each other, been able to find in Galileo their heart's desire." [14]

We might reasonably assume that the experimental activities of the Academy would be greatly stimulated by the work of Galileo, the more so because both its founder and his brother had been great admirers of that famous man and enjoyed his conversation during his declining years, and because Vincenzio Viviani, one of the two leading spirits of the Academy, had been Galileo's last pupil. In point of fact this influence does not seem to have been overwhelming; Galileo is cited only six times in the *Saggi*, compared to four references to Robert Boyle and four to the freethinking French cleric Gassendi out of a total of twenty-four citations altogether. As we shall see in chapter 4, the experiments made by the Academy that were directly inspired by Galileo were among those that the master suggested but clearly did not attempt himself. It is entirely misleading to say, as does Gaetano Pieraccini,[15] that the Academy "followed [Galileo's] method of research and of criticism."

But Galileo, however you consider him, was by no means the most avid experimenter in the period before the foundation of the Academy; indeed, his particular genius dictated that he should assign to experimentation a subsidiary role. There are a great many examples of this in his works, the best known and most striking of which may well be the occasion in the *Dialogo* (1632) during the discussion of what happens if a stone is dropped near the mast of a moving ship, when he makes Salviati say, "Without making the experiment I am sure that the result will be as I am telling you, because it must necessarily come out this way." [16] Other, lesser men were not as sure of

---

[12] See e.g., A. Koyré, *Études galiléennes*, 3 vols. (Paris, 1939).

[13] The discussion shows no signs of coming to a conclusion; see e.g., the twenty-three essays in *Galileo, Man of Science* (New York and London, 1967).

[14] A. C. Crombie, *Actes VIIIᵉ Int. Cong. Hist. Sci.* (Florence, 1956), III, 1090.

[15] Pieraccini, *La stirpe de' Medici di Cafaggiolo*, 3 vols. (Florence, 1924, 1925), II, 603.

[16] "Io senza esperienza son sicuro che l'effetto seguirà come vi dico, perchè così è necessario, che segua." (*G.OP.* VII, 170).

being right and, in consequence, took pains to make experiments as carefully as they could. One of these was the Minorite priest Marin Mersenne, famous as the correspondent of almost every contemporary natural philosopher of any consequence.[17] His many experiments in mechanics, pneumatics, magnetism, and other subjects have been discussed in a very interesting way by Robert Lenoble.[18] His experiments on falling bodies were also discussed by A. Koyré,[19] who, moreover, drew the attention of scholars[20] to the more important experiments of Father Giambattista Riccioli, a Jesuit astronomer whose experimental work is largely unrecognized because of his anti-Copernican stand, but even more because it is buried in the 1,504 folio pages of his enormous *Almagestum Novum*, published at Bologna in 1651, a volume that would daunt almost anyone but a Koyré. Riccioli, assisted by a team of Jesuit scientists, made a remarkably ingenious series of experiments on pendulums and on falling bodies in order to test the relations asserted by Galileo.[21] As far as I can tell, our academicians paid no attention to these experiments.

The Academy did read Gassendi,[22] taking from him some experiments on sound and one on freezing. They were probably attracted to Gassendi because by the standards of the time he had a very liberal outlook,[23] as opposed to Riccioli, who was an arch-conservative.

In Tuscany itself there was no particularly strong experimental tradition, although some important experiments were made there during the first half of the seventeenth century—one need only adduce the celebrated experiment with a glass tube and some mercury made by Viviani, at Torricelli's suggestion, in 1644—but such efforts, especially where Galileo's influence was strong, were conditioned by his attitude. As Moscovici has pointed out with great clarity, the major preoccupation at that time was not to *measure* effects but to *decide* between hypotheses.[24]

This naturally determined the kind of experiments that were performed. The Accademia del Cimento did indeed make experiments

[17] Most of his letters have been published in *Correspondance du P. Marin Mersenne, religieux minime. Commencée par Mme Paul Tannery, publiée et annotée par Cornelis de Waard*, 10 vols. Paris, 1933–67, in progress).

[18] Lenoble, *Mersenne ou la naissance du mécanisme* (Paris, 1943), *passim*.

[19] Koyré, *Proc. Amer. Philos. Soc.* 97 (1953): 222–37, and elsewhere.

[20] *Ibid.*

[21] *See* Koyré, *ibid.*, for the very interesting details.

[22] *Opera omnia*, etc., ed. H. L. Habert de Montmor and F. Henri, 6 vols. (Lyons, 1658–75).

[23] On this see J. S. Spink, *French Free-thought from Gassendi to Voltaire* (London, 1960), chapters 1 and 6.

[24] Serge Moscovici, *L'expérience du mouvement. Jean-Baptiste Baliani disciple et critique de Galilée* (Paris, 1967), p. 116.

with each of these purposes; but the more I study its work, both published and unpublished, the clearer it becomes that the mainspring of its effort was the desire to discredit the physics and astronomy of the peripatetics. In this respect they were faithful disciples of Galileo. This obsession sometimes led the Academicians into blind alleys and concealed from them their own metaphysical assumptions and, to use a military metaphor, allowed their strategy to be dictated by the enemy, a cardinal blunder in any conflict.

I must not end this introduction without a remark on the very delicate religious situation in which Italians found themselves after the condemnation of Galileo, a situation that we can appreciate in the twentieth century only by a strong effort of the imagination, or perhaps by a sidelong glance at contemporary totalitarian states. With the exception of a small number of clerics like Riccioli, who could write about astronomy and physics with safety because of their known and declared anti-Copernican stand, few people were anxious to expose themselves to the condemnation of the Church, and these sciences languished. The safe discipline was pure mathematics, and indeed there was a brief flowering of that discipline—we need only mention Cavalieri and Torricelli, both of whom died, too soon, in 1647. The Accademia del Cimento, under princely protection, corporately braved the storm for a few brief years. Not long afterwards, under the deadening piety of Cosimo III, the curtain came down on the natural sciences in Tuscany. The chief practitioners, men like Borelli and Malpighi, had already gone, or gone back, to other parts of Italy.

3. EARLY ORGANIZATIONS DEALING WITH EXPERIMENTAL SCIENCE

Apart from the groups that gathered around the famous philosophers of antiquity, the learned society is a product of Europe since the so-called revival of learning, and Italy is certainly its birthplace. According to Michele Maylender,[25] the very earliest one was probably the Accademia Aldina, founded in Venice about 1490 or 1495 by Aldo Manuzio, the celebrated proprietor of the Aldine Press. It claimed to be interested in both literary and scientific matters—especially medicine—and its founder must certainly have been thinking of the Academy of Plato, but it had no apparent effect on the progress of science. The same might be said of the Accademia degli Affidati, which was founded in Bologna in 1548 and lasted about thirty years.[26]

An Academy concerning which we know little, but which must have

[25] Maylender, *Storie delle accademie d'Italia*, 5 vols. (Bologna, 1927–30), I, 125–30. This invaluable work deals concisely with several hundred "Academies" devoted to literary and scientific activities.

[26] *Ibid.*, I, 72–82.

been a small and private affair, met at the house of Giovanni Battista della Porta in Naples for a few years beginning about 1560, until it was broken up because Della Porta was accused of meddling with the black arts, a charge which he almost encouraged by writing on "Natural Magic." In the preface to the 1589 edition of his *Magia naturalis* he refers to this gathering: "Nor did my house ever lack an Academy of inquiring men who worked strenuously at these investigations and experiments."[27] We can only assume that this little Academy, which was called the Academia Secretorum Naturae, worked at the sort of experiments described in Della Porta's book.

Of much more importance was the celebrated Accademia dei Lincei (lynxes), which, although it did not confine itself entirely to experimental science, was the true precursor of the Cimento.[28] The Accademia dei Lincei was the creation of one very young man, Prince Federico Cesi, then Marchese di Monticello. When he founded it in Rome in 1603 he was only seventeen; he drew up the rules, which were detailed and specific, paid the bills, ran it according to his own ideas, and defended it valiantly through a great deal of persecution until it finally ceased to exist in 1629. Cesi himself died in the following year.

From 1603 to 1608 there were four members, Cesi, Johannes Eck, Francesco Stelluti, and Anastasio de Filiis, all young men. De Filiis died in 1608, and in 1610 the first new member was added, none other than the aging Della Porta, whom Cesi had met in 1604 at Naples. Gabrieli tells us that this friendship with Della Porta greatly enlarged Cesi's horizons. Then on April 25, 1611, Galileo became a member of the little group, and this naturally drew the attention of the learned world to the Lincei, the more so because the great man seemed proud to use the title of Lincean, both in his letters and on the title pages of most of his subsequent books. I share the opinion of Professor Drake that this action of Galileo's is a mystery, for he towered above the other members, and was well aware of the fact.

Besides Galileo, four other new members were added in 1611, and ten more in 1612. From this time until about 1625 the Academy was very active. According to Professor Drake, its chief contribution to science was as a rapid and effective means of communication. It had no

[27] Preface, fol. a3v.

[28] The most authoritative modern account of the Lincei is by a secretary of the revived Accademia dei Lincei (the national scientific society of Italy), Giuseppe Gabrieli. This runs to 1,446 pages, consists mainly of letters, but Gabrieli's introduction and comments are invaluable. See *Mem. R. Acc. Naz, dei Lincei, cl. sc. morale,* ser. 6, vol. 7, parts 1–3 (1938–42). A charming and concise history, unfortunately not documented, has been given by Professor Stillman Drake, *Science* 151 (1966): 1194–1200.

regular publications of its own, although it published several books written by members, including two by Galileo. It was not by any means devoted to experimental science alone, but by reason of its activity and the celebrity of a few of its members it cannot be neglected in a discussion of the academies that preceded the Accademia del Cimento.

The Societas Ereunetica was established at Rostock on the Baltic in 1622 by Joachim Jungius, a remarkable man who was probably influenced by the *Reipublicae Christianopolitanae descriptio* of Johann Valentin Andreae,[29] a Utopian work that so greatly resembles the *New Atlantis* of Francis Bacon that some transmission of ideas seems certain.[30] The "laws" of the Societas Ereunetica called for the prosecution of experimental science, but had very distinct theological, and especially anti-Jesuit, overtones.[31] In any event, the social conditions—the Thirty Years' War, for example—were against it, and nothing seems to have been heard of it after about 1624.

In 1633 an academy was founded in Paris, open to all "bon esprits" who wanted "to discuss in public the most interesting questions in physics, moral philosophy, mathematics, and other disciplines."[32] The founder was the physician Théophraste Renaudot, who in 1629 had opened at his house a "Bureau d'Adresse" that performed some of the functions of today's columns of classified advertising. Meetings were held each Monday at two o'clock, and the language was exclusively French. From the published records of the meetings,[33] it appears that the papers on scientific subjects were of very little value. One of the rules was complete anonymity of publication. But there was much opposition from the University of Paris, and the last session was held on September 1, 1642.[34]

It is not surprising that scientific meetings were also held at Marin Mersenne's quarters, beginning about 1635, and later at the houses of

[29] (Strasbourg, 1619). Translated with a valuable introduction by F. E. Held, *Christianopolis, an Ideal State of the Seventeenth Century* (New York, 1916).

[30] Held concludes that Bacon had read the *Christianopolis*. I shall not enter into this question here.

[31] See G. E. Guhrauer, *Joachim Jungius und sein Zeitalter*, etc. (Stuttgart and Tübingen, 1850), pp. 69–70.

[32] See G. Bigourdan, *Compt. Rend.* 163 (1916): 937–43.

[33] *Première* [. . . *quatriesme*] *centurie des questions traitées ez conférences du Bureau d'addresse*, etc., 4 vols. (Paris, 1634–41). These were re-edited by Eusèbe Renaudot and republished in 4 octavo volumes in Paris, 1665–66, and in Lyons in 6 duodecimos in 1666.

[34] The proceedings were even translated into English by G. Havers and J. Davies, and published in two folio volumes in London in 1664 and 1665.

Etienne Pascal, Le Pailleur, and others.[35] This entirely informal group, which had no rules and no publications, should scarcely be considered an academy at all. Its importance lies in the fame of those who attended the meetings, a list that included the Pascals, father and son, Roberval, Desargues, Gassendi, and Hobbes. Mersenne died in 1648, Le Pailleur in 1654, and after the death of the latter the meetings continued at the house of Henri Louis Habert de Montmor. We shall hear more about the Montmor Academy in chapter 6.[36]

One more Italian academy should be mentioned. In or about 1650 the Accademia degli Investiganti was founded at Naples by Tommaso Cornelio of Roveto di Cosenza and Leonardo di Capua.[37] "For research on things philosophical and natural, Cornelio . . . assembled an Academy of learned men in his own house." Maylender names about a score of these academicians, and in the list we find Gassendi (presumably a correspondent), Giovanni Alfonso Borelli, and Sebastiano Bartolo. There was much opposition from the conservatives, and in 1656 the plague forced the suspension of all meetings. It had various later reincarnations, but nothing was printed in its name, and its fame is only that of some of its members.

In Germany the Collegium Naturae Curiosorum was founded in January 1652 at Schweinfurt by Dr. Lorenz Bausch.[38] This was entirely a medical society, as can be deduced from a long list of the members by date of election. It was officially recognized by the Emperor; hence the pompous title. In 1670 it began to publish its *Miscellanea Curiosa*.

I would suggest that of all these academies and societies the only two that are in the least likely to have influenced the founding of the Accademia del Cimento are the Accademia dei Lincei and the Accademia degli Investiganti. In chapter 3 I shall give reasons for thinking that even these had very little influence.

4. SOURCES OF INFORMATION

(A) PRIMARY SOURCES

By far the greatest part of the first-hand information about the Accademia del Cimento is contained in the remarkable collection of

---

[35] G. Bigourdan, *Compt. Rend.* 164 (1917): 129–30.

[36] For an admirable account of these early French Scientific Academies, see Pierre Gauja, "Les origines de l'Académie des Sciences de Paris," *in* Institut de France, Académie des Sciences, *Troisième Centenaire 1666–1966* (Paris, 1967), pp. 1–51.

[37] Maylender, III, 367–69.

[38] See A. E. Büchner, *Academiae Sacri Romani Imperii Leopoldino-Carolinae naturae curiosorum historia* (Halle, 1755).

manuscripts known as the Collezione Galileiana in the Biblioteca Nazionale Centrale at Florence. Of these there are 347 volumes, divided into five divisions and appendix: the predecessors of Galileo (10 volumes); Galileo's works and letters (89 volumes); his contemporaries (11 volumes); his "disciples" (148 volumes, of which 24 refer to Evangelista Torricelli and no less than 104 to Vincenzio Viviani); those that came after him (the "Posteriori," 49 volumes); and finally the "Appendix" in 40 volumes, consisting of manuscripts that came into the collection after the arrangement had been made. Nowadays each volume of this great corpus is generally referred to by its number in the entire series, as for example "Gal. 293." In this book the italic letter *G.* will be used as an abbreviation, so that *G.* 293 will signify volume 293 in this collection.

Apart from the correspondence of Viviani, which extends through the period of the Academy, most of the relevant manuscripts will be found in the series *G.* 259 to *G.* 286, and *G.* 312 to *G.* 315 in the Appendix, the latter group being copies of letters between Prince Leopold and various people.

In the room for the reading of manuscripts at the Biblioteca Nazionale there is a manuscript catalogue of this collection in 3 volumes, by Carlo and Favaro.[39] Although this is not indexed, it is invaluable, particularly when the correspondence must be consulted.

The "Galileo" manuscripts have had their vicissitudes, and it is remarkable that so much has survived, especially of the letters. The sort of thing that must have happened distressingly often is well illustrated in a story told by Giovanni Targioni Tozzetti in his celebrated *Notizie.*[40] Early in the eighteenth century a Dr. Giovanni Lami, taking some friends home to lunch at his house, stopped at a market to buy some mortadella. One of the friends was the historian G. B. Nelli, and when the sausage was unwrapped Nelli noticed that the wrapper was a letter of Galileo's, which he rescued, afterwards going back to the market and subsequently tracing further manuscripts.[41]

The chance that associated Targioni Tozzetti with the manuscripts of the Accademia del Cimento is so fortunate for posterity, and so interesting in itself, that I shall translate most of his account of the affair, noting only that the first sentence of the following passage is no idle boast.

One reads here and there a few notices about this celebrated Accademia del Cimento, but I flatter myself that I am the only one who can give

---

[39] This was given to *FBNC* in 1923 by Antonio Favaro's son Giuseppe.
[40] See below, p. 350 and list of abbreviations, p. xii. The three volumes of this work will be referred to as *TT* 1, *TT* 2, and *TT* 3.
[41] *TT* 1, 124–25.

very copious, circumstantial, and authentic information about it, thanks to the following piece of good fortune. You should know, then, that when the Cavalier Carlo Giuseppe Segni, son of Senator Alessandro and the last of his family, passed into eternal rest, the heirs named in his will, fearing to receive more loss than gain, repudiated the inheritance. Under the provisions of our laws, this devolved on the Royal Treasury, and it was appropriately disposed of by the ministers. Among the items of the inheritance, there were many printed books that were sold at auction on the advice of experienced booksellers. There were also a very large number of manuscripts in great confusion, and I was appointed by my brother-in-law the auditor Ippolito Scaramucci, Under-secretary of the Treasury, to sort and examine them.

In two mornings I executed this commission with great pleasure. When I had set apart the books and papers concerning the interests of the Segni family, and those belonging to various administrations and tribunals that had been confided to the direction of Senator Alessandro Segni, who died in 1697, I set aside those that were of concern entirely to science and various kinds of scholarship. These contained: I. Forty-three folio manuscripts and 33 in quarto or smaller, from various centuries, but for the most part from the great age of the Tuscan tongue, which had been used as authorities in the *Vocabolario della Crusca*.[42] Others were on history, poetry, and various disciplines. II. Six volumes containing various diaries and memoirs of the Accademia della Crusca, of which Senator Alessandro Segni had been secretary. These were subsequently given to that Academy. III. Four bundles of papers belonging to Prince Cardinal Leopold's secretariat, of which the same Segni used to be superintendent. I could not examine these in my own way, and I do not know whether it was so, but probably there would have been a few important philosophical notes among them. IV. Sixteen volumes or bundles of various sizes containing writings belonging to the Accademia del Cimento. Of the larger, bound volumes, one was the Diary of the Academy; . . . three were various drafts of the *Saggi d'esperienze naturali* of that Academy; two thinner ones contained water-color drawings and pen-and-ink figures of the instruments for the experiments, on the lines of those published in the *Saggi*, and several that have remained unpublished. There were four others of a few pages, bound in cardboard, the contents of which I shall indicate in the proper place; and finally there were six bundles of papers, tied with string and of various sizes, but not thicker than the width of four fingers.

Of these 16 manuscript items, I said in my report that I was unable to say anything certain about their value without making careful comparisons with the works printed in the *Saggi*, so as to make sure what and how much remained in addition to, or different from, that already printed. Therefore, so that I might make such a comparison conveniently and with care, all these writings were sent to me and I gave a receipt for them. Thus for some months I was able to examine them one by one at my pleasure. This occupation, which I can assert to have been one of the most joyful of my

---

[42] The great Italian dictionary that seems always to be in progress.

life, very soon made me aware that in this pile of papers there was concealed a precious treasure of physical information.

So I resolved to reduce them methodically to certain determined and distinct categories, the better to distinguish the published ones from the unpublished. In order not to change the form of the original manuscripts that had been confided to me, I had copies made in separate items and loose sheets, with great care and at no little cost, so that I could transpose them and bring them into a methodical series. I also had copies taken of the figures.

It was fortunate for me that these copies were completed by three different copyists in a few months, because I received an unexpected note from the Under-secretary of the Treasury, my brother-in-law Scaramucci, in the following terms: "Signor Cavaliere A . . . wishes to see those papers concerning the Accademia del Cimento that were found in the Segni inheritance, and that I should give them to him. I therefore beg you to favor me with them so that I can take them to him. I remain, etc. From my house, July 29, 1760." I immediately got together all these manuscripts, tied them up and sealed them, and sent them to my brother-in-law, who then gave them into the care of that gentleman, who enjoyed great authority in the Government at that time.

Nobody inquired again about these manuscripts for $19\frac{1}{2}$ years, that is, until it was seen in the draft of this work of mine that the writings of the Cimento belonged to the Royal Treasury. When they were not found in its archives, I was asked about it and in consequence I had to produce the note referred to above in order to clear myself.[43] Then the search was directed to the house of that gentleman, who had died several years before, and it was learned that he made a choice of the most important, among which was the Diary of the Academy, and gave them to a priest to copy, who died a little while later. Therefore the heirs, who had made no inquiry into these manuscripts and thought they were the priest's, sold them, with some other rubbish that he had left. It has not been possible to find out what happened to them.

Thus were lost these precious originals; thus, through my good luck, the copies that I took the trouble to make in time stand in their places. All the other volumes and papers of lesser importance, which had remained in that gentleman's house, were given back, and placed, as they deserved to be, in the Old Royal Secretariat, in the Department of the Secretariat of Prince Cardinal Leopold. There, according to regulations, immediately after the death of the Cardinal, Senator Alessandro Segni ought to have been compelled to deposit every one of the papers belonging to the Secretariat of that Prince, confided to his supervision. I am not sure how the affair turned out; it is certain that a large part of the letters and other papers of Prince Cardinal Leopold are preserved in the Old Royal Secretariat . . . . But it is also certain that many others remained in the house of Senator Segni,

---

[43] It must be stated that Targioni Tozzetti had been busy with many entirely different things during this long interval.

and later came to the Treasury, as I said. Segni may also have consigned to the Royal Secretariat only the papers that belonged more properly to the interests of his Prince, so that those concerned with studies and literary matters, and most of all those relating to experimental physics, which was in bad odor during the reign of Cosimo III, were left in his hands. At the least, either in one place or another, what belonged to the Accademia del Cimento might have been preserved or buried all together in one place until our day! In this way many papers, which have been missing ever since the Segni bequest devolved on the Royal Treasury, would not have been lost.[44]

Targioni Tozzetti goes on to list the papers in the condition in which they existed when he was writing, but this would be no use to us now. Suffice it to say that the papers in question fill the twelve volumes of "Galileo" manuscripts now numbered G. 259 to G. 270. The scientific letters to and from Prince Leopold do not belong to this group; as Targioni Tozzetti says, they were in the Old Royal Secretariat (now the Archivio di Stato), from whence they came to the Central Library. There is still a good deal in the Archivio di Stato, but it takes some ingenuity to find it. There can be few places in which the correspondence of the ruling class is better preserved than in Florence.

There are a relatively small number of pertinent manuscripts, mainly letters, in several of the other libraries of the city. There used to be four volumes bearing directly on the activities of the Academy in the Archives of the Ginori-Conti family, but these have now been dispersed by sale. There are also some interesting letters in London— particularly at the Royal Society—in Paris, and no doubt elsewhere. The surviving correspondence of the individual members of the Academy—other than that with Prince Leopold—is lost or widely scattered, with the exception of that of Vincenzio Viviani which, as I have said, is mainly in Florence.

(B) PRINTED SOURCES

The Accademia del Cimento published only one work, *Saggi di naturali esperienze fatte nell'Accademia del Cimento sotto la protezione del Serenissimo Principe Leopoldo di Toscana e descritte dal segretario di essa Accademia* (Firenze: Giuseppe Cocchini, 1667). This volume is discussed and translated in chapter 4, and a bibliography of its many editions and translations is presented in appendix A. Naturally it is essential, but it gives us almost no information about the actual working of the Academy and no references whatsoever to manuscript material.

[44] *TT* 1, 373–76.

The really essential printed source, without which a book such as the present one would be many times as difficult to write, is the work by Giovanni Targioni Tozzetti[45] already referred to as the *Notizie* and which I shall designate by the symbol *TT*. The full title of this very rare book of almost 2,000 pages, in 3 volumes divided into 4, is as follows:

Notizie degli aggrandimenti delle scienze fisiche accaduti in Toscano nel corso di anni LX. del secolo XVII. Raccolte dal Dottor Gio. Targioni Tozzetti. In Firenze MDCCLXXX. Con licenza dei Superiori. Si vende da Giuseppe Bouchard Libraio in Mercato Nuovo.

There is another edition bearing the same date and almost identical, but with a different publisher, and a title page that is much more fully descriptive of the contents of the book:

Atti e memorie inedite dell'Accademia del Cimento e notizie aneddote dei progressi delle scienze in Toscana contenenti secondo l'ordine delle materie, e dei tempi memorie, esperienze, osservazioni, scoperte, e la rinnovazione della fisica celeste e terrestre, cominciando da Galileo Galilei, fino a Francesco Redi, ed a Vincenzio Viviani inclusive. Pubblicate dal Dottore Gio. Targioni Tozzetti in Firenze MDCCLXXX. Con licenza dei Superiori. Si vende da Giuseppe Tofani Stampatore, e da Luigi Carlieri Librajo.

This edition seems to be even rarer than the other. I cannot account for the duplication.

The *Notizie* is a difficult work to use and badly needs an analytic index. There is a short manuscript index bound at the end of the last volume of the copy in the Sala da consultazione of the National Library in Florence, but this is entirely inadequate. Nevertheless, for the study of the Accademia del Cimento the work is immensely useful. Not only is it a guide to what was going on in Tuscan science before 1657 but it includes a heavily annotated edition of the *Saggi*, the annotations, with a few exceptions, being complete, dated paragraphs from the diary of which Targioni Tozzetti had had a copy made (now G. 262). The citations are generally exact, if we forgive the eighteenth-century printers their usual cavalier treatment of punctuation marks and capital letters, but a few of the dates are given incorrectly.

Having done this for the experiments reported in the *Saggi*, Targioni Tozzetti goes on to make a classified list of the ones that were not, with dated extracts as before. He then quotes numerous letters and other documents, the letters mostly from printed collections,[46] the other documents from the papers of the Academy.

---

[45] For biographical details see Appendix B.

[46] Especially A. Fabroni, ed., *Lettere inedite d'uomini illustri*, 2 vols. (Florence, 1773 and 1775). I have given this very useful work the symbol *LIUI*.

These documents are not referenced by Targioni Tozzetti, and a guide to them is to be found in the first volume of a planned 21-volume extension of the National Edition of Galileo's works. This grand design, which was to involve the publication and annotation of almost all the last three divisions of the "Galileo" manuscripts, was interrupted by World War II; but one extremely useful volume was produced by Giorgio Abetti and Pietro Pagnini under great difficulties in 1942, as a memorial to the 300th anniversary of the death of Galileo. This has the following title: *Le Opere dei discepoli de Galileo Galilei, edizione nazionale; Vol. I, L'Accademia del Cimento, Parte prima* (Firenze, 1942). In style and format it is uniform with the National Edition of Galileo's works, and it contains a reprinting (not in facsimile) of the *editio princeps* of the *Saggi*, preceded by a 75-page introduction of very great value, and followed by 227 pages of extracts from the manuscripts G. 259 to G. 270 inclusive.[47] This well-chosen selection includes the first draft of much of the *Saggi*, as submitted to the members for approval, and the comments of three of the members—Borelli, Rinaldini, and Viviani—on it, as well as some of the variants, tried or only suggested, on the published experiments, and finally some of the experiments that were published only by Targioni Tozzetti. All this is done with exemplary scholarship. Nothing has been copied from the diaries, as it was intended to produce two further volumes about these. The net result is that the provenance of many documents quoted by Targioni Tozzetti can be traced by the use of *DIS*,[48] while the others, from the diaries, are dated. A study of both *DIS* and *TT* is the best possible guide to the manuscripts.

Some readers may wonder at my apparent neglect of the *Storia del metodo sperimentale in Italia* of Raffaello Caverni.[49] The reason is that Caverni, writing from a strangely anti-Galileian standpoint, very often let his imagination carry him far beyond his sources.

Useful for orientation, and very pleasant to read, is the report of a tercentenary celebration held at Pisa on June 19, 1957.[50] Finally, concerning the instruments of the Accademia del Cimento, there is the beautifully illustrated appendix by Professor Maria Luisa Bonelli at the end of the 1957 edition of the *Saggi*.[51]

---

[47] For references to this work I have used the symbol *DIS*.
[48] Though the only index is a summary index of names, as it had been the intention to publish a full analytic index in the 21st volume.
[49] Published in 5 volumes (Florence, 1891–98).
[50] *Celebrazione della Accademia del Cimento nel Tricentenario della fondazione, Domus Galilaeana, 19 Giugno 1957* (Pisa, 1958).
[51] See appendix A, no. 13.

2.

THE FOUNDERS AND MEMBERS OF THE ACADEMY

1. GRAND DUKE FERDINAND AND PRINCE LEOPOLD OF TUSCANY

It is often stated that the Accademia del Cimento was founded by Prince Leopold of Tuscany. This statement is too simple and categorical. It overlooks the fact that Prince Leopold's elder brother, Ferdinand II, fifth Grand Duke of Tuscany, was just as keenly interested in the new science as Leopold and was probably an even more able experimenter who employed much of his scanty leisure in scientific pursuits for at least a decade before anything was heard of the Academy directed by his brother. The character and accomplishments of Ferdinand II are therefore of interest.

His father, Cosimo II de' Medici, had set out to follow the Medici tradition of patronage of the arts and sciences.[1] As a youth he had had instruction from Galileo during several summers beginning in 1605, and on July 10, 1610, five months after he had come to the throne at the age of nineteen, Cosimo recalled Galileo into his service from Padua[2] and, in the eleven years that remained to him, welcomed numerous scientific men to his court, both from Italy and from abroad. Nevertheless—to quote from the rhetoric of Giovanni Targioni Tozzetti—Cosimo's main gift to Tuscan science was:

two famous and incomparable protectors and promoters of good physics. Tuscany glories in having been endowed with its princes, the sons of Cosimo, especially His Serene Highness the Grand Duke Ferdinand II, and the most Serene and Reverend Prince Cardinal Leopold. The lofty virtues of these two heroes have merited the grateful veneration of the Tuscans, and have

---

[1] A readable account of the Medici family is to be found in G. F. Young, *The Medici*, 2 vols. (London, 1909; 2nd ed., 1911). Young's chief source for our period was Riguccio Galluzzi, *Istoria del Granducata di Toscana sotto il governo della Casa Medici*, 5 vols., 4° (Florence, 1781); also 8 vols., 8° (Leghorn, 1781). Galluzzi was immensely erudite, but gave no references to his sources. G. M. Bianchini, *Dei Gran Duchi di Toscana della reale Casa de' Medici*, etc. (Venice, 1741), is impossibly eulogistic.

[2] On Cosimo's interest in science, see *TT* 1, 7–92 *passim*.

received immense praise from the most cultivated nations. Indeed they will live forever in the memory of posterity, as long as some knowledge of true philosophy subsists among mankind.[3]

Cosimo II and his wife Maria Maddalena had eight children, and it may be well to list them here:

Cristiana (1609–32)
Ferdinando (1610–70)
Giovanni Carlo (1611–63); made a cardinal in 1644.[4]
Margherita (1612–79); married in 1628 to Odoardo Farnese, Duke of Parma.
Mattias (1613–67)
Francesco (1614–34)
Anna (1616–76); married Archduke Ferdinand of Austria in 1646.
Leopoldo (1617–75); made a cardinal in 1667.

In spite of the ecclesiastical influence that was strong in the court of Cosimo II, Maria Maddalena saw to it that her sons were given an excellent education, including a good deal of the science of the time. Mattias and Francesco seem not to have had scientific interests, but Giovanni Carlo is known to have had the Torricellian experiment made before him as early as February, 1645,[5] and to have been visited by Marin Mersenne in Rome.

Ferdinand, who came to be known as the peacemaker, was of a gentle and affectionate disposition that rendered him somewhat less than effective in dealing with the disturbed politics of the period. For the first seven years after Cosimo's death, Tuscany was governed by regents, namely, Ferdinand's grandmother, Christina of Lorraine, and his mother, Maria Maddalena of Austria. The excessive piety of these women, particularly the former, led to a dangerous increase in the power and insolence of the clergy, especially after the election of the Barberini pope Urban VIII, who reigned from 1623 until 1644. Ferdinand took charge of the government on July 14, 1628, but Christina survived until 1636, and her bigotry must have contributed greatly to Ferdinand's inability to shake himself free from ecclesiastical domination.

[3] *TT* 1, 93. In spite of this there is still no biography of Leopold, as far as I have been able to ascertain; Gaetano Pieraccini, *La stirpe de' Medici di Cafaggiolo*, 3 vols. (Florence 1924–25), II, 603–32, is useful.

[4] Not to be confused with Cardinal Carlo de' Medici (1596–1666), brother to Cosimo II, and hence an uncle of Ferdinand II and Leopold. He was made a cardinal in 1615.

[5] Athanasius Kircher, *Musurgia universalis sive ars magna consoni et dissoni in X libros digesta*, etc. (Rome, 1650), p. 11.

His failure to do this was strikingly illustrated during the plague of 1630, when he did everything he could to help the populace, and decreed strict sanitary measures for everyone. But the pope threatened to excommunicate Ferdinand's officers for having dared to enforce these rules on the priests and the monks, and the Grand Duke permitted them to be obliged to beg pardon on their knees for this offense against the privileges of the Church.

In this same epidemic, Ferdinand redeemed this moral cowardice by his physical courage, riding out into the city with his brothers—Leopold was twelve at the time!—to cheer up the populace and to supervise measures of relief.[6]

Christina, to whom any disobedience to a pope was a mortal sin, was still alive when Galileo was allowed to go to Rome to answer the charges of the Holy Office. On these grounds Young[7] excuses Ferdinand's failure to defend Galileo.

In 1632 the old Duke of Urbino died and the papal troops at once occupied that city. Although Ferdinand had a good claim to the Duchy of Urbino, he was unable to enforce it, and the result was an undeclared war between the Barberini and the Medici. Indeed the war became a shooting war in 1642, when Ferdinand joined the Venetian Republic and the Duke of Modena in an alliance for the defence of the territories of his brother-in-law Odoardo Farnese, Duke of Parma. The hatred of the Medici shown by Urban VIII, whose main endeavor was to enrich his family in whatever way he could, lasted until his death. He "made every priest and monk in Tuscany an enemy of the Government."[8] It must have been a great relief to Ferdinand when in 1644 Urban VIII died, and was succeeded by Innocent X, who reigned until 1655. Innocent X was friendly to the Medici family, and immediately gave a cardinal's hat to Ferdinand's eldest brother, Giovanni Carlo. He was followed by the Chigi pope Alexander VII, a patron of learning and of art, who died on May 22, 1667. After Urban's death Ferdinand had no trouble from Rome.

An innovation that Ferdinand introduced into the government of Tuscany was to associate his three surviving brothers with him in running the state—a sort of despotism by committee. Mattias was given almost complete control over military affairs; Giovanni Carlo over finance; and Leopold over political matters. The truth seems to be that the Grand Duke was not very fond of the business of governing

[6] Galluzzi, *Istoria* (octavo edition), VI, 30.

[7] Young, *The Medici*, II, 404–6. The bigotry of its womenfolk undoubtedly contributed greatly to the decline of the Medici family. See Harold Acton, *The Last Medici* (London, 1958).

[8] Young, *The Medici*, II, 404.

and preferred his scientific experiments and other amusements. There can seldom have been a more benevolent despot.

Many travellers attest to his generosity and personal charm. I shall quote only one, the Englishman Richard Lassels, who made the Grand Tour in the 1660's:

> The Duke himself also who makes this Court, makes it a fine Court. His extraordinary civility to strangers, made us think our selves at home there. He is now above fifty, and hath an Austrian look and lip, which his mother Magdalena of Austria Sister to the Emperour Ferdinand II. lent him. He admits willingly of the visits of strangers, if they be men of condition; and he receives them in the midst of his audience chamber standing; and will not discourse with them, till they be covered too. Its impossible to depart from him disgusted, because he pays your visit with as much wit as civility: and having enterteined you in his chamber with wise discourse, he will entertein you in your owne chamber too with a *regalo* of dainty meats, and wines, which he will be sure to send you.[9]

Ferdinand made an unfortunate marriage to Vittoria della Rovere, to whom he had been betrothed while still a child. There seems to be no doubt that her excessive piety ruined the character of their eldest son, whose education she superintended and who became Cosimo III.[10] From about 1642 to 1659 Ferdinand and Vittoria lived entirely apart, but were reconciled in 1659, and this resulted in the advent of their second son Francesco Maria. But there seems to have been no real affection, on her part at least, and it is recorded that in Ferdinand's last illness she visited him only once, and then only at his request.[11]

As I have said, Ferdinand found the conversation of learned men much more interesting than the business of government. It would be outside the scope of this book to go into detail about his benefactions to such people, and the reader must be referred to the first volume of Targioni Tozzetti's *Notizie* for a great deal of well-documented information. Let it suffice to point out that no considerations of nationality were allowed to interfere with his judgment. His liberality in these matters greatly increased the prestige of the University of Pisa,[12] soon, alas! to relapse into the most unenlightened ecclesiastical control under his son, Cosimo III.

Ferdinand's own scientific activity seems to have begun before Gali-

[9] Lassels, *The voyage of Italy, or a Compleat Journey through Italy*, etc. [op. posth., ed. S. Wilson], 2 vols. (Paris, 1670), I, 217.

[10] See Acton, *The Last Medici*, and also Young, *The Medici*, for the very sad details.

[11] Gaetano Pieraccini, *La stirpe de' Medici di Cafaggiolo*, II, 508, quotes documentary evidence of this.

[12] See Bianchini, *Dei Gran Duchi*, pp. 100–8.

leo's death in 1642. His thermometer, that makes use of glass balls floating in spirit[13] was almost certainly in use by 1641, and his hydrometers cannot have been much later. Although his quarrel with the Barberini was at its height in the early 1640's, he managed to do, and promote, scientific work.[14] There is abundant evidence that Torricelli (until his early death in 1647), Viviani, and others were associated with this activity, and some writers[15] would like us to believe that a formal academy was set up, meeting in the Pitti Palace, as the Accademia del Cimento did later. We must not dismiss this possibility out of hand, for it would not have had to be very rigidly organized to be as formal as its successor. It is only because no connected minutes have come down to us from this period that we may wonder whether there was any organization at all. Unfortunately only a few of the experiments can be dated; the earliest in this category may be those on natural freezing, made on December 30, 1648.[16]

The Grand Duke invented several instruments of general utility in the experiments of the period, namely the sealed spirit-in-glass thermometer,[17] a condensation hygrometer,[18] and several hydrometers,[19] both weighing and volumetric. The condensation hygrometer, probably invented in 1655, was at first called the "stillatorio."[20]

A number of physical experiments were done by or for the Grand Duke in the years preceding 1657; notably experiments on the speed of sound, made by Borelli and Viviani in October 1656.[21] This interest in physics was maintained into the period that marked the Accademia del Cimento, and I shall suggest later that in 1657 two sets of experiments were often conducted in the palace at the same time. Nevertheless, in the following years his activity in science declined, as his brother Prince Leopold took over the encouragement of scientific work in Tuscany.

Rafaello Caverni denies almost all merit to Ferdinand in scientific

[13] See p. 98 below.

[14] Galuzzi, *Istoria*, VI, 281. Regarding the Barberini, see pp. 19, 21.

[15] Young, *The Medici*, II, 437. Stefano Fermi, *Lorenzo Magalotti, scienziato e letterato, 1637–1712*, etc. (Piacenza, 1903), pp. 77–78. Young's account of the supposed academy would have Niccolò Aggiunti in it, but he had died in 1635, which reduces our confidence in the story.

[16] G. 259, 6r.

[17] G. 269, 227r. See also my *A History of the Thermometer*, etc. (Baltimore, 1966), pp. 27–35; and p. 93 below.

[18] See p. 99 below.

[19] G. 269, 227r–29v.

[20] G. 259, 15v.

[21] Viviani reported on these, probably to Michelangelo Ricci, at some later date (G. 268, 155r–58v; *DIS*, 449–52).

research and the invention of instruments, hinting that the things attributed to him were all done by sycophantic courtiers.[22] But in my opinion there is more than enough evidence to render the probability of this extremely small.

Ferdinand was also interested in astronomy and in biological questions. Targioni Tozzetti quotes Filippo Baldinucci to the effect that in about the year 1642 the Grand Duke had several noted painters try to make drawings of the moon, using a large telescope made by Galileo.[23] He also attracted famous anatomists to his court, among them Marcello Malpighi, whom he appointed to the chair of theoretical medicine at Pisa,[24] Francesco Folli of Poppi, his own physician Francesco Redi, and, in 1666, Nicolaus Steen (Steno).[25]

Of all Ferdinand's brothers, Leopold was the most brilliant and the most closely associated with him. They lived in perfect harmony and had the same tastes. When Leopold had been made a cardinal and spent some time in Rome during the spring of 1668, the two brothers kept up a constant and affectionate correspondence,[26] in which it is evident how much they missed each other's company. As time went on, Leopold conducted more and more of the scientific correspondence, especially with foreigners. Ferdinand's favorite study was experimental physics, while Leopold's was astronomy.

Leopold was born in Florence on November 6, 1617, and died there on November 10, 1675. In his youth he had Iacopo Soldani for his tutor, a firm adherent of the "new philosophy" as taught by Galileo. Later he studied hard at physics and mathematics under Famiano Michelini, who is "justly revered as one of the brightest lights of Italian hydraulics,"[27] and who later became lecturer in mathematics at the University of Pisa. It is, therefore, not surprising that as a young man Leopold busied himself with that endemic Italian problem, the regulation of rivers.[28]

At the age of nineteen Leopold was sent to Sienna as governor, in place of his brother Mattias, and held the post until 1641, and again from October 1643 until the end of 1644. He seems to have ruled wisely and well. The rest of his life, except for two visits to Rome in

[22] Caverni, *Storia del metodo sperimentale in Italia*, 5 vols. (Florence, 1891–98), I, 186. Re Caverni, see p. 15 above.

[23] *TT* 1, 250–51. For Ferdinand's astronomical interests, see *TT* 1, 251 ff.

[24] Malpighi, *Opera postuma . . . quibus praefixa est ejusdem vita à seipso scripta* (London, 1697), p. 2.

[25] Lorenzo Magalotti to Alessandro Segni, Aug. 24, 1666. Printed by Fernando Massai, *Rivista delle Biblioteche e degli Archivi* 29 (1918): 41–42.

[26] *FAS*, filza Medici del Principato 5508.

[27] *G.OP.*, XX, 494.

[28] *TT* 1, 372.

1668 and in 1670, was spent in the valley of the Arno with the court, at Florence for the most part, but often at Pisa, Artimino, and other places nearby.

There can be no doubt whatever that he was much more than a spectator in the scientific activities going on at the court. A remarkable example of his acumen is his response to a suggestion of Borelli about the cause of the fall of the mercury in the Torricellian tube in wet weather and its rise when the skies are clear. This was, in fact, the hypothesis that the winds might pile the air up in some places, taking it from others, an idea that was about a century and a half ahead of its time.[29] Nor was he unwilling to suffer discomfort in the cause of science, as is shown in a letter to his secretary Alessandro Segni, dated February 5, 1666, in which he expresses his annoyance at the "gentlemen who are helping with the Dictionary"—i.e., the *Vocabolario della Crusca*.[30]

I have provided in the usual room at the Sanità for those gentlemen who are helping with the Dictionary, so that they will lack the excuse of a cold room. And I do not believe that any of them would have passed several evenings, as I have done, observing the freezing of water at midnight in the open air. As to this, I realize that it is necessary to make many experiments, because in the few that I have made this year, unexpected and astonishing things have been recognized, which we have not yet begun to explain.[31]

Leopold was on very friendly terms with his secretaries, Alessandro Segni and (later) Lorenzo Magalotti, and wrote them lively, cordial, and familiar letters, at the same time insisting that they should carry out his often detailed instructions. Camerani published a set of instructions for the journey of Segni and Riccardi, 1665–66, which take up nearly four pages of 8-point type. They concern the people they are to greet and the books they are to buy, and end as follows: "Make friends with nobody while you are travelling, not even when you have stopped in cities; and if you deem some inhabitant or other person worthy of your conversation, you may invite him once to lunch or dinner, but never make friends with anyone."[32]

Not content with science, Leopold patronized the arts and collected paintings, sculpture, and objets d'art; in fact a fair proportion of the treasures that we admire today in Florence were assembled by Fer-

---

[29] I have dealt with this episode in *A History of the Theories of Rain* (London, 1965; New York, 1966), pp. 66–67.

[30] This letter was published by Sergio Camerani, *Arch. Stor. Ital.* (1939, part I), p. 37. At this time Segni was in Paris.

[31] On these experiments see also p. 199 below.

[32] *Arch. Stor. Ital.* (1939, part I), pp. 32–35.

dinand II and his youngest brother. He also collected a splendid library, a catalogue of which is to be found in the State Archives in Florence.[33] This shows the breadth of his interests; but it is remarkable that the catalogue contains few scientific books of the seventeenth century. It also seems that he was very generous in lending books from his library, for in this catalogue there is a long list of missing volumes.[34]

The reader should be warned against the facile conclusion that the contents of this library only show what was presented to the prince by virtue of his high position. A fair proportion of his correspondence with the Tuscan agents abroad is devoted to orders for specific works that could not be obtained in Florence, and these cover almost the whole field of knowledge.

One of Leopold's interests that may surprise us is his attraction towards Jansenism, the Catholic revivalist movement that was finally destroyed by Louis XIV and the Pope in 1713. Regarding the beliefs of the Jansenists, it will be sufficient to say here that they demanded a more personal and less authoritarian kind of Catholicism, a point of view that naturally brought them into violent conflict with the Jesuits. Bronowski and Mazlish[35] make the interesting point that they believed in simplicity of style, just as did the young Royal Society.[36] They claim that the Jansenist influence was strong in the group that collected around Father Mersenne in Paris,[37] and compare this to the Puritan influence in the group or groups from which the Royal Society took its rise. It is not unreasonable to suppose that in another place or time, Leopold would have been a distinguished freethinker.

Distinguished he most certainly was, among the princes of his age and his country. He deserves study. In his book about the dynasty,[38] Young remarks on the strange absence of a biography of the last member of the family who showed its characteristic and exceptional ability.

Leopold was painted several times by Sustermans, the "official painter" of the Medici at this epoch; but I have chosen to reproduce (Plate 1) the excellent canvas by G. B. Gaulli, called Baciccio, in the Uffizi Gallery, which shows him as cardinal, perhaps sad and certainly in poor health and prematurely old. It was painted in 1670, or thereabouts. There is also a portrait bust by an unknown sculptor in the

[33] *FAS*, filza Medici del Principato 5575a.

[34] See also Pieraccini, *La stirpe*, etc., II, 615.

[35] J. Bronowski and B. Mazlish, *The Western Intellectual Tradition from Leonardo to Hegel* (London, 1960), p. 258, note 9.

[36] Thomas Sprat, *The History of the Royal Society of London, for the Improving of Natural Knowledge* (London, 1667), p. 113.

[37] See p. 8 above.

[38] Young, *The Medici*, II, 435.

Uffizi, not without merit but emphasizing the enormous nose and the somewhat prognathous jaw to an extent that, as Antonio Archi remarks,[39] causes one to suspect an attempt to caricature.

PLATE 1.   Cardinal Leopold de' Medici, from the portrait by G. B. Gaulli (*Uffizi Gallery, Florence, by permission*).

---

[39] A. Archi, *Il tramonto dei principati in Italia* (Rocco San Casciano, 1962), p. 165.

Leopold's tomb, in the crypt of the Medici Chapel in Florence, is covered with a stone bearing the following inscription:

H•S•E
LEOPOLDVS
COSMI•II•M•D•ETR•QVARTI•F
SENARVM•PRAEFECTVRA
FLORENTI•AETATE•PERFVNCTVS
DEIN•CARD•S•ECCL•ROM
QVO•AVCTORE•ET•AVSPICE
SODALES•FORENTINI•AB•EXPERIMENTO
AD•PHYSICAS•DISCIPLINAS•PROVEHENDAS
ANNO•MDCLVII
PRIMO•OMNIVM•CONSPIRAVERE
O•IIII•IDVS•NOV•A•MDCLXXV
QVVM•ESSET•ANNOR•LVIIII
SACRAE•PVRPVRAE•HONORE•VSVS•ANN•VII

It is interesting that his successors should in this way have recognized his Academy as his greatest achievement.

2. THE ACADEMICIANS

The Accademia del Cimento was conducted in such a private and informal way that it is even difficult to be sure that we know the names of all who should be considered to have belonged to it. In this it contrasts with the first Accademia dei Lincei, where Federico Cesi delighted in rules and regulations.[40] The only thing we can do, as Targioni Tozzetti points out,[41] is to restrict ourselves to those who are named in the diaries or other writings of the Academy, and on this basis he provides the following list:

Giovanni Alfonso Borelli (1608–79)
Candido del Buono (1618–76)
Paolo del Buono (1625–59)
Alessandro Marsili (1601–70)
Francesco Redi (1626–97)
Carlo Rinaldini (1615–98)
Antonio Uliva (d. 1668)
Vincenzio Viviani (1622–1703)

---

[40] See G. Gabrieli, "Carteggio Linceo della vecchia Accademia di Federico Cesi (1603–1630)," *Mem. R. Acc. Naz., Lincei, Classe sc. mor.*, ser. 6, vol. 7 (1938–42), parts 1 to 3; especially pp. 229–30.
[41] *TT* 1, 418–19.

and as secretary,

Alessandro Segni (1633–97)

replaced on May 20, 1660, by

Lorenzo Magalotti (1637–1712)

The name of Carlo Dati (1619–76) does not appear in this list, although it is certain that he proposed a number of experiments that were performed by the Academy[42] and was privy to the discussions about the appearance of Saturn that went on in 1660.[43] On the other hand Paolo del Buono, who was in Austria and Poland from about 1655 until his early death in 1659, ought surely to be considered one of the correspondents, among Ottavio Falconieri and Michelangelo Ricci in Rome, Melchisadec Thevenot in Paris, and perhaps Geminiano Montanari in Bologna; but Vincenzio Antinori states without documentation that Prince Leopold wrote to Paolo in Vienna, expressly appointing him a member.[44] It is certain, at any rate, that Paolo took the Academy very seriously and wrote to Borelli, probably in September, 1657, proposing laws for it. Borelli's answer[45] exists; on October 10 he wrote, "First, about our Academy . . . I wish that the laws that you imagined might be established in it, but the worst of it is, that only disorder is to be found there . . ." and a great deal more.

This letter, written at a moment when Borelli was angry with one of his fellow Academicians, expresses as much about the character of its writer as about the Academy, and will serve to introduce some notes about Giovanni Alfonso Borelli, who was not only first alphabetically among the Academicians, but, as far as experimental science is concerned, easily its most active and distinguished member.

Giovanni Alfonso Borelli was born at Castelnuovo in Naples on January 28, 1608. He was the son of Michele Alonzo, a Spanish soldier, and an Italian woman, Laura Porello.[46] Baptized Giovanni Francesco Antonio, he later adopted the Italian form of his father's surname as a middle name, dropped the "Antonio," and changed his surname to Borelli, a corruption, or perhaps an upgrading, of Porello. As a young man he went to Rome and studied under Benedetto Castelli. Torricelli and he were fellow students. From 1635 to 1656 he occupied the chair of mathematics at the University of Messina and was then ap-

---

[42] *TT* 1, 447.

[43] See chapter 5.

[44] Antinori, in his introduction to the 1841 edition of the *Saggi*, p. 78.

[45] *LIUI*, I, 94. Quoted in *TT* 1, 440–42.

[46] For a short, fairly well documented life see Gustavo Barbensi, *Borelli* (Trieste, 1947). Borelli, probably because he had been treated very well in Messina, sometimes claimed to have been born there.

pointed by Ferdinand II to that position at the University of Pisa, where he remained until 1667, taking a prominent part in the experiments of the Accademia del Cimento and making or superintending the observations of weather at Pisa as part of Ferdinand's extensive meteorological network.[47] On his arrival at Pisa he did not want to give an inaugural lecture, being anxious to "get on with the job."[48] Nevertheless, he was obliged to give the lecture, but it was not a success and got little applause, apparently because of his inelegant style.[49] His regular lectures seem to have attracted good audiences,[50] so it must be assumed that his students soon came to esteem his great learning and quick intelligence. He was extremely versatile, writing on mathematics, astronomy, physics, hydraulics, and—though he was not a medical man[51]—physiology and medicine. The best known of his works are the *De vi percussionis* (Bologna, 1667), the *De motionibus naturalibus a gravitate pendentibus* (Reggio Giulio, 1670), and the posthumous *De motu animalium*, 2 vols. (Rome, 1680). It is pleasant to know that in the midst of this enormous activity he could occasionally relax; in a letter to Prince Leopold, in May 1665, we find him recommending a comedy by Susini.[52]

Unfortunately, his great intelligence was accompanied by an intolerance of criticism and a proud and quarrelsome disposition. His ill-nature seems to have been legendary. We find Magalotti writing to Viviani from Rome on December 4, 1661, that "Borelli . . . has it for a maxim that one cannot be a man and not be evil."[53] Again, Magalotti writes to Falconieri on December 1, 1665, and refers to a book by "Bellini, a young man of 22 or 23, very clever and studious. He is a pupil of Borelli, but up to now none of his master's ill-nature has stuck to him."[54] Even his constitution was thought to be unusual, for it was reported that a high-cut diamond that Leopold used to wear in a finger-ring would not become electrified when stroked on Borelli's skin, but became strongly so when stroked on the skin of many other persons.[55]

[47] Angelo Fabroni, *Vitae Italorum doctrina excellentium, qui saeculis XVII & XVIII floruerunt*, 20 vols. (Pisa, 1778–1805), I, 236. See also *TT* 1, 208–9.

[48] Filippo Magalotti (the "bursar" at Pisa) to Prince Leopold, March 8, 1656; in *LIUI*, I, 85–87.

[49] F. Magalotti to Leopold, March 20, 1656 (*ibid.*, I, 87–89). In *DIS*, 21, note 1, it is stated that his inaugural lecture could not be completed because his audience were so annoyed at his delivery.

[50] Borelli to Leopold, April 12, 1656 (*ibid.*, I, 90–92). The autograph original is in G. 275, 42r–v.

[51] It has been suggested, and denied, that he took a medical degree at Rome in his youth.

[52] G. 277, 179r.

Naturally Borelli did not get along with all his colleagues in the Academy. Apart from his spiny character, it is more than likely that his frenetic energy annoyed them by making them feel that they were almost standing still most of the time. He always knew the answers, or managed to give the impression that he did. He seemed particularly obnoxious to Vincenzio Viviani, a competent mathematician who made a good deal of capital out of having been Galileo's last pupil, but who was plagued by ill-health and was less energetic. I shall come back to the relations between these men in a later chapter.

In fairness to Borelli, it should be noted that the Roman mathematician and churchman Michelangelo Ricci esteemed him and, on April 4, 1667, wrote to Prince Leopold that he was sorry that Borelli had left Rome on his way to Messina. "He is a great scholar, and a good old friend of mine; and I don't like to have him so far away. It is almost like losing him, because of the difficulty of correspondence."[56] His mutually stimulating friendship with the anatomist Marcello Malpighi shines through the many letters preserved in the library of Bologna University. He was also very conscientious, as is shown by a letter written on March 26, 1663, to an unidentified priest. After the priest had left Pisa, Borelli thought of a defect in an experiment that they had seen performed for the Prince and carefully explained what had been wrong with it.[57] Nor was he ungrateful, for long after he had left Tuscany, upon receiving the congratulations of Prince Leopold, now a cardinal, on his *De motionibus naturalibus a gravitate pendentibus* (Regio Giulio, 1670), he began a letter[58] thus:

I received the most blessed letter from Your Most Reverend Highness, in which you condescend to thank me for what Your Highness has, one might say, generously given me, inasmuch as no small part of the notions and speculations contained in that book of mine had their basis in the experiments made in Your Highness' academy, without which I should not have been able to complete such a work.

For one reason or another, Borelli became disenchanted with the service of the Tuscan court and, in March, 1667, asked leave to return to Messina, which was granted. He was greeted there with honor, but in 1674, after an uprising in the city, was accused of sedition and obliged to take refuge in Rome, where he was received into the Acad-

[53] In *Delle lettere familiare del Conte Lorenzo Magalotti*, etc., 2 vols. (Florence, 1769), I, 5.
[54] *Ibid.*, pp. 148–49.
[55] *TT* 2, 552.
[56] *LIUI*, II, 156.
[57] *G.* 276, 188r.
[58] Borelli to Leopold, from Messina, April 13, 1671 (*G.* 279, 49r).

emy of Queen Christina of Sweden, who had become a Catholic and was living there. Finally, because of a theft by one of his servants, he had to retire to teaching in a religious academy. He died on December 31, 1679.[59]

Two brothers, Candido and Paolo del Buono, were members of the Accademia del Cimento. Little seems to be known of the life of Candido, except that he was born in Florence on July 22, 1618, and died at Campoli, where he was the parish priest, on September 19, 1676.[60] He appears to have learned mechanics from Galileo[61] and to have suggested several experiments to the Academy. On the strength of a categorical statement by Magalotti, Targioni Tozzetti ascribes to him the design of an ingenious mounting for a very long telescope made by Campani,[62] but there were four del Buono brothers, and Prince Leopold sent a drawing of this mounting to Christiaan Huygens on June 15, 1661, saying that "a way of making a telescope occurred to Antonio Maria, brother of the late Paolo del Buono" and that it is very elegant and easy to operate.[63] The social position of the del Buono brothers is obscure, the editors of Huygens' Oeuvres stating that Antonio Maria was an instrumentmaker. We shall hear of him again on page 258.

Rather more is known about Paolo del Buono. Born in Florence on October 26, 1625, he appears to have been a pupil of Galileo near the end of that great man's life. The evidence for this is contained in a letter from the mathematician Antonio Nardi to Galileo, dated from Rome, Sept. 7, 1641. After setting out a theorem in geometry, Nardi writes, "You will excuse these trifles of mine, which I have the impudence to send you, not to occupy your own time, but to give a little amusement to Messrs Paolo del Buono and Viviani, to whom I beg you to give my regards."[64] He later studied with Famiano Michelini at Pisa, where he obtained a degree in 1649. In 1655 he went across the Alps and entered the service of the Emperor Ferdinand III, becoming his "Master of the Mint" and doubling as a mining engineer. As far as is known he never returned to Italy, but

[59] The circumstances of his death are given in some detail by Giovanni Viovannozzi in Atti Pontif. Accad. Romana dei Nuovi Lincei 72 (1919): 81–82. For a list of his published works see DIS, 21–22, note 1.

[60] DIS, 20, note.

[61] TT 1, 435.

[62] TT 1, 436; it is described and illustrated in TT 2, 799–800 and plate IX.

[63] Oeuvres complètes de Christiaan Huygens publiées par la Société Hollandaise des Sciences, 22 vols. (The Hague, 1888–1950), III, 130. The folding plate agrees with Targioni Tozzetti's plate IX.

[64] G.OP. XVIII, 352.

died at the Polish court in the latter part of 1659.[65] It is, therefore, surprising that he is considered a full member of an Academy, none of whose meetings he can have attended. Several of his letters, both to the Grand Duke and to Leopold, have been preserved,[66] as well as a copy by Boulliau of an undated copy of a long letter from Paolo to Prince Leopold, which on internal evidence must have been written in late August or September 1657.[67] I very much doubt whether this letter, the original of which has not been found, ever reached its destination; which is a pity, because it contains a clear description of the experiment which, in other hands, led to Boyle's law, but which Paolo could not make for want of the necessary glassware.[68] One gets the impression that Paolo del Buono might have been of great use to the Academy and might have lived longer if he had come back to Florence.

Carlo Roberto di Cammillo Dati, a Florentine gentleman, was born in 1619 and lived until 1676. He was better known as a literary man than as a scientist, but his knowledge of at least some science is attested to by the little book that he published pseudonymously in 1663 in defense of the memory of Evangelista Torricelli, which is noteworthy for the first publication of the famous letters of 1644 describing the Torricellian experiment.[69] He was reader in humane studies at the Studio in Florence and an officer of the Accademia della Crusca. His influence in the Accademia del Cimento was slight, though he may have been one of those who examined drafts of the *Saggi*.

Lorenzo Magalotti was born in Rome of noble Florentine parents on December 13, 1637, and died on March 2, 1712, in Florence, the last survivor of the Academy.[70] Between the ages of thirteen and eighteen he studied at a Jesuit seminary in Rome, having as his master Antonio Uliva, later professor of medicine at Pisa, where Magalotti

[65] Our evidence for this is in a letter written to Prince Leopold by the French astronomer Ismael Boulliau (Bullialdus), who had relations with Warsaw. On December 19, 1659, he writes of Paolo del Buono as "nuper defunctus in Poloniae Regis Aula." The letter is printed in *LIUI*, I, 198–202. The original is in G. 275, 169r–70r.

[66] G. 275, 38r–39v, 78r–79v; G. 285, 20r–21r; G. 286, 24r–25r, 26r–27r, 30r, 31r, 32r.

[67] *BN*, fonds français, 13039, fol. 126r–40v.

[68] I have published this document, with comment, in *Archive for History of the Exact Sciences* 6 (1969): 1–28.

[69] *Lettere a Filaleti di Timauro Antiate della vera storia della cicloide e della famosissima esperienza dell'argento vivo* (Firenze, 1663).

[70] See Stefano Fermi, *Lorenzo Magalotti, scienziato e letterato, 1637–1712*, etc. (Piacenza, 1903).

went in 1656 to study law. He kept at this for only sixteen weeks, according to his own account, and left the law because of his interest in science, studying for three years under Vincenzio Viviani, who became his lifelong friend. The young man must have been extraordinarily intelligent and quick to learn, and undoubtedly displayed a virtuosity that was unusual even in those days. As he matured, however, it became clear that his real love was literature, and he grew to be very critical in matters of style.

When the Accademia del Cimento began meeting, in 1657, with Alessandro Segni as its secretary, Magalotti was a very learned young man of nineteen. He took a small part in the activities of the Academy until May 20, 1660, when, on the advice, it is said,[71] of Viviani, Leopold replaced Segni with Magalotti as its secretary, a choice that has been adversely criticised, and certainly led to a disastrously long delay in the publication of the *Saggi*.[72] A few months before the completion of the printing, Magalotti was sent by the Grand Duke, together with Paolo Falconieri, on a long voyage in northern Europe, including England, where they attended some meetings of the Royal Society and formally presented a copy of the *Saggi* on Leopold's behalf at the meeting on March 12, 1668.[73] They returned to Paris, but not before visiting Robert Boyle at Oxford, where they were most cordially received. Ferdinand II ordered them back from Paris to accompany his son Cosimo (later the Grand Duke Cosimo III) on a long journey around Europe; and from this time until about 1678 Magalotti was employed on diplomatic missions. In 1689 Cosimo III made him his Third Counsellor of State.

Lorenzo Magalotti seems to have been a most attractive man who, although immensely erudite, enjoyed life to the full. Fabroni tells us that besides being very good-looking, he was an excellent dancer and horseman, so that he seemed born to be a courtier.[74] His letters to Prince Leopold, then a cardinal, from England and France in 1668, are delightful. I shall quote from one of these, written from Paris on June 29 of that year:

At the Royal Academy [of Sciences], which meets on Saturdays at the house of Monsieur Carcavi, His Majesty's librarian, I have found nobody who has offered to introduce me, and I have not recommended myself for admission at all, my ambition being extremely moderate in that direction.

[71] Girolamo Tiraboschi, *Storia della letteratura italiana*, etc., 9 vols. in 20 (Florence, 1805–13), VIII, 244.

[72] See p. 294 below.

[73] Thomas Birch, *History of the Royal Society*, 4 vols. (London, 1756), II, 256. See also chapter 6 below.

[74] *Delle lettere familiari*, vol. I, p. XIX.

I could never tell Your Highness how prejudicial it is for a man of fashion from that side of the mountains to pass for a philosopher and mathematician. The ladies at once believe that he must be enamored of the moon, or Venus, or some silly thing like that. To show that this is true, Dr Bernadin Guasconi,[75] who is no fool, began to be jealous of me in the house of a certain widow in London, where he had introduced me. He merely told the lady that I was a philosopher, and it was as so much poison to me, for from that time on I was considered a platonic lover, and in consequence I could bring nothing to a conclusion other than to admire *the high first cause* in her external beauty. I know that with your infinite tact Your Highness will not be scandalized at all by this too circumspect warning of mine, if perchance the devotions of Rome have not made you take a more rigorous line towards the fragility of others.[76]

Apart from his work as secretary, which involved the writing of the *Saggi*—an occupation to which he did not take kindly, as we shall see[77]—his contribution to the experiments of the Academy was sporadic and rather dilettante. His natural ebullience, which was probably interpreted as superficiality, is illustrated by an undated note, most likely to Viviani, about an attempt that he had made to "refract cold" through a glass lens.[78] At the top of the sheet he has written in enormous fancy letters, "E viva gl"ATOMI FRIGORIFICI"—long live the atoms of cold! Nevertheless, he seems to have had a sound knowledge of the science of the time, even though he advanced it hardly at all.[79]

Alessandro Marsili was born in Sienna in 1601 and died there in 1670. He took a degree in law in 1622 and in philosophy a year later, and afterwards occupied chairs of logic and of philosophy at the University of Sienna. In 1638 he became professor of philosophy at Pisa. It appears that Galileo thought highly of him, for he praised him extravagantly in a letter to Leopold dated March 31, 1640;[80] Galileo had evidently been influential in getting him the chair at Pisa.[81] It seems that they had been together a great deal when Galileo had passed some time in Sienna in 1633, and it is clear that Marsili had

[75] A Florentine who had settled in London as Bernard Gascoigne, and was knighted and elected F.R.S.

[76] G. 278, 185v–186r. Published in *LIUI*, I, 308, in which the italics were added by the editor.

[77] In chapter 4.

[78] Florence, Bibl. Riccardiana, cod. 2487, fol. 7r. The Academy did a rather similar experiment in September 1660.

[79] For an interesting appraisal of Magalotti's complex character, see Emilio Cecchi in *Il sei-settecento* (Florence, 1956), pp. 147–66.

[80] *G.OP.*, VIII, 542.

[81] *Ibid.*, 496.

been a good listener. Writing to Galileo on October 11, 1636, he "confessed" having learned more in those few months with Galileo than in all his studies with others, etc.[82] His membership in the Accademia del Cimento may be explained, at least partly, by this recommendation by Galileo, for in spite of all he had learned, Marsili was in no way freed from his peripatetic prejudices. Borelli loathed him and could have been referring to no other of the Academicians when he wrote of a "rotten and mouldy peripatetic" among their number.[83] And because, as Borelli added, "the bottle pours out the wine that it is filled with," Marsili's contribution to the Academy was unimportant.

Francesco Redi, chief physician to Ferdinand II and later to Cosimo III, was a distinguished poet as well as a medical man and zoologist of great repute. He was born in Arezzo in 1626 and died at Pisa in 1694. Targioni Tozzetti thought that there was some doubt that Redi was really a member of the Academy,[84] but we have two published letters that make it seem probable. The earlier of these was written to Michele Ermini on April 25, 1659: "I wanted to come today to wish you a good journey, but it wasn't possible, because today the usual meeting of the Accademia del Cimento was held, so that I am taking this way out, wishing it by letter...."[85] The trouble with this is that its date is in the long period between September 1658 and May 1660, during which no meetings were recorded, or at least no records of them have survived, as they would have had they been entered in the great diary, of which we have a copy.[86]

On May 9, 1660, he wrote to Carlo Dati, also mentioning the Academy—"Nell'Accademia del Cimento si lavora"—[87] but this was still in the long holiday. We must reserve judgment, with the remark that he would certainly have been an able and useful member.[88]

This description also fits Carlo Rinaldini (spelled Renaldini in the titles of his books, all of which are in Latin), born in Ancona in 1615 to a family that originated in Sienna. After serving as a military engineer in the forces of Urban VIII and Innocent X, he became senior professor of philosophy at Pisa, and also taught mathematics to the young prince, Cosimo. In 1667, on the pretext that the climate of

[82] *Ibid.*, 502.

[83] Borelli to Paolo del Buono, Oct. 10, 1657, quoted in *DIS*, 20, note 2. The letter is also in *LIUI*, I, 94–100.

[84] *TT* 1, 450–51.

[85] *Opere di Francesco Redi gentiluomo aretino*, etc., 7 vols. (2nd ed.; Naples, 1740–41), IV, 22.

[86] *G.* 262. See p. 11 above.

[87] *Opere di Redi*, IV, 22.

[88] See also p. 259 below.

Pisa did not suit him, he obtained permission to depart and took up the chair of philosophy at Padua in December of that year at about twice his former salary, which may explain his preferences as to climate. He stayed there thirty years, and retired to Ancona shortly before his death in 1698.

Rinaldini wrote several ponderous books in Latin, including a "huge and frightening tome"[89] called *Philosophia rationalis, naturalis, atque moralis* [etc., etc.], which has 1,088 numbered folio pages and many unnumbered ones. This was to have been only "Part I," and was published at Padua in 1681. "I have not," says Targioni Tozzetti, "succeeded in finding in it any notice of what Rinaldini had been doing while he was in Tuscany."[90] His books were quickly forgotten, but it should be put to his credit that he was the first to have the courage to lecture at Pisa on the works of Galileo, which had been condemned by the Holy Office, and on the atomistic philosophy of Gassendi, which was suspect because of its derivation from Epicurus.[91] To the Academy he proposed an experiment to determine whether heat diffuses spherically. I have shown elsewhere[92] that his interpretation of this experiment makes him the discoverer of convection in air. He was also one of those chosen to examine the draft of the *Saggi*.[93]

Alessandro Segni was born in Florence in 1633 and died there in 1697. As he was Prince Leopold's secretary in 1657, he was naturally pressed into service to record the minutes of the meetings of the Academy. But although he had been a pupil of Torricelli, he preferred literary studies and made no ascertainable contribution to the Academy's scientific activities. In 1662 he is found in the "Civil List" of the Grand Duke with the title of "Librarian in the service of Prince Cosimo."[94] Later he travelled extensively in Leopold's service, and in 1674 became the Prince-Cardinal's chamberlain and superintendent of his secretariat, which explains how the papers of the Academy came into his possession, as we saw in chapter 1.

The strangest of the Academicians was Antonio Uliva (or Oliva), "a man," says Tiraboschi, "to tell the truth, not very worthy to belong to that famous circle."[95] He was a native of Reggio di Calabria in the far south of Italy, and the date of his birth is unknown. Early in life

[89] *TT* 1, 347.
[90] *Ibid.*
[91] On Gassendi see J. S. Spink, *French Free Thought from Gassendi to Descartes* (London, 1960), *passim*.
[92] *Physis* 10 (1968): 299–305.
[93] See p. 68 below.
[94] *TT* 1, 447.
[95] *Storia della letteratura italiana*, ed. cit., vol. VIII, p. 246.

he held the post of theologian to Cardinal Francesco Barberini, but was dismissed because of his irregular habits, and went back to Calabria and became an outlaw, passing some time in prison. On his release he went to Tuscany, and in 1663, having ingratiated himself with the Grand Duke and with Leopold, was actually made professor of medicine at Pisa. He was apparently liked at court on account of his quick intelligence, but he appears to have been entirely undisciplined.

He seems to have accomplished nothing at the Accademia del Cimento, apart from assisting Borelli with a few experiments,[96] though he announced that he was working on a book on the nature of fluids, and Leopold asked Viviani to sacrifice a work of his own on the subject, so as not to enter into competition with Uliva.[97] It appears from an undated letter from Viviani to Dati that Leopold was in deadly earnest about this.[98] Nothing of the treatise has been found except a set of not very illuminating headings,[99] and it is only conjectured that these are by Uliva. Yet Leopold, writing to Ricci, said that he himself had seen a large part of Uliva's treatise.[100]

In 1667, for reasons that are not clear, Uliva left Tuscany and went to Rome to practice medicine, but was arrested for conduct that seems to have been scandalous even in seventeenth-century Italy, and was tortured. Fearing worse, he threw himself from a window and died of his injuries.

Finally, there was Vincenzio Viviani, whose conduct, unlike that of Uliva, was exemplary. He was born in Florence in 1622 and began to study mathematics at the age of sixteen.[101] In 1639 he went to help the aging and blind Galileo, and ever afterwards referred to himself as "Galileo's last pupil." His devotion to the memory of the great man largely directed the course of Viviani's life until his death in Florence in 1703. While he was with Galileo he got to know Torricelli and later became his pupil; he was, in fact, the "laboratory assistant" in the famous experiment that led to the invention of the barometer.

[96] See for example G. 277, 63r (Borelli to Prince Leopold, December 22, 1664).

[97] *TT* 1, 427.

[98] *Ibid.*, 434–35. The original is in the library of Pisa University, ms. 70, fol. 164r–65r.

[99] G. 268, 173–74.

[100] *LIUI*, II, 130. The letter is undated, but was evidently written in June 1665.

[101] See Fontenelle's "Éloge," read to the Académie des Sciences in Paris on April 2, 1704, in *Hist. Acad. R. Sci. Paris* (1703): 137–48. This volume was printed only in 1706.

At this time Viviani steeped himself in the mathematics of antiquity and formed the idea of restoring, or imagining, the lost five books of Aristaeus the elder on conic sections, and also the lost books of Apollonius of Perga. For fifteen years this occupied what leisure he could find from his employments at the Court, which were rather onerous. In 1656 he was surprised to hear that Borelli had discovered, in the library of the Grand Duke, an Arabic manuscript entitled *Apollonii Pergaei Conicorum libri octo*, and this caused him to take steps to prove that he had never seen this and knew no Arabic. Although Borelli got permission to have this translated by the Roman scholar Abramo Ecchellense, it was not published until 1661.[102] Viviani's "divination" appeared in 1659,[103] was found to be more than just guesswork, and received great praise.

During the next few years Viviani was employed by the Grand Duke as "consulting engineer" on the never-ending problems of flood control, until in January 1666 he was excused from this and given a small salary so that he could concentrate on his studies.[104] But before this, through the influence of the Italophile writer Jean Chapelain,[105] both he and Carlo Dati had been given pensions by Louis XIV. The official letter from the king's minister Colbert to Viviani, transmitting a letter of credit, is dated November 14, 1664,[106] and its conclusion contains a broad hint of the real reason for such pensions: "And if you will strive to give the public something to the glory of such a great prince, although you may be pledged to this by any other motive than self-interest, it will be a certain means of obliging him to continue [the payments] in the future."

Louis had to wait a long time for this contribution to his glory; Dati, at the suggestion of Segni, dedicated his *Vite dei pittori antichi* to him in 1667, but Viviani did not manage anything until 1701, when he published his guess at the lost work of Aristaeus.[107] Nevertheless,

[102] *Apollonii Pergaei conicorum lib. V VI VII paraphraste Abalphato Aspahanensi nunc primum editi*, etc. (Florence, 1661).

[103] Viviani, *De maximis et minimis geometrica divinatio in quintum Conicorum Apollonii Pergaei adhuc desideratum*, etc. (Florence, 1659).

[104] Magalotti to Segni, from Florence, January 15, 1666, published by F. Massai, *Riv. delle Biblioteche e degli Archivi* 28 (1917): 134–38. But it appears that Viviani did not get the salary until the 15th of May; the decree is in *LIUI*, II, 23.

[105] This is clear from the correspondence between Marucelli, the Grand Duke's resident in Paris, and Viviani. See, e.g., G. 162, 147r–48r; see also *Lettres de Jean Chapelain*, ed. Ph. Tamizey de Larroque, 2 vols. (Paris, 1880 and 1883), II, *passim*.

[106] G. 162, 227r–28r.

[107] *De locis solidis secundo divinatio geometrica in quinque libris iniuria temporum amissos Aristaei senioris geometrae* (Florence, 1701).

the king continued his pension for years, many of which Viviani, plagued by poor health, seems to have passed in collecting materials for a life of his adored Galileo. In 1696 he was made a foreign member of the Royal Society, and in 1699 *associé étranger* of the Académie des Sciences. At this time he was offered the post of chief astronomer in Paris, which he wisely declined, as he did a similar offer from King Casimir of Poland.[108]

Viviani seems to have been a quiet and courteous man, magnanimous and generous and unwilling to pick a quarrel. At least in his earlier years he must have been a free-thinker, if we can believe the traveller Balthasar de Monconys, who met him on November 6, 1646, and recorded in his *Voyages* that "Viviani . . . told me his opinion of the sun, which he believed to be a fixed star, the necessity of all things, the non-existence of evil,[109] the participation of the Universal Soul, the conservation of all things."[110] Nevertheless, he was fundamentally an antiquarian. In the words of Luigi Tenca,[111] "He lived *outside of his own epoch*; he understood the great men of the golden age of Greek civilization as if he had known them personally; he did not understand and *was not conscious of* the new infinitesimal analysis that had its two illustrious protagonists in his own day, Isaac Newton and Gottfried Leibniz. It is a strange spiritual phenomenon."

As to his work in the Accademia del Cimento, there is no doubt whatever that Viviani and Borelli were responsible for by far the greatest part of the solid accomplishments of the Academy. It is never easy to decide between the claims of these two men, so utterly different in temperament, and it is more than likely that on many occasions, thinking about the same problems, they came up with the same answers. Viviani knew very well that it would not do to be a shrinking violet, not in the Tuscan court, and there exists a sheet of paper in his execrable but unmistakable handwriting[112] in which he flatly claims about a score of instruments and experimental ideas, more than half of which appear in the *Saggi*.[113] In various places in G. 268, especially fol. 5r–6r and 43r–v (*DIS*, 398 and 408), he claims dozens of others, as he does in G. 263 (*DIS*, 396–97). The historian G. B. Nelli asserted that he could prove from the documents that most of the electrical ex-

[108] See Pierfrancesco Tocci, in *Le vite degli Arcadi illustri, Parte Prima* (Rome, 1708), pp. 123–34.

[109] *la nullité du mal*; perhaps he was denying "original sin."

[110] Monconys, *Voyages*, 2 vols. (Lyons, 1666), I, 129.

[111] L. Tenca, *Rendiconti Ist. Lombardo Sci. e Lett.* 90 (1956): 109. The italics are from the original.

[112] G. 269, 259r–v.

[113] I have tried to mention all these claims in the notes to the translation of the *Saggi* in chapter 4.

periments made by the Academy were devised by Viviani,[114] and I have the impression that he was right. Viviani's "laboratory notes" and rough sketches in G. 269 show a very active mind. He even produced a list of twenty-four experiments to be performed in a "great vacuum," into which he thought a man might be introduced without danger![115] If these had been carried out, he would have been a pioneer in "space research," but fortunately they were not. Add to all this that he was a good mathematician and astronomer[116] and it will be clear that he was a tower of strength to the Accademia del Cimento.

In addition to those we have mentioned, who, with the possible exception of the excellent Redi, were regular members of the Academy, there were a few people outside Tuscany whose correspondence with Prince Leopold and some of the others was so voluminous that they ought probably to be considered corresponding members. These were Michelangelo Ricci and Ottavio Falconieri in Rome, Melchisadek Thevenot in Paris, and perhaps Geminiano Montanari in Bologna. Falconieri seems to have been sent across the Alps "in servizio del Accademia filosofica"[117] as early as 1658. These are the four mentioned by Targioni Tozzetti.[118] If volume of correspondence is any criterion, the Paris astronomer Ismael Boulliau ought to head the list, for there are about forty letters between him and Prince Leopold during the years 1657–67 in the "Galileo" manuscripts, and more in Paris. Both he and Thevenot, as well as Ricci, will receive further mention in chapter 6.

It is now time to consider as a whole this strangely assorted Academy, headed by a prince whose genuine love of the new knowledge is entirely beyond doubt, and to speculate—for one can do no more—about what led to his choice of those who might be admitted to this privileged circle. His choice, or that of the Grand Duke, for I think there is reason to believe that a member of the Accademia del Cimento had to be *persona grata* to both the brothers.

The nine included one really first-class man, Borelli; another, Viviani, of great competence, but less energy; and perhaps a third, the medical man Redi, whose name has survived both as a biologist and

---

[114] G. B. C. Nelli, *Saggio di storia letteraria fiorentina del secolo XVII* (Lucca, 1759), pp. 110–11, note.

[115] G. 269, 263r–v.

[116] Nelli, *Saggio*, pp. 111–12.

[117] O. Falconieri to Leopold, from Brussels, November 2, 1658, G. 275, 124r; November 30, 1658, *ibid.*, 129r.

[118] *TT* 1, 453–58.

as a poet. This was a trio of which any prince might be proud. Then there was Rinaldini, a man of no great originality or force, and the del Buono brothers, the more competent of whom was the absentee member Paolo, whose early death may have been more of a loss to science and engineering than was realized at the time. Besides the secretary of the moment—first Segni and then the engaging Magalotti—the Academy was completed by Dati, who was a literary man more devoted to Tuscany than to science, the peripatetic Marsili, who if not quite as "rotten and mouldy" as Borelli would have us believe was nevertheless a strange choice, and finally the extraordinary adventurer and libertine Antonio Uliva.

Another surprising circumstance is that only six of the ten, even including the secretary, resided all the year round in Florence during the life of the Academy. This may well explain why Viviani's name occurs so frequently in the manuscripts, for I cannot escape the impression that for the most part he directed the day-to-day operations of the laboratory. However, Borelli and Rinaldini regularly came to Florence in the summer, and were almost certainly there when the Academy opened, for their presence is mentioned in the diary on June 21, 1657.[119]

I think that there is a reasonable explanation for the presence of Marsili the peripatetic, if we remember the character Simplicio in Galileo's *Dialogo sopra i due massimi sistemi del mondo*. He was the Aristotelian on whom the others sharpened their wits. May we not suppose that Leopold and his brother remembered this device of their revered master and decided that it would be well to have a Simplicio at hand?

It is more difficult to explain the presence of Uliva. It is true that Ferdinand and others of his family have been accused of indulgence in habits of a sort that Uliva, to judge by his later record, might well have abetted. Giovanni Carlo was certainly a notorious womanizer and, according to a manuscript in the Moreniana Library published in the nineteenth century by F. Orlando and G. Baccini,[120] Ferdinand himself was a homosexual, his favorite being the page Bruto Molara. We may take comfort from the observation that Leopold seems not to have been tainted by this environment, for even the author of this scurrilous account acknowledges that he "can say nothing about the Cardinal except that he was a gentleman full of clemency and goodness, and so lived his whole life and so died."[121]

[119] G. 262, 3v. See also pp. 53 and 359.
[120] F. Orlando and G. Baccini, *Vita di Ferdinando II Granduca di Toscana* (Florence, 1886) [*Bibliotechina Grassoccia*].
[121] *Ibid.*, p. 28.

3.

THE ESTABLISHMENT AND OPERATION
OF THE ACADEMY

1. INTRODUCTION

It was briefly noted in chapter 2 that a great deal of scientific work was being carried on for more than a decade before the year 1657 by the Grand Duke and under his auspices. This work was relatively unorganized and sporadic, and little of the evidence that has come down to us can be dated, but a few dates can be established, the earliest being the invention of a kind of thermometer consisting of a number of hollow glass balls, of slightly different weight-to-volume ratio, floating in a tube of spirit. These date at least from 1641.[1] Five years later the traveller Balthasar de Monconys, in Florence on November 7, 1648, described this sort of thermometer as well as two others that do not seem to have been described anywhere else.[2]

On December 30, 1648, the natural freezing of water was carefully observed,[3] and on February 23, 1654, the freezing of various liquids was watched on a night when the temperature went down to about $-9°C$ (7⅔ degrees on the scale of the 50-degree Florentine thermometer).[4] In 1644 Ferdinand had experts brought from Egypt to demonstrate, in the Boboli Gardens, the artificial incubation of hens' eggs.[5]

There are two incomplete and somewhat random accounts of the experiments made by the Grand Duke during this period. One, which still exists in manuscript,[6] is entitled "Register of various experiments made and observed by the Most Serene Grand Duke Ferdinand II,

---

[1] See p. 98 below.

[2] *Journal des voyages de M. de Monconys*, 3 vols. (Lyons, 1665–66), I, 130. See my *A History of the Thermometer* (Baltimore, 1966), p. 29.

[3] G. 259, fol. 6r.

[4] *Ibid.*, 7r. See also p. 96 below.

[5] *TT* 1, 150, with references.

[6] G. 259, 10r–17v. This is apparently a copy made by Viviani. Targioni Tozzetti (*TT* 1, 151) tells us that G. B. Nelli had the original.

and collected by Paolo Minacci for his own amusement."[7] The other was published, almost in full, and is headed "Note of experiments made by the Most Serene Grand Duke of Tuscany."[8] About two dozen of the eighty-six numbered experiments and observations in the latter are represented in the former, although not usually in exactly the same words. The experiments are on a great many different subjects. Of the few that are dated, all those in the Minacci account were made in 1655; but, besides these, the other list includes experiments made during the period in which the Accademia del Cimento was in operation. However, of the fourteen dates concerned, eight have no recorded meeting of the Academy, and of the remaining six only two show the experiments noted by the unknown writer. The entries for these two agree word for word with the diary G. 262, and are also found in G. 260; but in neither codex are there the Roman numerals printed by Targioni Tozzetti. On the other hand a series of these experiments, some dated in July and September 1657, have evidently been copied, though without their Roman numbers, into a gathering identified by the letter *K* and now found in G. 260.[9] This brings up the subject of the very interesting lack of correspondence between some of the various diaries or journal books that have come down to us, especially from the last months of the year 1657.

2. THE DIARIES THAT HAVE SURVIVED

As we saw in chapter 1, the "great diary" of the Accademia del Cimento was lost, but not before Giovanni Targioni Tozzetti had had a copy made, which is known to us as the manuscript G. 262. Everything that I have been able to deduce convinces me that this is probably a very nearly exact copy of the original, apart from punctuation and capitals, perhaps; but it must be emphasized that the latter was a somewhat conventional representation of the actual experiments, set down at not less than two removes from the original laboratory notes. My confidence in the value of G. 262 is greatly strengthened by the existence, in G. 260, of an earlier version of most of the diary, some of it—after May 1660—in the unmistakable handwriting of Lorenzo Magalotti,[10] which agrees almost exactly with the version in G. 262, as does a diary in the handwriting of Vincenzio Viviani covering the

[7] Minacci (1625–95) of Sienna seems to have been chiefly a writer of stories (*Grande dizionario enciclopedico* [Turin, 1958], s.v.).

[8] *TT* 2, 163–82.

[9] *TT* 2, 176–80; G. 260, 5v–9v.

[10] It is strange that the earlier part is not in the hand of Segni, but in that of an amanuensis.

period June 19 to December 4, 1657, in which, however, all 141 experiments are numbered serially.[11] This is in such careful handwriting—for Viviani—that one may suppose that it was intended for presentation. Thus we can, I think, be quite sure that we have the "official report" of the experiments made in the Accademia del Cimento from June 19, 1657, until March 5, 1667.

But there are two other documents that do not fit neatly into this scheme, although they refer to the period after June 19, 1657. One is a gathering marked "D" consisting of rough notes in the writing of Rinaldini[12] and headed by Magalotti with the words "Distesi d'Esperienze fatte dal Sig. Dottor Rinaldini l' 8bre [i.e., Ottobre] dell'anno 1657."[13] The other is much more important, and occupies all 67 numbered leaves of G. 261, with a covering leaf marked "C" that carries a baroque banner with the word "Sperienze," supported in space by two *putti*. It describes 288 dated and numbered experiments, made between June 19, 1657 and January 23, 1658, which are listed by title in chronological order on folios 1 to 6, and described in detail on folios 32 to 67. The remaining leaves, 7 to 31, are devoted to figures, excellently drawn, illustrating those experiments for which figures are desirable. The text is all in the same handwriting, plainly not that of a professional secretary, although I hesitate to suggest who the writer might have been. I get the impression that he did not always entirely understand what he saw. His orthography is sometimes a little wilful or archaic (for example *bocha* and *attachata*), and the word *fistola* is used throughout instead of the more usual *canna* to denote a tube.

For the present purpose the important feature of this document is its lack of correspondence with the group of journals that agree with G. 262. The differences are of several kinds, and I have thought it worth while to present a table of headings as appendix C, so that the reader may judge for himself. I think that he will find it difficult to escape the conclusion that there were two sets of experiments being made during this period.

Referring to the two journals simply as "G. 261" and "G. 262," it will first be noted that the former describes almost twice as many experiments as the latter. Viviani has saved us the trouble of counting those in G. 262. Secondly, G. 261 refers to no less than twenty-eight days for which nothing whatever is recorded in G. 262, including eight

---

[11] G. 260, 226r–81r, on the rectos only. There is a verbatim copy of the first three days of this in G. 263, 154v, 155r, and 156r. I cannot identify the handwriting.

[12] G. 260, 34r–49r.

[13] *Ibid.*, fol. 34r. Actually they run from June to October.

days during the seven months following December 22, 1657, when the Accademia del Cimento is presumed to have been on leave.[14] On the other hand G. 262 includes 11 days that are not represented in G. 261 at all.

Of the days that the two diaries both report there are only two (November 5 and December 3), and perhaps a third (October 13), in which they have all their experiments in common. On twenty-five others they share one or two experiments, seldom reported in the same terms, but occasionally with identical data. Many, perhaps most, of the experiments that are reported in both diaries are displaced by less than a week. A fair number are displaced by only one day, and we may suppose that these experiments were made in one place and carried to another overnight; but there is a possibility that some of the descriptions are of the same experiment, one being misdated. It is almost certain that an experiment with a brass ball, described in G. 261 on October 16 and in G. 262 on October 17, is but one experiment, because both accounts refer to a suspected crack in the metal, and the trial destroyed the ball. The resemblance between the two sets of experiments is greater at some periods than at other times; for example when work started again on December 3 they were identical for one day.

Further notes will be found in appendix C, but I wish to remark here that most of the experiments by Rinaldini that can be identified correspond to those described in G. 261, not those in G. 262. There are also descriptions of experiments made in Pisa and Leghorn, two of which, one on "positive levity" and the other on projectiles, found their way into the Saggi, although they are not represented in G. 262. On the other hand there are excellent things in G. 261, the omission of which from the Saggi is puzzling; for example, two optical experiments, one (no. 168) to find out whether a beam of sunlight is refracted on passing into a vacuum, the other (no. 178) on refraction by a surface of water.

It is interesting to compare the three diaries—Viviani's in G. 260, that in G. 261, and that in G. 262—on the occasions when they were describing the same experiments. Sentence for sentence, that of Viviani almost always follows the "official" version closely, but it has been rewritten rather than copied and sometimes includes additional items of information. Apart from this there would be no reason to choose one as the original rather than the other, and, indeed, we might suppose that Viviani would be more likely than Segni to be able to write

[14] On fol. 50v of G. 262 there is a heading "Riaprimento dell' Accademia fatto il dì 23. Luglio 1658."

down an intelligible account of the proceedings. A search through the huge mass of Viviani's papers might decide this point.

Turning to G. 261, we find the explanations briefer and more matter-of-fact, except where numerical data appear, as often happens. The experiments, even those made on the same day in the two "laboratories," often differ in detail, or one may have an additional feature lacking in the other. I shall illustrate by translating the two accounts of an experiment performed on June 25, 1657:

[From G. 262, 5v–6r]

It was investigated whether what Gassendi wrote is true, that if a quantity of common salt is put on a plate of ice that is placed on a table, then after the plate has remained there a little time, it will become so strongly attached to the table that it cannot be detached without great force. This was found to be true; and it was also found that the water that had run on to the table was salty. It was also seen that the side of the ice turned towards the table, that is to say the side below the one covered with salt, was rendered opaque by a white cloud composed of various particles of salt. At last, if after the salt had been there we looked at the light through the ice, this was seen to be minutely roughened, and as if prettily engraved with a diamond point; in short, like those drinking glasses that are usually called "iced" (diacciati), from the resemblance that they have to ice. Finally it was observed that it was impossible to detach such a plate by pushing it horizontally, but on the contrary it was done more easily by lifting it forcibly in the perpendicular.

[From G. 261, 33r]

19. A test was made with a plate of ice placed on a table, with some common salt put on top of the ice, to see whether it stuck to the table. It stuck marvellously well; and if this was done with niter in the same way it never would stick.

The contrast between these two accounts seems to me to be a certain indication that two sets of experiments were actually being done concurrently at the Court.

It would also appear that the Grand Duke had firm control of the services of the celebrated glassblower Alamanni, for we find Prince Leopold writing to Borelli on December 15, 1657—a week after the Grand Duke and most of the Court had gone to Pisa—

Meanwhile let me know what instruments or other things you think I should bring with me to Pisa, and tell me whether you want us to make some observations with them here first, as long as we don't have to have instruments made by the glassblower, who is setting off in your direction tomorrow. Also prepare for the observations that have to be made at Leghorn, and tell me if anything is needed for these also.[15]

---

[15] G. 282, 3v (draft).

3. The "Founding" of the Accademia del Cimento

I have dealt with the journal that forms the manuscript G. 261 at some length because I think it casts a little light on what really took place on June 19, 1657, and discloses a rather extraordinary situation that persisted for the remainder of the year. There can be no possible doubt that at least during that period two separate "research laboratories" of some kind were in operation at the Tuscan Court, one almost certainly centered on the Grand Duke and the other on Prince Leopold. On the evidence of the two diaries, the former was more active and continued during periods when the Court went to Artimino in September and to Pisa and Leghorn at the end of the year. The personnel of the two was at least partly the same; Rinaldini was clearly involved in both, and on August 7 Prince Leopold, Rinaldini, and Uliva made observations that are recorded in G. 261, while on the same day an entirely different group of experiments was being made in the other "laboratory" that is commonly identified with the Accademia del Cimento. On the following day nothing is recorded from this quarter, but in G. 261 we find the names of Prince Leopold, Rinaldini, Uliva, one Falconetti, and Segni. On August 9 Borelli and Magalotti joined this group, while again an entirely different experiment was recorded in G. 262.

What, then, happened on June 19? The problem is made no easier by the fact that the two diaries begin simultaneously on that day, so that we are tempted to conclude that the Grand Duke and his brother each "founded" an experimental Academy. Nor are we helped much by the surviving correspondence; indeed there is little between those mainly concerned during the summer of 1657. It is possible to deduce from a letter written by Borelli to Viviani on June 4 that Borelli and Rinaldini left Pisa for Florence on June 9 ("We shall leave on Saturday after lunch"),[16] presumably on the orders of the Grand Duke or Prince Leopold. They evidently stayed until the end of October,[17] and with Viviani were undoubtedly the committee of experts on which everything depended. It is probably significant that during the extensive periods when Viviani was away from Florence in 1660, 1663, and 1665 no meetings of the Accademia del Cimento are reported.[18]

There is evidence that the interest of the two princes in natural science was on the increase in 1656. Marcello Malpighi, who took up his chair at Pisa in that year, noted this, and believed that the Ac-

---

[16] G. 254, 58r.

[17] See p. 310 below.

[18] The dates can be deduced from the drafts of his correspondence, especially in G. 157.

cademia del Cimento was the result of this interest.[19] There is a further indication that before the end of 1656 Leopold was thinking of organizing something better than the sporadic experimentation that had been going on for years at the Court, for he commissioned Rinaldini to make a list of things worthy of investigation. Rinaldini, who was erudite in the fashion of his time and rather a pedant, searched all the books he could lay hands on and sent a list of these books to the Prince from Pisa on November 6, 1656.[20] On November 15 he sent a longer list,[21] which has been reprinted by the editors of *DIS*[22] and will be discussed later. Rinaldini's selection of topics has not survived, but the mere list of books shows that Leopold's anticipated program was extensive.

All this leads me to believe that by this time Leopold had formed the design of doing some serious experimentation, as a contribution to extending the work of Galileo in the liberation of the human mind from the scholastic fetters that were only too obvious in Tuscany. Even in Pisa, in spite of the Grand Duke's efforts, "almost all academic teaching was still in the hands of the Peripatetics."[23] The necessity of organizing experiments and taking notes on them must have occurred to one of the brothers, and this procedure was begun on June 19, 1657. We can only guess at the reason for the surprising duplication of experiments disclosed in the various diaries. It is very unlikely that there was any dispute between the Grand Duke and the Prince. It may be that there was a sort of friendly competition that was abandoned at the end of the year.

If the experiments were going to advance the cause of science they would need to be published, and I shall suggest in the next chapter that Leopold had plans for this at least as early as October 1657, ten years before they finally bore fruit.

If this interpretation is correct, there seems no reason to believe that the example of the Accademia dei Lincei,[24] which had already been extinct for twenty-eight years, had anything to do with the foundation of the Accademia del Cimento. As to the Accademia degli Investiganti of Naples,[25] there is every likelihood that news of it had reached Florence before 1657, but I cannot document this, and without firm

---

[19] Malpighi, *Opera postuma . . . quibus praefixa est ejusdem vita à seipso scripta* (London, 1697), p. 2.
[20] *G.* 275, 44r is the covering letter.
[21] *G.* 275, 48r is the covering letter, the list is on fol. 49r–v.
[22] *DIS.*, 51–52.
[23] *Ibid.*, p. 7.
[24] See p. 7 above.
[25] See p. 9 above.

evidence I do not think we need suppose any influence from that quarter.

It is necessary to deny the statement sometimes seen in general works that Leopold brought nine scholars together into a formal association, creating an entirely new laboratory of experimental physics and chemistry. The laboratory was there already and needed only some planning and the taking of notes. I have even seen the assertion that Leopold was *elected* president at the first meeting, thereby fulfilling one of his virtuous ambitions.[26] Such a statement betrays a lack of historical perspective; seventeenth-century Courts did not operate in that way. In later sections of this chapter I shall examine how the Court of Tuscany conducted its Academy; but first I must deal with a theory about the establishment of the Accademia del Cimento espoused by the historian Riguccio Galluzzi, who believed that it "was undoubtedly stimulated (*senza dubbio promosso*) by the example of the [Imperial] Court of Vienna."[27] As Keeper of the Tuscan State Archives, Galluzzi had seen the documents that I am about to cite, but I must contend that his interpretation of them is quite unjustifiable. Let us examine them.

In Vienna at that time the Empress Eleanora, who was a Gonzaga of Mantua, was very fond of Italian literature, and this taste was shared by the Emperor Ferdinand III and the Archduke Leopold Wilhelm. It is a matter of record that in December 1656 the Archduke instituted an Academy of Belles Lettres composed of ten members, every one an Italian, under his direction. This Academy, however, seems scarcely to have survived the Emperor of the Holy Roman Empire, who died on April 7, 1657, less than four months later.

In order to find out what sort of an Academy this was, we need only go to the Archivio di Stato in Florence and consult the official reports of the Tuscan Resident at the Imperial Court, Felice Marchetti, and his letters to Giovan Battista Gondi, one of the secretaries of the Grand Duke Ferdinand II.[28]

On December 30, 1656, the Resident can scarcely conceal his pleasure that

The Archduke has desired that I too shall be numbered among the academicians elected by His Highness. I have not been able to disobey his orders, the more so as I know the high regard that the Emperor and the Empress have for me. There are ten academicians, and they are as follows:

---

[26] Pieraccini, *La stirpe de' Medici de Cafaggiolo*, 3 vols. (Florence, 1924, 1925), II, 603.

[27] R. Galluzzi, *Istoria del Granducato di Toscana sotto il governo della Casa Medici*, 4 vols. (Florence, 1781), IV, 129.

[28] *FAS*, filza Medici del Principato 4401. The leaves are not numbered.

Count Raimondo Montecuccoli, General of His Imperial Majesty's cavalry; 2. Marquis General Mattei, Commander of the Archduke's cavalry; 3. Count Francesco Piccolomini of Aragona; 4. Marshal D. Giberto Pio of Savoy; 5. Baron Orazio Buccellieni, Colonel first class in the Regiment of Vienna; 6. Baron and Colonel Mattias Vertemuti; 7. Abbé Spinola; 8. Count Francesco d'Elci; 9. The noble Francesco Zorzi of Venice; 10. Marchetti, Resident of his Serene Highness of Tuscany.

There were many other people who could have been appointed to this academy, but His Highness said that their Majesties and he had thought that the assembly would have more dignity if none were admitted but those of eminent and distinguished nobility; so that I did not esteem it to be to my disadvantage to be recognized by all the Court, in this way too, as not unworthy to enter this rank.

I have spoken to you about this matter by way of an extra, not having any other news worth imparting to you.

This was a letter to Gondi; but the second paragraph of the official report of the same date is about the Academy.

His Highness the Archduke, busy with good works as usual, has succeeded in having some Italian noblemen chosen, in order to form an Academy of Belles-lettres in the Imperial palace. Already there have been ten elected for this purpose, which their Majesties support with much applause; in fact it has been resolved that the first function will be held on the first Sunday of the year in the Emperor's own apartment.[29]

In subsequent reports and letters to Gondi we get further glimpses. The academy met every Sunday. A table was provided, with a red cloth, and candelabra so that the academicians could see to read what they had composed. The report on January 13, 1657 tells us that the sessions ended with amateur music-making by the academicians, "which was a marvellous success, seeing that they are not professionals." A week later Gondi was told that the academy continues, but the papal nuncio, the Venetian ambassador, and the ambassador from Spain "are offended at not being named." It was still meeting in February, but on the 10th Marchetti reports that a ball at the palace of Marshal Traun seemed to have been of much greater interest. And on April 7, 1657 he records the death of the Emperor. The world was thrown into some commotion, and Marchetti's correspondence becomes more and more political, and often in cipher.

I submit that this rather absurd academy, picked by an Italophile Court from military men and diplomats (with one abbé) can scarcely have had any other effect at the Court of Tuscany than to produce a little amusement. It had nothing whatever to do with science, experimental or otherwise. Nor did the Tuscan princes need to be told how

[29] *Ibid.*

to set up an academy; in 1638 they had re-started the Accademia Platonica, originated by their ancestor Lorenzo the Magnificent in the fifteenth century;[30] and Leopold protected, and sometimes egged on,[31] the perennial Accademia della Crusca. The whole theory smells of *post hoc, ergo propter hoc*, but it gave Galluzzi one more chance to display his really extraordinary erudition.

It should also be recorded that Rinaldini, in a letter to Ricci written in February 1667,[32] attributed to himself the merit of persuading Prince Leopold to undertake experiments on natural objects and said that this resulted in the founding of the Academy. Abetti remarks dryly that Rinaldini "was probably not over-modest." On the other hand, it seems generally to be agreed that Viviani must have had a good deal to do with it, though I have found nobody who is able to document the assertion.

4. THE NAME "DEL CIMENTO"

I do not recall having seen it pointed out that the name of the Academy is in reality a *double entendre* that was entirely appropriate at the time. In a twentieth-century Italian dictionary we read: "cimento, *s.m.* 1. (*ant.*) analisi scientifica, che si faceva con una mistura a base di sali e altre sostanze, per saggiare metalli preziosi; esperienza, saggio 2. rischio, pericolo; prova ardua." [33] So soon after the death of Galileo, with the Counter-Reformation in full swing, the activities of the Academy were indeed risky and needed all the protection that the prince could bestow. Nevertheless, the translation "Academy of Experiment" probably expresses the idea that the experimenters themselves wished to convey.

It is of some interest to try to determine the date on which the name "del Cimento" began to be used. Targioni Tozzetti thought that it was seen only when the *Saggi* was published, for he had not found it among the papers,[34] but the editors of *DIS* point out[35] that it does occur in G. 267. The Academy was usually referred to as "Prince Leopold's Academy" in the years after 1657. However, as the reader will remember from chapter 2, Francesco Redi wrote two letters nam-

---

[30] G. Bianchini, *Dei Gran Duchi di Toscana della reale Casa de' Medici*, etc. (Venice, 1741), p. 93; see *TT* 1, 98.

[31] See p. 23 above.

[32] Cited without further reference by Georgio Abetti in *Celebrazione della Accademia del Cimento nel tricentenario della fondazione* (Pisa, 1957), p. 4.

[33] *Dizionario Garzanti della lingua italiana* (Milan, 1965), p. 367. "cimento, *n.m.* 1. (*obsolete*) scientific analysis, made with a mixture based on salts and other substances, for assaying precious metals; experiment, test 2. risk, danger; arduous ordeal."

[34] *TT* 1, 412.

[35] *DIS.*, 19, note 1; also p. 64.

ing the Accademia del Cimento, one to Michele Ermini on April 25, 1659 and the other to Carlo Dati on May 9, 1660. It was there noted that no meeting of the Academy is recorded on either date; what is more, both dates come in the long hiatus between September 7, 1658, and May 20, 1660, during which nothing whatever was reported. Eleven days after the second letter, on May 20, 1660, the Academy was reopened with Magalotti as secretary, who recorded the event in his own writing in these words:

May 20.
The Academy was reopened, and on that morning all the papers belonging to it were handed over to me, Lorenzo Magalotti, including diaries, as well as experiments proposed, and to be made, and at the same time, various matters concerning the resumption of the Academy were decided.[36]

It is therefore impossible to believe that what we know as the Accademia del Cimento had a meeting on May 9, 1660. Let us look at more of the letter that Redi wrote on that date:

They are working in the Accademia del Cimento, and you may be assured that My Lord the Grand Duke is most enthusiastic about it; one might say very much indeed. All scholars are greatly obliged to this great Prince. On his orders I am working at many things, but particularly at artificial salts obtained from the ashes of wood, grass, and fruits, and I have already made some fine discoveries, which will come to light at the proper time.[37]

That these artificial salts were being worked on under the direction of the Grand Duke and not in the Academy of Prince Leopold is clear from a letter from Borelli to Malpighi, dated June 12, 1660, found in the library of Bologna University and paraphrased by Adelmann.[38] In this letter, after stating how happy he was that the Academy was again functioning, and then dealing with several other subjects, Borelli tells Malpighi that the Grand Duke was having the shapes of the salts of various vegetable and mineral materials studied.

The conclusion that I draw from the two letters written by Redi is perhaps novel, but, I think, inescapable. It is simply this, that in 1659 and 1660 it was the Grand Duke's experimental group, not Prince Leopold's, that Redi called the Accademia del Cimento. It is possible that Redi was alone in using the name at that period, and that he may indeed have suggested it. In any event, it is certain that nearly thirty

[36] G. 260, 99r. There is a small sheet, fol. 92, with an exact copy in another hand; it is also copied exactly in G. 262, 66r.

[37] Opere di Francesco Redi, gentiluomo aretino, etc., 2nd ed., 7 vols. (Naples, 1740, 1741), IV, 19. Targioni Tozzetti (TT 1, 451) quotes only the first sentence of this extract.

[38] Howard B. Adelmann, Marcello Malpighi and the Evolution of Embryology, 5 vols. (Ithaca, N.Y., 1966), I, 170.

years later he boasted to the Marchese Bartolommeo Verzoni of Prato that he glorified "in having been one of the first founders of the famous Tuscan Accademia del Cimento."[39] Should we conclude that as far as Redi was concerned all the experimenting at the Court had been part of the activity of the Academy? I am willing to concede that a mystery remains here.

The editors of *DIS* have attempted to establish that the use of the designation "del Cimento" for the Prince's Academy dates only from the summer of 1667.[40] But we owe to Ferdinando Massai the publication in 1917 of a letter dated January 15, 1666 from Magalotti to Alessandro Segni;[41] a letter in which Magalotti cites the title of the *Saggi* in almost its final form. Segni was in Paris at the time, and it is therefore not surprising that Melchisadek Thevenot used the term in a letter written to Prince Leopold on October 16, 1666.[42]

5. The Motto "Provando e riprovando"

We do not know which of the academicians suggested the motto "Provando e riprovando" that appears in the engraved decoration on the title-page of the *Saggi*, and again in the preface. It is reasonable to suppose that whoever was responsible remembered the verses of Dante:

> Quel sol che pria d'amor mi scaldò il petto
> di bella verità m'avea scoverto
> provando e riprovando, il dolce aspetto.[43]

But, as Bruno Nardi reminds us in a delightful article,[44] this was simply a verbal memory, for the context makes it quite clear that the

---

[39] Redi to Verzoni, September 5, 1686, in *Opere di Francesco Redi*, 5 vols. (Florence), V (1727), 232.

[40] *DIS*, 64–65.

[41] One of a number of letters published by Ferdinando Massai, in *Rivista delle Biblioteche e degli Archivi* 28 (1917): 134–38. An extensive extract is translated on p. 74 below.

[42] G. 277, 347r–48v. Printed, with a French translation (Thevenot wrote in Italian) by R. M. McKeon, *Rev. Hist. Sci.* 18 (1965): 1–6.

[43] Paradiso, III, 1–3.

> that sun which first had warmed my heart with love
> Had now, by argument and refutation
> Revealed to me the lovely face of truth.

—*The Divine Comedy . . . a new Translation into English Blank Verse by Lawrence Grant White* (New York, 1948), p. 132. Copyright by Pantheon Books. Quoted by permission.

[44] "Significato del motto 'Provando e riprovando'," in *Celebrazione della Accademia del Cimento nel tricentenario della fondazione* (Pisa, 1957), pp. 71–79.

poet intended the other meaning of *riprovando*, that is to say confuting, or refuting. Nevertheless, it is clear that what the academicians —or most probably their secretary Magalotti—intended to convey was the idea "by trial and error" or "testing and re-testing." The metaphor, used in the preface to the *Saggi*, of the heap of jewels that have come out of their setting[45] would make the first of these translations the more appropriate. Nardi reminds us that "In the same way, we all know those highly educational toys with which children practice reconstructing shapes and drawings that are cut into small parts and mixed up in great confusion."[46] The attentive reader of the *Saggi* will not find this metaphor of the jig-saw puzzle too far-fetched.

6. THE OPERATION OF THE ACCADEMIA DEL CIMENTO

Probably because of its limited and relatively well-defined objectives, the operation of the Accademia del Cimento was quite unlike that of the multitude of other academies in which Italy abounded. In these— if we except that official guardian of the Tuscan language, the Accademia della Crusca—the main, and often the only, activities were the reading of literary compositions and the occasional subsidizing of publications.[47] Even the Accademia dei Lincei, although its interests were largely scientific, had operated in this way.[48] Usually these activities were governed by a formal constitution and bylaws. But nobody has ever discovered any evidence of formal rules governing the Accademia del Cimento, and it had no officers other than a secretary. It had, however, something much more valuable than rules—a well-equipped laboratory; not to mention an apparently inexhaustible source of apparatus and materials. It is refreshing to read how they "piled fifty plates of gold one upon the other, [and] saw a needle placed on the top plate obey the motion of a magnet moved close to the under surface of the bottom one."[49]

We get surprisingly little information about how the Academy was actually run. An entry on June 21, 1657—the third day of its official career—is almost unique: "It was decided that Dr. Rinaldini, Borelli, and Uliva should meet at the Palace at 21 o'clock[50] every day to discuss the experiments to be made one after another on the following

---

[45] See p. 90 below.

[46] Nardi, "Significato del motto . . . ," p. 71.

[47] The standard work on these academies is by Michele Maylender, *Storie delle accademie d'Italia*, 5 vols. (Bologna, 1927–30).

[48] See p. 7.

[49] *Saggi*, 1667 ed., p. CCXX.

[50] The day began one hour after sunset.

day, and to give the necessary orders."[51] This arrangement cannot have persisted, however, for Borelli and Rinaldini were back at Pisa by the end of October, frequently suggesting experiments in letters to the Prince. Uliva's activity in the Academy is not well documented, and it is very likely that the burden fell on the already overloaded shoulders of Viviani. But like Targioni Tozzetti, I can only record the impression derived from a perusal of the papers, that "this Academy had neither laws nor any sort of constitution, but everything depended on the commands or agreement of Prince Leopold, and on the activity and goodwill of the Academicians."[52] There seems to have been no order of precedence among the members, and each one was allowed freely to propose experiments and to discuss those that had been done. Occasionally, Leopold did indeed give particular tasks to one or other of his Academicians.

The fact that the publication that eventually resulted from the activities of the Academy is completely anonymous should not cause us to lose sight of the fact that the authors of a great many of the experiments, published as well as unpublished, are mentioned in the diaries or easily ascertainable from the other papers. Indeed, the requirement of strict anonymity seems to have grown slowly, for occasional names occur in the first existing draft of the *Saggi*, probably written in 1661.[53] It cannot be ascribed to a surplus of modesty, but arose from prudence, for the new natural philosophy was even more hateful to the Church than it had been in Galileo's day. On the same cautious sentiment must be blamed the almost total failure in the *Saggi* to declare in favor of any hypothesis, although the purpose of a given experiment might be obvious.

It is not quite clear whether meetings of the Academy as such were ever held outside of Florence, even when the Court moved to Pisa or Leghorn or Artimino, as it often did. G. B. Nelli thought not, basing his opinion on the fact that some of the members had employment in Florence, especially Viviani, who was a lecturer at the University of Florence, and also chief engineer at the Office of River Control.[54] From Viviani's correspondence[55] it is possible to date his longer absences from Florence, and it is noteworthy that no meetings of the Academy are recorded at these times. Borelli and Rinaldini seem to have spent most of the summer and early autumn of every year in

[51] G. 260, 229r; also G. 262, 3v.
[52] *TT* 1, 412.
[53] See also p. 69 below.
[54] Nelli, *Saggio di storia letteraria fiorentina del secolo XVII* (Lucca, 1759), p. 83, note 4.
[55] Especially in G. 157.

Florence, and most of the meetings in the active years 1657 to 1660 were held during such periods. Experiments were certainly made at Pisa and Leghorn in the presence of the Grand Duke and his brother, as in January 1658, when investigations of air pressure, freezing, and "positive levity" were reported by Rinaldini and by Borelli in letters to Viviani.[56] There were no recorded meetings of the Accademia del Cimento between December 22, 1657 and July 23, 1658.

Evidence that out-of-town scientific activities were not considered to be meetings of the Academy is contained in the diaries. After a reference to two experiments made on September 17, 1657, the account continues as follows:

Until September 23 the time was occupied in observing the intersections and expansions made by the circles produced by the fall of projectiles into water, without anything certain being collected or established about it for the time being.

Meanwhile news came from the Court, which was[57] at Artimino, that in confirmation of what was written in France[58] to give validity to the pressure of the air as the true cause of the elevation of the mercury, it was observed that in going down from the top of the palace at Artimino to the base of the mountain . . . the mercury in the instrument known as the hydrostathmion[59] continually rose during the descent. . . . On going up again it fell enough to come to its usual height at the top. The same experiment is being made on the hill of S. Giusto, and on the tower of the Palazzo Vecchio in Florence. . . .

Then on the 24th, with the departure of Prince Leopold for Artimino, the Academy was given leave until October 3, which was the first day it met after His Highness' return.[60]

There follow details of experiments with a water barometer set up at Artimino on September 28, 1657. This shows conclusively that the work at Artimino was not considered part of the activity of the Academy. However, experiments first made in the Academy were sometimes repeated before Ferdinand, as on June 28, 1660, when "After lunch . . . the experiment against positive levity was made before the Grand Duke, and also that of boiling with ice."[61]

[56] G. 283, passim.

[57] la quale si trovava. The imperfect tense tends to confirm the impression that the diary was "written up" later. Viviani's dairy (G. 260, 264r) uses the same words.

[58] They read about this in Jean Pecquet's Experimenta nova anatomica, etc. (Paris, 1651). See p. 129 below.

[59] Apparently the instrument of fig. XIV, p. LVIII, of the Saggi; really a barometer with an unusual cistern.

[60] G. 262, 34v–35r; also in Viviani's version, G. 260, 264r.

[61] G. 262, 83r; TT 2, 580.

It was mentioned in the third section of this chapter that in 1656 Rinaldini had been instructed to make notes on a large number of books, presumably with the idea of using them to establish a program for the Academy. The notes have been lost, but three lists of the books exist, and if he read them all he must have been a busy man. In his covering letter of November 6[62] he refers to them as books that he had used and was using "in reference to experimental matters," but he cast his net so wide that I am obliged to assume that he was simply trying to seem more learned than he was. There is the usual list of classical *opera omnia*—Plutarch, Galen, Seneca, Pliny, Plato, Plotinus, and of course Aristotle and all the commentaries Rinaldini could find, though what use he made of these last is hard to imagine. There are a score of fifteenth- and sixteenth-century philosophers, now mere names. He had fished in the more turbid streams of his own age—Robert Fludd, Athanasius Kircher, Paracelsus. On the other hand, he listed the works of Gassendi, Mersenne, Descartes, Galileo (of course), Emmanuel Maignan, François d'Aguillon, Giambattista della Porta. There are a great many more; but one omission is surely surprising and, I believe, significant. The name of Francis Bacon does not appear. I shall defer a discussion of the so-called Baconianism of the Academy until later.

There exists another catalogue by one "Bernardo Paludano," presumably the Flemish traveller Bernard ten Broecke,[63] but, if this ascription is correct, it can have nothing to do with the Academy, for its author died in 1633. It names seventeen works on minerals, metals, and natural history, and it is known that the princes were enthusiastic about collecting minerals.

7. THE ROLE OF PRINCE LEOPOLD

It would be easy, but quite wrong, to make the facile assumption that the Accademia del Cimento was merely one of the diversions of these Medici princes, on a par with hawking or the pleasures of the bedchamber. There is too much evidence of serious purpose in their scientific activities for such an interpretation to be defensible. It is therefore of interest to inquire into Leopold's relation to the daily work of his Academy.

The way Leopold saw himself is neatly summarized by Magalotti in a letter written to Ottavio Falconieri on July 29, 1664, in which he says that Leopold "likes to act as an Academician, and not as a Prince. He is content to play the second role only on occasions when there is a question of expense, generously supplying the needs of the Acad-

[62] G. 275, 44r.
[63] G. 268, 1r–v; *DIS*, 52.

emy."[64] And, indeed, Leopold did make his own contribution to the researchers. For example, Borelli assures us that the Prince was the first to discover that the volume of a glass flask increases when it is put in hot water and decreases when it is put in ice,[65] an observation that Borelli supported by the ingenious experiment illustrated in fig. XII and described on page CLXXXXII of the *Saggi*. This observation, which really consisted in noticing the small, quick rise in the level of the liquid in a long-necked flask as soon as this was put into ice, fascinated Leopold and led to the seemingly endless series of experiments on natural freezing that take up twenty pages of the *Saggi* and evidently bored some of the Academicians, among them Viviani, in whose copy of the diary, in place of the details reported in G. 262, we read: "On November 4, 5, 6, 7, 8, 10, 1657, the observations of the freezing of water were repeated several times, without finding anything different from what was found on previous days."[66]

Leopold's interest in the process of freezing seems to have endured for years, and he made elaborate experiments on natural freezing in the winter months, even as late as 1667.[67] After entertaining friends at dinner on cold, clear evenings, he would invite them to witness these experiments. Such an occasion, in January 1666, is described by Magalotti in a charming letter to Alessandro Segni written on the fifteenth of that month.[68]

The opinion of Magalotti, cited on page 56, as to the role of the Prince must not be taken too literally, however, for it is clear that Leopold had no patience with mere "pottering about." In 1660 Magalotti had to have the experiments prepared that were to be made every day in the Prince's presence,[69] and he found this a somewhat onerous task. By 1662 Leopold was anxious to get something definite for publication and began to put on the pressure. In the diary for July 31, 1662, we read: "The Academy met at Sig. Lorenzo Magalotti's house, about repeating some experiments that appeared most necessary to the finishing of the work that is to be printed. All of these, when they have been made easy by practice, have to be done again in the presence of His Highness."[70] Indeed, a great many of the earlier experiments were repeated in the ensuing months. The occasion for the

[64] *Delle lettere familiari del Conte Lorenzo Magalotti e di altri insigni uomini a lui scritti*, 2 vols. (Florence, 1769), I, 86.

[65] Borelli, *De vi percussionis liber . . .* (Bologna, 1667), p. 236.

[66] G. 260, 279r.

[67] See p. 199 below.

[68] Published by F. Massai in *Rivista delle Biblioteche e delle Archivi* 28 (1917): 134–38.

[69] Magalotti to Segni, October 26, 1660, in *Lettere familiari*, I, 67.

[70] G. 262, 132r–v; *TT* 1, 411.

meeting described in the above entry was a handwritten note that Magalotti had just received (Plate 2):

Sig. Lorenzo: Remember to adhere strictly to what I told you, to bring into the experiments only those who will work continuously, because too much time is being lost and I am anxious to get on with it. And to those Aca-

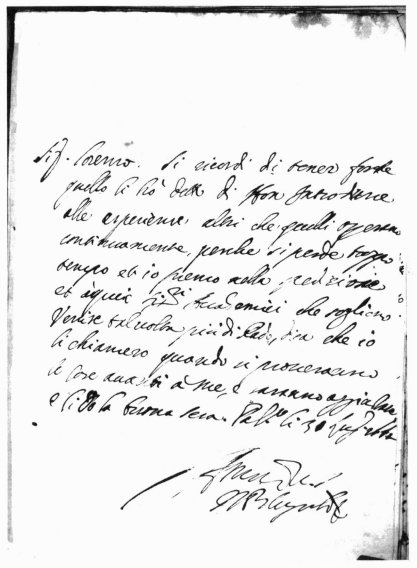

PLATE 2.   A letter to Magalotti in the handwriting of Prince Leopold
(*Biblioteca Mediceo-Laurenziana*).

demicians who sometimes are in the habit of coming less often, say that I shall call them when the things are put in order and demonstrated before me; and I wish you good evening.

From the Palace, July 30, 1662[71]

As the usual close of such notes of Leopold's that I have seen is "and may God prosper you," I think we may conclude that he was somewhat annoyed on this occasion. In any case, he was certainly giving orders and playing the prince rather than the academician, but we ought not to conclude that the others resented this or found it at all unnatural. After all, they were courtiers and knew the rules of the game.

One of the rules, among friends, was to praise each other to the Prince, even at the cost of some hypocrisy. An obvious instance of this is in a letter written by Rinaldini at Pisa to Viviani at Florence on January 9, 1658, a date when the Court was at Pisa. It should be explained that Viviani had invented several instruments, really rather freakish air thermometers, with the idea of measuring small variations in the density of the air. Four of these got into the *Saggi*,[72] but they were really not very useful, as Borelli and Rinaldini, at least, knew very well, and the latter wrote: "I have tried on the Campanile your instrument for weighing the air, but I have had difficulty with heat and cold [i.e., with the temperature]. So you ought to give some thought to this. I have praised the invention and I have had occasion to speak about you to Prince Leopold with such esteem that you could not wish for more. I have done this with pleasure and shall do it again, knowing I speak the truth."[73]

Another rule was to show extreme deference, even to despots as benevolent as Leopold and Ferdinand, and even if your scientific attainments were of the highest order; even if you were Borelli, in fact. In March 1660 there was a most interesting difference of opinion between him and the two Princes regarding the reason for the high barometer in fine weather, and on this occasion, as I have shown elsewhere,[74] these noblemen were much more nearly right than was Borelli. However, the latter defended his theory and criticized that of the Princes at some length.[75] Then he wondered if he had been too enthusiastic, and his letter ends:

[71] Florence, Bibl. Mediceo-Laurenziana, MS. Ashburnham 1818, fol. 9r. This codex contains a number of such short notes to Magalotti in the handwriting of the Prince.

[72] See p. 131 below.

[73] G. 283, 33r.

[74] Middleton, *A History of the Theories of Rain* (London, 1965), pp. 67–69.

[75] Borelli to Prince Leopold, from Pisa, March 16, 1660, G. 276, 12r–15r.

Now I most humbly beg Your Highness to represent to the Grand Duke the very great caution with which I have spoken about this event, and which I am not giving up even now; so that I have not meant to say anything dogmatically, but merely as a conjecture that may be exposed as fallacious. And therefore I beg you in all humility to receive the things I have said, not to make them known, or to oblige me to support them, but as being confused and imperfect, and communicated privately to Your Highness. I am flattered by the infinite clemency with which you are wont to receive my fantasies, and excuse my defects. And in conclusion I pay my most humble respects, as is my duty.[76]

Borelli was probably in no danger of princely displeasure on this occasion, for the Prince knew his value. Lesser lights, not in the Academy, such as the hydraulic engineer Famiano Michelini, were treated simply as civil servants and expected to produce, as we are told by Angelo Fabroni in a footnote to a letter from Michelini to Leopold dated February 3, 1662, complaining about his poverty.[77] Michelini, says Fabroni, "never got round to publishing any of the many things that he said he had found; and the eagerness of the Medici Princes to see useful and glorious works come into being under the protection of their favor was the reason that they demanded of Michelini, even with threats, the performance of his promises." Michelini seems to have got busy after this, and his *Trattato della direzione de' fiumi* was published at Florence in 1664.

But even, or perhaps especially, people like Borelli were glad to have the protection and diplomatic assistance of the Prince. In 1657 Borelli had blundered into giving serious offense to Viviani in a discussion concerning the thermal expansion of solids.[78] On November 28 he wrote a letter to Viviani,[79] the first part of which dealt with the expansion problem. He continued:

They are saying here that the judge Vettori is coming to Pisa. From over there [i.e. in Florence] you will be able to tell better than we can whether this is true, because if he should not come to Pisa I am obliged to beg you to request him to give a commission to some doctor of Pisa to look over my book, so as to grant the license to print it. And if it should even seem necessary to you to make use of the authority of His Highness the Prince in this affair, I also beg you to say a word to him about it, so that with his authority he may make this business easy.[80]

---

[76] *Ibid.*, 14v–15r.

[77] *LIUI* 1, 174.

[78] See p. 311 below.

[79] G. 283, 22r–23r.

[80] *Ibid.*, 22r. Alessandro Vettori was "Auditore (judge) of the Knights of St. Stephen."

The book must have been his *Euclides restitutus,* published at Pisa in the following year. Even though this was a text on geometry, the ecclesiastical authorities seem to have been somewhat difficult about it. There is much evidence of the struggle between the "new men" implanted by the Grand Duke at Pisa and the old-line scholastics who predominated in the University there.

8. THE INTERRUPTIONS IN THE WORK OF THE ACADEMY

The meetings of the Academy were not held at all continuously between the terminal dates on which they are recorded, June 19, 1657, and March 5, 1667. Its activity seems to have come in bursts, interrupted for various reasons. I shall examine a few of the interruptions, to show what a marginal activity the Academy really was, in spite of the genuine interest of the Princes in the new science.

The Academy had been in operation only a few months when, on November 10, 1657, it adjourned until December 3. In G. 262 this is not explained; but on November 17 Viviani wrote to Rinaldini that the Academy had been dismissed "for many days, in view of the arrival of this very eminent person" (*quest'Eminen.mo*). He heard that the visitor would leave "next Monday."[81] This would have been November 19. In G. 261, however, we learn who the visitor was, together with the very significant fact that both academies, Ferdinand's and Leopold's,[82] stopped work on November 10 and started on December 3; for at the end of the entry for November 10 there is a note reading "for the coming of Cardinal Antonio Barberini we remained until December 3 without doing any more." [83]

Leopold's Academy shut up again on December 22, when the Court went to Pisa, but the other one continued at Pisa and Leghorn until about January 23, 1658.[84] After that we hear nothing certain about Ferdinand's; and about Leopold's nothing until the next July 23, when the diary (G. 262) resumes with the bare heading, "Reopening of the Academy, done July 23, 1658."

After July 23 the meetings continued fairly steadily until September 7, and then, except for one undated meeting, there was a hiatus of twenty months until the formal reopening on May 20, 1660, when Magalotti was made secretary and all the papers were turned over to him.[85] One of the reasons for this long break emerges from a letter from Leopold to the Parisian astronomer Ismael Boulliau, dated

[81] G. 252, 31r.
[82] See p. 45 above.
[83] G. 261, 60r.
[84] See appendix C.
[85] See p. 51 above.

April 24, 1659 in which he writes, "Various circumstances have resulted in many of my Academicians having been and still being separated, in different places. Thus we have not applied ourselves for some time to the experiments and studies that were begun."[86] But during 1659 and 1660 Borelli and Rinaldini were at Pisa or Florence as usual. Viviani undoubtedly travelled about a good deal on his engineering work and was certainly away for most of the spring of 1660. Prince Leopold's health was perhaps in one of its more fragile phases.[87] After the reopening on May 20, 1660, there ensued a very active three months, with no less than seventy-eight recorded meetings by September 1. The entry for this date ends with the following passage, which occurs both in Magalotti's writing in G. 260 and in the copy of the diary, G. 262:

Then the remainder of this month [and of the next, October, until the 25th],[88] although the Academy met several times, this was only in order to make observations of Saturn,[89] and about the dispute with Father Fabri, and finally to examine the propositions of the commentaries that were sent to Holland,[90] as appears in the book.[91]

In the month of October the holidays of the Academy ran on until the 25th when it opened again.[92]

Indeed it opened again, bursting by repeated freezings a golden ball filled with water. If one just looked at the headings in G. 262, one might suppose that the reopening was only for one day, for the next entry is headed "October 19, 1661."[93] But on folio 110v, without even a paragraph to itself, is concealed the following: "On the following days, until November 15th, the two following experiments were worked on, of which, because various difficulties were encountered which several times caused us fruitless labor, it will suffice to give an account in the form in which they were successfully performed." These experiments were on the weight of air, and its weight relative to water. At the end of the long account we read, "The Academy was again given leave, and was recalled on October 19, 1661."[94] It seems clear that the Academy had a good many meetings that are not dignified by separate entries in the diaries.

[86] The draft is in G. 282, 10r–13r.
[87] Pieraccini, *La stirpe* . . . , II, 622.
[88] The passage in brackets is in G. 260, but is cancelled.
[89] See p. 258 below.
[90] To Christiaan Huygens.
[91] These documents, or copies of them, are in G. 271.
[92] G. 260, 136v; *verbatim* in G. 262, 109v–10r.
[93] G. 262, 114r. The entry for October 25, 1660, is not in G. 260.
[94] G. 262, 114r.

The long break of a year in 1660 to 1661 can be explained only in part. Leopold's health was always somewhat precarious, but 1661 does not seem to have been a bad year.[95] However, it contained one event that eventually shook Florence to its foundations, the wedding of the heir-apparent Prince Cosimo and Marguerite Louise, daughter of Gaston of Orleans, a marriage that turned out disastrously for the Medici dynasty. Such a wedding was not a matter of a week's festivity. The Princess arrived in Florence on June 15, 1661, and the wedding was on June 20. But as early as April 21 we find Leopold excusing himself in a letter to the Parisian dilettante Melchisadec Thevenot. After thanking him for a list of experiments performed in Paris and another of some done in London, the Prince continues:

I am very sorry indeed not to be able now to send in this manner the recompense that I owe you for your kindness, as I should have wished to do, because we are all busy on the preparations for the reception of Her Highness the bride, and for the festivities to be arranged for the wedding. For the time being, both I and these Academicians have given up studies of this kind. But, *given that this function comes to an end,* we shall again take up the thread of the experiments, and then I hope to be able to correspond with you . . .[96]

The phrase I have emphasized suggests that the Prince was finding the wedding a bore already, so that the preparations may have been going on for some time before April 21. On the first of June he had to apologize to the great Dutch astronomer Christiaan Huygens in rather the same vein: "For now I am not telling you my other opinion, or that of my Academicians, about this, even though there may be much to say on many subjects, because they, as well as I myself, are busy in the service of the Grand Duke."[97] Acton tells us that the public celebrations continued until at least the middle of July.[98]

After the Academy reassembled on October 19, there was another spate of activity, thirty-seven meetings, until January 18, 1662, when there began a seven-month recess, for which I have no explanation. I shall show in chapter 5 that it was during this period that Leopold must have received from Robert Boyle a copy of the Latin edition of his *New Experiments Physico-Mechanicall.*[99] We have seen[100] that

[95] Pieraccini (*La stirpe* . . . II, 622) finds few medical notations for that year.

[96] The draft is in G. 282, 50r–v. My italics.

[97] *Oeuvres complètes de Christiaan Huygens, publiées par la Société Hollandaise des Sciences,* 22 vols. (The Hague, 1888–1950), III, 274–75.

[98] Harold Acton, *The Last Medici* (2nd ed.; London, 1958), p. 83.

[99] See p. 263 below.

[100] P. 57 above.

in response to the Prince's directions the Academy next met on July 31, 1662. They worked through August and part of September, mainly repeating and improving experiments made earlier.[101]

This almost completes the recorded experimental work of the Academy, except for a period at the end of 1666 and the beginning of 1667 when some unusually cold weather led to elaborate experiments on natural freezing, and there were a few on amalgamation and on heat that seem to have been made at the instance of Rinaldini. In this long interval there was an attempt to find a finite speed for light, made on July 24 and 25, 1663,[102] and a few botanical experiments in March and April of the following year.[103] In 1664 and 1665, however, there were many astronomical observations which will be mentioned in chapter 5.

Now Magalotti was away from Florence a good deal in 1663 and 1664, and Targioni Tozzetti speculates that meetings may have been held when Magalotti was not on hand to record them.[104] We just do not know, although there is a slight possibility that the answer might be found in the many kilometers of shelves of the *Archivio di Stato*. I am inclined to think that Targioni Tozzetti's suggestion is unlikely to be true, both on the general grounds of what I know about the documents, and because it is surely reasonable to ask what on earth would have prevented the Prince from drafting one of the other Academicians as temporary secretary?

In conclusion, I do not think that we should believe that the somewhat sporadic meetings of the Academy indicate that the institution was merely a pastime of Leopold's, to be called together when he was in the mood and dismissed whenever it suited him. Apart from his delicate health he had, we must remember, a considerable part in the government of Tuscany. The most valuable Academicians were sometimes absent from Florence for long periods. Furthermore, what is not generally recognized is that the curiously limited program that the Academy seems to have set itself was substantially complete by the end of 1662, an assertion that I shall try to document in the following chapter.

[101] As far as the diary G. 262 is concerned, this activity ended on September 9; but there are rough notes in G. 260, 218r and 219r, referring to experiments on the 11th, 12th, and 14th. Targioni Tozzetti's reference (*TT* 2, 423) to a meeting on September 20 is probably a mistake for September 2.

[102] G. 262, 155r–60r.

[103] G. 268, 137r–44r. These are not mentioned in G. 262.

[104] *TT* 1, 411.

4.

THE *SAGGI DI NATURALI ESPERIENZE*

1. Introduction

The Accademia del Cimento has a place in the general history of the physical sciences almost entirely because of its one publication, the *Saggi di naturali esperienze fatte nell' Accademia del Cimento.* Apart from a study of the surviving manuscripts at Florence, the only way that the student can learn much about the other activities of the Academy is through the very rare *Notizie* of Giovanni Targioni Tozzetti, the two thousand pages of which are mentioned more often than they are read, or by a study of the work referred to here as *DIS.*[1]

Its remarkable singleness of purpose and its pointed avoidance of speculation give the *Saggi* its importance and make it unique in its time, quite apart from its interest as the first report from what we should now call a research laboratory. It is not, as sometimes stated, the first book entirely devoted to experimental research, an honor that belongs to Robert Boyle's *New Experiments Physico-Mechanicall* (Oxford, 1660).

Apart from a new translation into English, the first since 1684, this chapter will be devoted to a brief exposition of what can be deduced about its writing and the printing and distribution of the first edition. Its reception by the European scientific community will be discussed in chapter 8.

A bibliography of the *Saggi* will be found in appendix A.

2. The Compilation and Printing of the *Saggi*

The idea of publishing the results of the efforts of the Academy certainly was in the mind of Prince Leopold shortly after the date in June 1657 generally given for its founding. The earliest evidence of this that I have discovered is contained in a letter that he wrote to the French astronomer Ismael Boulliau from Florence on October 13, 1657:

[1] See p. xii above.

As I hinted to you, I had resolved to send you further letters in which I was to tell you about an idea of mine, and a work that I have in hand, which (if I succeed in bringing it to a conclusion, and if I am not mistaken) ought in some ways to be of no small use to the Republic of Letters, in which you hold one of the principal posts. I should have asked you for your counsel and aid, and I will do so.[2]

It is difficult to imagine what else this could refer to.

By 1660 others were suggesting publication. Michelangelo Ricci wrote to Leopold from Rome on December 13 of that year:

I have had the good fortune to hear, from the most noble and learned Lorenzo Magalotti, much about the experiments that are being made from day to day in your Highness' Academy, worthy of being printed so that the world may benefit from them, and return the applause that is merited by the talent and diligence of these gentlemen, and first of all by the magnanimity of Your Highness.[3]

Leopold must have recognized the wide interest in the doings of his Academy. On December 3, 1660, we find him writing to Melchisadec Thevenot in Paris:

You must not be surprised at my reply reaching you so late, for this has happened only because I wished to be able to accompany it with an account of some experiments that we have made in this Academy of ours. But as Signor Lorenzo Magalotti, its secretary, has been allowed to go to Rome, and as I myself am greatly occupied and distracted from study, what with the new baby of my sister-in-law the Grand Duchess,[4] and the Prince[5] having caught the smallpox, I have not been able to satisfy myself as I wished.[6]

The result of many such hints and much speculation was that by about 1662 the "book of experiments" was momentarily expected throughout the learned world. On March 29, 1662, the Dutch philologist Nicholas Heinsius wrote to Christiaan Huygens from Stockholm, "You will perhaps have learned from the most Serene Prince of Tuscany, Leopold de' Medici, himself, that he has already committed his natural observations to the press."[7] On November 21, 1662, Ricci told

---

[2] *BN, fonds français*, ms. 13039, fol. 71r. There is an enormous correspondence between Boulliau and Leopold, chiefly astronomical, of which I have read only a small part.

[3] *LIUI*, II, 110.

[4] Francesco Maria was born to Ferdinand II and Vittoria della Rovere on November 12, 1660.

[5] Prince Cosimo, the heir apparent.

[6] G. 282, fol. 45r.

[7] *Oeuvres complètes de Christiaan Huygens publiées par la Société Hollandaise des Sciences*, 22 vols. (The Hague, 1888–1950), IV, 101. I have not found the letter from Leopold to Heinsius that led to this remark.

Leopold that Thevenot had informed him that "Huygens, and the gentlemen in Paris and in England, are waiting very impatiently (*con sommo desiderio*) for the book of the experiments made in your Highness' Academy."[8] As it turned out, they had to wait more than five years, and we must now look into the reasons for this long delay.

An examination of the diary shows that apart from some of the experiments on freezing, one pneumatic experiment, and the attempt to measure the speed of light, all the experiments described in the *Saggi*, and nearly all that were made and not published, had been made by the end of 1662, or indeed by the beginning of that year. Nothing was recorded between January 18 and July 31, 1662, and on the latter date: "The Academy assembled in the house of Signor Lorenzo Magalotti for the purpose of repeating some experiments that seemed most necessary for the perfection of the work that is to be printed. When these become easy by practice, they all have to be made again in the presence of his Highness."[9] In fact, with the above-mentioned exceptions, almost all the experiments after that date were repetitions.

This extract makes two important points: first, that by this time it had been definitely decided to print the book, and second, that the Prince meant to be very careful that whatever experiments went into it would really work. It is also certain that before the middle of 1662 the writing of the text had begun and that early in that year some, at least, of the illustrations had been drawn, for we find Leopold writing to Magalotti from Pisa in his own scarcely legible hand that "it is felt to be advisable that until I get back to Florence no orders whatever should be given to have any work done on the illustrations. Only Signor Falconieri could be allowed to see a sample of them, for the purpose of seeing whether any of them conform to the way in which the copper plates would need to be engraved for our requirements."[10]

Not only did Leopold assume control over the contents of the proposed book, but Magalotti's successive drafts had to be submitted to the other Academicians for approval or suggestions. There were at least five drafts, and portions of each have been preserved among the manuscripts in *FBNC*.[11] The most interesting is the one to be found in G. 266, fol. 30r–82v, and G. 265, fol. 33r–50v, which corresponds to

[8] *LIUI*, II, 111.

[9] G. 262, 132r–v. I am inclined to think that this reactivation was sparked off by the advent of the Latin edition of Boyle's *New Experiments*. See p. 263 below.

[10] Florence, Bibl. Laurenziana, ms. Ashburnham 1818, fol. 10r. This is dated "8 April 1662."

[11] See Abbreviations, p. xii.

the experiments in the *Saggi* between pages XXVI and CLXXXXIV. It is interesting because this draft was one of those submitted to the Academicians, and the remarks of Borelli, Rinaldini, and Viviani have been preserved and can be compared with it, for its pages were numbered. The editors of *DIS*[11] have done us the service of publishing this draft, which they refer to as Codex A, in full,[12] as well as the remarks of the three referees.[13] They have published a good proportion of the other drafts, also variants of some of the experiments and many other matters that are to be found in the manuscripts G. 263 to G. 270. All this occupies about 170 large quarto pages of rather fine print, and my readers will understand the impossibility of dealing with more than a little of it here.[14]

The three Academicians who examined the draft each made a large number of suggestions for the improvement of the descriptions of the experiments. Many of these ideas were adopted. They also noted many experiments that should, in their opinion, have been included, and some that ought to be made. In addition to this technical criticism, Borelli, though apparently not the others, kept a watchful eye open for violations of the unwritten laws by which the Academy was supposed to be governed. For example, "I should think that it would agree better with our terms of reference to write historically, rather than in the form of an argument."[15] And again, regarding one of the experiments on the pressure of the air (p. LIV of the *Saggi*): "I raise the point whether all the subsequent discussion might well be omitted, as it does not bring in experiments, but opinions and counter-arguments to the things that might be pointed out against the pressure of fluids."[16] The draft looks innocuous enough nowadays.

Borelli also had a sharp eye for nonsense, and his remarks on the five kinds of air thermometers, invented by Viviani for measuring small differences in atmospheric pressure,[17] are scathing. In discussing Codex A he says at this point:

I submit for consideration that from an Academy set up by such a famous Prince for making experiments on natural things, and from the promise and title of this work, strangers eagerly expect to hear of experiments, and not at all their mode of operation with instruments so difficult to operate that we ourselves, with such advantages, have not been able to make them work nor derive any benefit from them. But if we must be seen to make experiments

---

[12] *DIS*, pp. 281–322.

[13] *Ibid.*, pp. 325–48.

[14] *DIS* is completely out of print, but the publishers inform me that a reissue is intended.

[15] G. 267, 18v; *DIS*, 330.

[16] G. 267, 18r; *DIS*, 329.

[17] The reader may consult pp. 131–36 below.

with them on the amount of compression of the air at various heights, and if this is impossible, and it still seems that we should record them to make clear the great industry that is displayed in the Academy, we ought to confess frankly that it has not been possible to make such instruments work.[18]

It may be supposed that Viviani did not care much for this sort of language.

It is possible to establish the fact that Codex A was written before July 31, 1662, indeed some time before, as Borelli, at least, must have made his report[19] by this time. Remarking on the experiments about the change of volume of liquids during freezing,[20] he writes as follows:

Such an Academy does not seem to be doing its duty if, into this account of freezing, it does not bring exact information of the times, accurately measured by means of pendulums, in which all the above-mentioned operations of contraction, expansion, etc., took place. Besides this, curiosity also needs to know what degrees of cold produce similar effects, and these may be measured exactly with very sensitive air, or spirit thermometers. The same particulars should be noted in the lists of observations of the freezing of wines, vinegars, etc.[21]

Now an examination of the diary G. 262 will show that these experiments on freezing were made in three groups: the first on five days in October and November 1657; the second on ten days in July and August 1660; and the third on fourteen days in July and August 1662. In neither of the first two groups are there any measurements of time or temperature, and, in fact, the ones reported in the *Saggi* come from the third group, which began on July 31, 1662, the day on which the Academy met at Magalotti's house after six months in which nothing was recorded. The passage quoted on page 57 is immediately followed by this: "They wished to see again the progress of freezing in the usual bulb, and also, by means of the pendulum, to observe the times at which the various changes of condensation and rarefaction take place in natural water while it is freezing."[22] They made the experiment then and there, also using a thermometer. Borelli had had his way. It is also highly probable that a letter written by Borelli to Leopold on March 27, 1662, beginning "According to Your Highness' orders I have considered most of the experiments, and have made what notes I have considered necessary,"[23] refers to this draft.

It is much more difficult to make an intelligent estimate of the date when Codex A was begun. One reason for this is that although the

---

[18] G. 267, 18v; *DIS*, 330.
[19] G. 267, 16r–23v; *DIS*, 327–33.
[20] *Saggi*, ed. princ., pp. CXXXXVII–CLXV.
[21] G. 267, 22r; *DIS*, 334.
[22] G. 262, 132v.
[23] G. 276, 160r.

Academy met on twenty-six occasions in 1661, the experiments they made, apart from a few on heats of solution, are among the many that are not dealt with in the *Saggi* at all. Thus, as far as the materials for this draft are concerned, it could have been written at any time after the end of October 1660.

At the end of 1662 and the beginning of 1663, Magalotti spent a long time in Rome and, evidently at Leopold's orders, discussed the draft at length with Michelangelo Ricci, who, as we saw in chapter 2, was a "corresponding member" of the Academy. On January 21, 1663, Magalotti wrote that he and Ricci had

already had many sessions about the book. I am delighted to be able to tell your Highness that you are very well served by Signor Michel Angelo, whom I find to be very different to what I had represented to you last week. He has undertaken the revision of this work with so much inclination and with such application, that I shall venture to say to your Highness that there is nothing more to be desired, and I am sure that after the work has passed through this very fine sieve, you will need nothing else, and I promise myself that what has pleased him will please others. What will turn out to be the most useful of all is the study that is being made of it in order to fashion the marginal notes, of which a very full index will be drawn up, so that the greatest simplicity and clearness will be added to the work. We are still busy with the things concerning quicksilver, from which Sig. Michel Angelo gets a marvellous delight, and I see that they put him in a good humor; but what makes me happiest is to see that he esteems certain things for which we should have the least regard; those that we think good seem very good to him, and the finest wonderful and stupendous. I mean that the multiplicity of things and their having become so familiar makes them seem to us of much less value than they are. Finally, we could not do any better; this is a man who understands without having seen, which is what the wise sagacity of your Highness so greatly desired. May it please God that he behaves in this way when he sees the parts about ice, which made us marvel. It seems like a thousand years to me.[24]

Indeed it appears to have been an ordeal to Magalotti, who was more of a poet than a scientist. Moreover, he was only twenty-five years old at this time, and a perfectionist. He was obliged to satisfy Borelli, Ricci, Rinaldini, and Viviani, and—which seems to have been harder —he had also to satisfy himself. I reproduce fol. 151 of G. 266 (Plate 3) as an example of his struggles, and there are dozens of others that could have been chosen, "with so many rearrangements, deletions, cross-references, changes, and bits of paper pasted on," writes Targioni Tozzetti, "that you could not believe it if you had not seen it."[25]

[24] *LIUI*, I, 292–93.
[25] *TT* 1, 417.

PLATE 3. A page of one of Magalotti's drafts of the *Saggi* (*Biblioteca Nazionale Centrale*).

"Thus (he goes on) I know that the excellent and so fluid sweep of the *Saggi*, which delights us, and seems as if it were dictated quite without interruption, almost as if he were talking, is a very difficult thing that deceives us by seeming easy (*è un facile difficilissimo, che inganna*), and cost Count Magalotti an immense amount of very tedious work, and a great deal of time."[26]

But there was another worry, not only for Magalotti but for Leopold and all the Academicians. Remembering what had happened to Galileo, they were in no mood to defy the Church, even now that Alexander VII occupied Saint Peter's chair. There is a good deal of evidence for their concern; for example, Leopold wrote to Magalotti from Pisa on March 25, 1664, "Sig. Lorenzo: I prefer that the manuscript should be sent on through Cardinal Rannucci. I warn you that nothing will be printed against his wishes."[27] But a more dramatic document can be found in the manuscript G. 266, of which fol. 160 *bis* is a small sheet of paper on which someone has written in a rough scrawl, "It is dangerous to make original conjectures, and I may have left the censor's office. So look at it again before giving it to the printer and also give it to me to do, although I am busy." In a later paragraph I shall make a guess as to who wrote this.

There is another draft of a great deal of the work, in which many of the experiments are described at much greater length.[28] This has led the editors of *DIS* to assign to it a later date than Draft A, and to call it Codex B.[29] It may well date from the last months of 1662 and be the one that Magalotti discussed with Ricci, for on March 5, 1663 Carlo Dati wrote to the Prince, "Magalotti tells me that he has the first treatise on the experiments (*il primo trattato delle esperienze*) ready to be given to the censor."[30] The last leaf of the draft contains the original autographs of the ecclesiastical censors. These are four in number, and the procedure was as follows: In May 1664 the manuscript was sent to the vicar-general of Florence, Vincenzio Bardi, so that it might receive the ecclesiastical permission. On May 12 Bardi passed it on to Francesco Ridolfo, who read it and on the 16th reported that it was very well worthy to be printed and contained nothing repugnant to the Catholic faith or to public morals. The Vicar-General then

---

[26] *Ibid.*

[27] Bibl. Laurenziana, ms. Ashburnham 1818, fol. 16r.

[28] G. 264, 94r–132v; G. 266, 107v–22v; G. 267, 11r–v. These leaves bear an old series of numbers, at first in pages and then in folios; folio 132 of G. 264 is numbered 55, and folio 11 of G. 267 is numbered 56.

[29] *DIS*, 61.

[30] G. 276, 186v. It is possible that the date in *stile commune* may really be 1664.

sent it on to the Chancellor of the Holy Office for Florence, who sent it to the censor, Lelio Mela, who read it and passed it on May 18; and so it went back to the office of the Vicar-General and finally received the *imprimatur* on July 31.[31] Meanwhile, the paper had even been ordered, for we find Carlo Dati assuring Leopold on February 1, 1664 that "four boxes of the best paper, and perhaps more, are already on hand for the printing of the book of experiments," and that he had advised Magalotti of this.[32]

Everything was now in the clear, and in July (apparently without waiting for the final license) sheets were being printed and sent to Cardinal Sforza Pallavicini, a great authority on the Tuscan language, for his opinion;[33] not, be it noted, for his opinion on its orthodoxy but on its style. Ottavio Falconieri reported that

In general [the Cardinal] highly praised the clarity and elegance of the style, and said only that he had observed a few little things that are somewhat affected. These are almost the same ones that I pointed out to you in the letter written to you by Burattino. He also asked me whether the work would contain a variety of things, etc. I answered that I believed so, but not being informed of the details, I went no further.[34]

It would seem that he had been given only the first sheets by this time. Magalotti dashed off a reply by return of post on July 29:

I am extremely annoyed, particularly by what you tell me you wrote to me about the affectations that you say you found in my book, which seem also to have offended the ear of Cardinal Pallavicino.[35] For the time being, I put off complaining about you, for I have seen how you come to the rescue of my impudence with your tactful liberality. But if you think you have been sparing of criticism, blaze away before I have to be offended at you; and above all tell me one by one the things that do not please you, either by an immoderate affectation of Tuscan purity, or by any other defect whatever. [A list of chapter headings follows.] But remember that they are samples, and in none of these subjects is experimenting as yet finished. I say finished, meaning what it is planned to do, for I know very well that what might be done is infinite.[36]

---

[31] The documents are printed in *DIS*, 276.

[32] *G.* 276, 184r.

[33] Falconieri to Magalotti, July 26, 1664, in A. Fabroni, ed., *Delle lettere familiari del Conte Lorenzo Magalotti e di altri insigni uomini a lui scritti*, 2 vols. (Florence, 1769), I, 83–85.

[34] *Ibid.* This is also printed in *TT* 1, 416, and in *DIS*, 26–27.

[35] In *G.OP.*, XX, 500–1, he is called Pallavicini, as are other members of his family.

[36] *Lettere familiari*, I, 85–86; also in *TT* 1, 416–17, and *DIS*, 27.

It cannot be too strongly emphasized that Magalotti was much more a poet than a scientific man, and we may be sure that slurs on his style would have affected him much more than the discovery of factual errors.

Now, although the printing must have been well advanced by the end of July 1664, it suddenly ceased, and, as far as I have been able to learn, there was no progress for over a year. There was, however, at least one more draft. Our copy of the 1664 draft, carefully written out by an amanuensis, is well provided with marginal notes and cancellations by Magalotti. Indeed in the manuscripts there are a great many partial drafts in various secretarial hands, but no clean copy of the whole work. Even the courageous editors of *DIS* have not entirely succeeded in reducing this chaos to order.[37]

On December 11, 1665, Alessandro Segni, who was in Paris, wrote to Magalotti that his book was anxiously awaited and that a title was even being attributed to it, *La fisica provata per l'esperienze*.[38] Magalotti put an end to speculation about the title and also showed how things stood, in the course of a long reply dated January 15, 1666.[39]

Your letter certainly stung me out of my laziness, as it immediately made me take up once more the interrupted printing of the book, so as not to interrupt it again until the work is entirely finished. . . . Meanwhile I am sending you the title established and confirmed by the Academicians, which will serve to correct the one that, you wrote, is being given to it where you are. *Saggi di naturali esperienze fatte nell'Accademia del Cimento sotto la protezione de Ser.mo P.pe Leopoldo di Toscana, descritte dal Segretario di Essa Accademia. A Ferdinando Secondo Gran Duca de Toscana.* Beneath this goes a tablet with the following coat-of-arms: A glass vessel of the kind used to try gold with aqua regia, with a mass of gold at the bottom that is dissolving in the acid, as will be recognized by experts from the shape of the bottle and the smoke that it is making, although there is no fire underneath it. The motto, taken from Dante, "Provando e riprovando." At the foot of the tablet is written "Cimento," which is the name.[40]

The title was thereafter altered only by removing the phrase referring to the Grand Duke; but the coat-of-arms was completely changed. Instead of the vessel in which a solution is being carried out we have three vessels on a table in front of a blazing fire, probably in reference to the assaying of gold by cupellation. The final coat-of-arms was apparently by Ciro Ferri, a painter of the school of Pietro da Cortona.

---

[37] See *DIS*, 67–71.

[38] Ferdinando Massai, *Rivista delle Biblioteche e degli Archivi* 28 (1917): 136, note 3.

[39] Printed by Massai, *ibid.*, pp. 134–38.

[40] *Ibid.*, pp. 135–36.

The authority for this attribution is a note in G. 267, "Show the Cimento to Ciro Ferro for making the drawing of the coat-of-arms."[41]

On March 30, 1666, we find Magalotti writing again to Ottavio Falconieri in a mood of unrelieved gloom:

I don't know what to tell you about the book. I am ashamed of myself; but it is my fault, or rather the fault of an unconquerable aversion that I have formed for it. It is only a few weeks since I have again taken up the printing, which has been in abeyance for sixteen months. To tell you the truth, I am very ill-satisfied with what I have written, and it is impossible for me to put my mind to it. You must believe that I consider this book as something that must entirely discredit me, and now you see whether I go slowly about it with good reason. But I swear to you that I will not now take my lips from this bitter cup until not a drop remains there.[42]

This has messianic overtones, probably as an over-compensation for his sins of omission. It was apparently the preface that gave him trouble, and by May 18 he has thought of an ingenious dodge, as he tells Falconieri:

Oh, what would I give to spend an evening with you, to get your advice about an idea that I have about a certain preface to the book of experiments, which has of necessity to be finished by the end of February [1667?]. Besides the dedicatory letter to the Grand Duke, and an address to the reader, in which I have to tell about the founding, the objects, and the principles of our Academy, I am considering that it might be ornamented by adding a discourse about experimental philosophy. And because it seems hard work for me to do this, I should like to see if it is possible to introduce gracefully one of those that have been done already—but enough! I will send it to you and tell you how I am thinking of tacking on something that may turn out well, so that you may help me to make easy a transition that, I suspect, is going to be very difficult for me.[43]

Fortunately, the preface that eventually appeared contained no such scissors-and-paste job, nor indeed any dissertation on experimental science. It does contain a good deal of very high-flown language; but that is another matter.

As far as the progress of the *Saggi* is concerned, we know that copies of a title page dated 1666 were printed in that year,[44] but little further progress seems to have been achieved. It is noteworthy that Magalotti had been given a deadline—far enough ahead, one would suppose— for the completion of the preface. But the deadline passed and still the work was not printed.

---

[41] Quoted in *DIS*, 274.
[42] *Lettere familiari*, I, 159; also in *TT* 1, 415; *DIS*, 26.
[43] *Ibid.*, I, 171–72; also in *TT* 1, 415–16; *DIS*, 26.
[44] See appendix A.

Finally the dedication to Ferdinand II, signed "Il Saggiato[45] Segretario," was dated July 14, 1667, and a license to print was obtained on October 11. But Magalotti had evidently had enough and was not even in Florence on the date of the dedication. He wrote to Ottavio Falconieri on July 23, from Venice:

> You have done very well to come to an understanding with Luigi Rucellai about the preface, because he and the Prior, (hush! hush!), will correct everything that needs it, without bothering the Prince about it. That's enough; I should not like to be obliged by Signor Michelagnolo[46] Ricci to recast my thoughts; for I would rather have it printed in Geneva.[47] Good heavens! this man ruins my temper; either the difficulty arises from the thing itself, or from ignorance, or from the subtlety of the censor [Father Sebastiano de Pietrasanta, an Observant friar]. If it is due to this, why can't he be made to see the point? Now let them do almost anything they want and make a mess of it in the end.[48]

This Sebastiano de Pietrasanta, the *revisore*, had been one of Viviani's teachers, and there is nothing to make us suppose that he was inimical to the Academy. I think that it is rather likely that it was he who wrote the brief warning referred to on page 72.

At any rate the *Saggi*, passed once again by the censors, was printed before the end of October 1667[49] and, thanks to an undated account written by Magalotti, we know some of the details.[50] It appears that 800 copies were run off, 50 on "royal paper," 25 on *carta detta de' frati* (friars' paper), and 725 on ordinary paper. The ones on royal paper had larger margins. Magalotti thought that the inside margins had been made too narrow and went on to complain of the printer Cocchini, who was not used to printing books with many illustrations and not very good at striking copper engravings. Some get misplaced, some inverted, and "It is certain that in order that these errors should seem fewer, Cocchini would have the defective leaves separated and dispersed among the good ones. Therefore Ciaccheri,[51] before putting the

---

[45] The members of Italian academies gave themselves fanciful names. "Il Saggiato" [the tried?] seems to have been that of Magalotti.

[46] There seem to be two spellings, Michelangelo and Michelagnolo, in contemporary letters.

[47] To escape from ecclesiastical control.

[48] *Lettere familiari*, I, 176; also in *TT* 1, 416; *DIS*, 26. The expression "he and the prior" (*egli ed il Priore*) is strange, because Luigi Rucellai seems to have been a prior. I do not know who the other man was.

[49] The evidence for this is in several minutes of letters in G. 282, uniformly dated "Ottobre 1667," from Leopold to various people, transmitting copies. A typical one, to the Abbé Séguin, is on fol. 161r.

[50] This document was in the Ginori Venturi archives, since dispersed; fortunately it was printed in *DIS*, 28–31.

[51] Apparently the name of the binder.

leaves in twos,[52] will have to give each leaf a quick examination, to see that the ornaments, the tail-pieces, and the copperplates are in their places,"[53] and throw out the defective sheets.

This document seems to have been written after the text had been printed and before the printing of the end matter, because Magalotti says:

In binding the books, immediately after the leaf where the two titles are [i.e., the half-title and the title], I should think that a portrait of the Grand Duke ought to be put, and right after that the letter of dedication and the preamble. In case it is decided to put in some poetry (and if it is not very fine it would perhaps be better to leave it out), then between Signor Carlo Dati and Signor Luigi Rucellai, they will be able to see what is the most proper place for it.

It seems to me that the index should be put at the end of the book, a copy of which you could begin to send as quickly as possible to the Abbé Falconieri, so that he and Sig. Michelangelo Ricci could begin to work at this index, as Sig. Michelangelo promised me in Rome a few years ago. . . .

The copperplate of the coat of arms that goes on the title-page has to be printed, and this will be in the hands of Signor Luigi Rucellai.

The portrait of His Highness [the Grand Duke] is in the hands of Sig. Paolo Falconieri, to whom it is absolutely necessary to send the paper so that he can have it printed in Rome. An engraving of such fineness must on no account be entrusted to the awkward skills of these men of ours. This could be done at any time.

In the meantime His Highness will be able to make up his mind about the quatrain that goes under this same portrait, and to order Signor Paolo or, if he is not in Rome, his brother [Ottavio] the Abbé, to have it engraved there.[54]

Nothing could make it clearer that such verses were anything but spontaneous demonstrations of esteem. It is probable that Leopold wrote the quatrain himself, for in a file containing a great number of the prince's poetical efforts there are no less than seven versions of it, including the one that was used.[55]

The final form of the title-page was clearly established before the end of 1666, as is evinced by the existence of many copies bearing that date.[56] It is also highly improbable that the "1666" title-pages were printed in 1667 with a wrong date and that the date was corrected during a run; for the type was entirely reset, which would not have been necessary in such an event. Beyond this I do not think we can go.

---

[52] They were printed singly but, for the most part, sewn in sections of two. See also appendix A.

[53] *DIS*, 28.

[54] *DIS*, 30.

[55] *FAS*, Miscellanea Medicea, filza 3, inserto 1, items 58–60.

[56] See appendix A.

The drawings of instruments and apparatus in the *Saggi* are distinguished by their functional simplicity and absence of all ornament, contrasting in this respect with the baroque love of decoration that pervaded the century. In particular, very few of them give any indication of how the delicate and sometimes heavy apparatus was actually supported. From the historian's point of view this is a pity. It might be supposed that the deficiency is repaired by a series of elegant pen-and-wash drawings that has been preserved, showing most of the instruments with elaborately decorated supports.[57] I reproduce one of these as Plate 4.[58] It is possible that these are by a well-known artist, Stefano della Bella.[59] The question that must occur to us regarding such figures is whether the ornamentation actually existed or was imagined by the artist. There is evidence for the latter alternative, as for example in Rinaldini's remarks on the figures. The second one displeased him for eight reasons, one of which was "because there are supports with puerile ornaments and things."[60] And in the suggestion of Magalotti, at the beginning of a series of remarks on the figures: "take away from the first one the decorations on the vessels." And again "The ornaments on the vessels must not get in the way of what has to be seen, and of the operation."[61]

A good many of the figures given to, or prepared by the engraver Modiana, plain this time, have also been preserved,[62] as has a collection of line-and-wash drawings of various instruments and apparatus, in blue and ochre.[63] Not all of these instruments are represented in the *Saggi*.

It seems most probable that the final sheets of the work, that is to say, the dedication, the preface, and the end matter, came from the press in October 1667 and that Leopold at once gave instructions for its distribution. It must be emphasized that none of the copies were made available through booksellers. All that were distributed were sent out as gifts to the friends and correspondents of the Prince, including several crowned heads, and most of the important scientific workers in Europe. Outside of Italy, copies were sent to the various Residents of the Grand Duke, usually in sheets, to be bound locally

[57] *G.* 270, fol. 17r–60r.

[58] *Ibid.*, fol. 24r.

[59] See *G.* 290, fol. 7r. Targioni Tozzetti also makes this ascription (*TT* 1, 381), but the editors of *DIS* (p. 66) question it. The Court diary of Settimani (*FAS*, Manoscritti, filza 136, fol. 229r) calls Della Bella "the celebrated engraver on copper" in reporting his death on July 23, 1664.

[60] *G.* 267, 39r. *DIS*, 345.

[61] *G.* 267, 43r. *DIS*, 347.

[62] *G.* 270, 5r–16r.

[63] *G.* 270, 115r–36r.

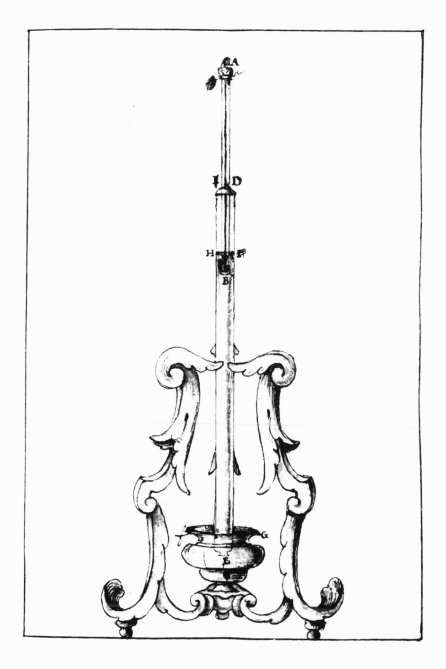

PLATE 4.   A pen-and-wash drawing of an apparatus
(*Biblioteca Nazionale Centrale*).

before presentation. Rather special arrangements, which will be discussed in chapter 6, were made for England.

As a result of this method of distribution, it was well into 1668 before most of the copies for other countries were received and acknowledged, and some did not reach their destinations until late in the summer. There were after-thoughts, of course; on June 5, 1668, Leopold wrote a charming letter to Pierre Petit in Paris, sending him the book.[64]

I have no means of estimating how many copies were sent out. Certainly the number that were acknowledged in letters that have been preserved falls far short of the 800 that were printed, but I do not feel constrained to pursue this matter further. Targioni Tozzetti notes that in 1737 "several bundles" of copies, many with badly printed sheets, were found in the Magliabecchi library.[65]

3. The Title of the *Saggi*

A few words are desirable on the significance to the Academicians themselves, or to Magalotti and to the princes, of the title of their book.[66] Readers of the English translation by Waller[67] will have concluded that *Saggi* must mean "essays." But the noun *saggio* has several meanings in Italian. It may indeed mean an essay; it may also mean an assay, as in testing the noble metals. But another meaning is common—a sample, a specimen, a "part or example of a whole for examination."[68] It should, however, be mentioned that a noted lexicographer has suggested a fourth interpretation of *saggio*: "a study, monograph, memoir, as evidence of researches, professed theories;"[69] but the force of this is considerably lessened by the fact that the passage cited in support of this definition is precisely "*Saggi di naturali esperienze*, di L. Magalotti." I hesitate to believe that Magalotti used this occasion to enlarge the Italian tongue.

The interpretation "examples" is strongly supported by the preface to the *Aggiunte* (additional material) collected from the documents of the Accademia by G. Gazzeri and appended to the third Florentine edition.[70] The preface begins as follows:

[64] G. 282, 165r.
[65] *TT* 1, 418.
[66] This note is borrowed from my paper in *British J. Hist. Sci.* 4 (1969): 283–86.
[67] *Essayes of Natural Experiments*, etc. (1684). See appendix A, no. 14.
[68] *parte o esemplare di un tutto da prendere in esame.* (Dizionario Garzanti della lingua Italiana, Milan, 1965).
[69] "Studio, monografia, memoria, come prova di indagini, teorie professate." Zingarelli, *Vocabolario della lingua Italiana* (Bologna, 1955).
[70] See appendix A, no. 10.

The title *Saggi di naturali esperienze* given by the academicians to the interesting book placed before us here is a sufficient announcement that the experiments described in it are a selection of a few among many. And indeed a very great number were made by the Academy between its foundation and its too speedy end.

I trust that it will be clear to the reader that I am not concerned with the word "essays" as it appeared to Waller's seventeenth-century compatriots, for a glance at the Oxford English Dictionary or even the Shorter Oxford will show that the meanings "assays" and "attempts" were all current at the time, and the word "of" in the title makes it hard to decide which of the last two Waller meant, the former seeming to demand "on" and the last to require "at." My question is whether *saggi* should be translated "essays" at all.

Further support for "examples" comes from the first sentence of the dedication to the Grand Duke Ferdinand II in the original: "Il pubblicar con le stampi i primi Saggi delle naturali esperienze, che per lo spazio di molti anni si sono fatte nella nostra Accademia...."[71] The reader is asked to note that in this extract *Saggi* is followed by *delle*, not by *di* as in the title. It seems to me that this makes the matter very clear. Waller translates the passage as follows: "The Publishing of these First Essays of Natural Experiments, which for many Years have been made in our Academy...." I incline, on considering this, to the belief that Waller interpreted "essays" as "short (literary) compositions." I am quite certain that the Academicians did not mean this. To emphasize the point, let me quote from Giovanni Targioni Tozzetti, who knew more about the Accademia del Cimento than anyone else before or since:

> ... It appears that the first idea of the Academy was to print all the observations and experiments that had been made up to that time ... but ... wishing to make a selection (*scelta*) of the better things under the name *Saggi*, they changed their first plan, and so many other very useful discoveries remained buried.[72]

The translation of *naturali* has bothered me for many years. I cannot imagine what a natural experiment might be; surely an experiment is essentially artificial, an interference with Nature in order to ask her a question. It is scarcely natural to put a lizard or a bird into a vacuum.

Nowadays we should say "scientific experiments." As a matter of fact, Waller would not have been an innovator if he had written this,

---

[71] "The printing of the first examples of the experiments in natural philosophy that have been made in our Academy during the space of many years...." Concerning the translation of *naturali* see below.

[72] *TT* 1, 425.

not by at least six years. I quote from the Oxford English Dictionary, s.v., "Scientific": "1678 Moxon *Mech Dyalling* 3 Scientifick Dyalists . . . have found out Rules to mark out the irregular motion of the Shaddow . . . And these Rules of adjusting the motion of the Shaddow to the motion of the Sun may be called Scientifick Dyalling." But still, he might have been less easily understood than if he had entitled his translation *Examples of Experiments in Natural Philosophy made in the Academy del Cimento. . . .* So doing, he would, I believe, have come very close to the meaning of the original. But it must be noted that he did not attempt to translate the word *cimento,* which in other contexts contains an element of risk or hazard. In this context, neither shall I.

4. The Style of the *Saggi*

We must turn to Italian critics for judgments on the literary style of the *Saggi.* Targioni Tozzetti, while he had no doubt whatever about the value of its contents, had reservations about the style and was also scathing about the printing of the *editio princeps.* He thought that the woodcut initials, the decorations, and the tailpieces were all too large—"suitable for a Choir Book." The spelling was "rather disgusting," and in fact little was right. Prince Leopold had spent a great deal on the book, but had been badly advised. Targioni Tozzetti believed that it would have been better if Viviani or Dati had been told to write it.[73]

In our own century Stefano Fermi expressed some reservations about the style of the *Saggi* in his biography of Magalotti.[74] He says that Magalotti, born in Rome, was forcing himself to write like a Tuscan, and the result was a certain affectation, a rather precious style. It is interesting that Fermi contrasts it with the style of Galileo, who, he says, brought about a revolution in literature as well as in science, introducing a new prose, easier, clearer, and more natural—the best prose of the seventeenth century.[75] A non-Italian can only record such an opinion; but, having attempted a translation of the *Saggi,* I feel constrained to say that there are two distinct styles in it. The actual record of the experiments is clear, concise, and easy to translate. On the other hand the introductions to the chapters[76] and the preface show the preciosity and striving for effect that Fermi deplores.

[73] *TT* 1, 417–18.

[74] S. Fermi, *Lorenzo Magalotti, scienziato e letterato, 1637–1712,* etc. (Piacenza, 1903), pp. 91–94.

[75] *Ibid.,* p. 94, note 3. This seems to be the general opinion at the present day.

[76] See for example the chapter on artificial freezing, p. CXXVII.

Examples
of Experiments
in Natural Philosophy

Made in the Academy
del Cimento
Under the Protection of the Most Serene
Prince Leopold of Tuscany
and Described by the Secretary of that Academy

Florence
Guiseppe Cocchini at the Sign of the Star, 1667.
With the permission of the Authorities

SAGGI
DI NATVRALI
ESPERIENZE
FATTE NELL' ACCADEMIA
DEL CIMENTO
SOTTO LA PROTEZIONE
DEL SERENISSIMO PRINCIPE
LEOPOLDO DI TOSCANA
E DESCRITTE DAL SEGRETARIO DI ESSA ACCADEMIA.

IN FIRENZE

Per Giuseppe Cocchini all' Insegna della Stella. MDCLXVII.
CON LICENZA DE SVPERIORI.

PLATE 5. The title page of the first edition of the *Saggi*.

TRANSLATOR'S NOTE

The *Saggi di naturali esperienze* has not hitherto been presented in a form that makes clear its relation to the science of its own day and the decades just before. The valuable edition of Giovanni Targioni Tozzetti (see appendix A, no. 8), which I have used a great deal, relates each experiment to the unpublished papers of the Academy, and the Latin translation of Musschenbroek (no. 15) comments on many of them from the standpoint of the early eighteenth century. In my notes to the present translation I have kept in mind two purposes: to establish the date or dates of each experiment, with appropriate references to the manuscripts and also to Targioni Tozzetti's *Notizie*; and to comment, where comment seemed to be called for, on the origins of some of the experiments and the reasons for some of the difficulties that led to inconclusive results. I have tried to give citations for the infrequent references to other authors and to supply modern metric equivalents for the Florentine weights and measures, based on the admirable account in *DIS*, pp. 74–75.

The beginnings of the pages in the original are given by Roman numerals in square brackets. The reader will notice that many of these are missing. This is because the plates and text in the original were numbered in the same series, and for the convenience of the reader the same plate was often repeated, sometimes several times. In this edition each plate is printed once, opposite the first reference to it. As the plates in the second edition of 1691 are clearer than those in the first, and were freed from some errors (see appendix A), those in this book have been reproduced from an excellent copy of the second edition in the library of the University of British Columbia. They are reduced to almost exactly two-thirds of their original linear dimensions.

The index, which by modern standards is of little use, has not been translated, but the *Saggi* has been included in the analytical index at the end of this volume.

The marginal rubrics in the original have been discarded with some regret, but such things do not lend themselves to modern methods of typesetting. However, it is doubtful whether they contain any information that is not in the text.

MOST SERENE LORD:

The printing of the first samples of the experiments in natural philosophy that have been made for many years in our Academy with the assistance and under the continued protection of the Most Serene Prince Leopold Your Highness' brother, will itself carry to those regions of the world in which virtue shines most brightly, new evidence of the great munificence of Your Highness and call back towards you with a new sense of gratitude the true lovers of the fine arts and the most noble sciences. We ourselves are the more greatly moved to a more devoted acknowledgment, for we have been so much nearer, enjoying the strong influence of your beneficent hand. Meanwhile, with the favor of your protection, with the stimulus of your ability and your own taste and inclination, and above all with the honor of your presence, sometimes coming into the Academy and sometimes calling it into your Royal apartments, you have given the Academy its reputation and its fervor and at the same time enlarged the progress of our studies.

These considerations very easily teach us our clear duty of consecrating to Your Highness' lofty name this first fruit of our labors, inasmuch as nothing can come from us in which Your Highness might have a greater part and which it would thus be more proper to offer you or which would come nearer to meriting the good fortune of your generous approval. Truly, for the superabundance of so many and so remarkable favors we feel no greater desire than to see ourselves thus strictly bound to Your Highness; not because we willingly carry the burden of such precious and valued obligations, but because we should wish only to be able to offer you something that is not your own, so that we might at least flatter ourselves that we had brought you some slight recompense and expressed some thanks of our own choice to Your Highness that would not be entirely your own or from necessity. But now we are perforce content to have in our hearts such just and proper sentiments, since the fruit of these new philosophical speculations is so strongly rooted in Your Highness' protection that not only what our Academy produces today but everything that matures in the most famous schools of Europe, and that will come to light in succeeding ages, will likewise properly be due to Your Highness as the gift of your beneficence. For as long as sun, stars, and planets shine and the heavens endure, there will remain the glorious memory of one who contributed so much with the power of his most happy favors to such new and stupendous discoveries and to the opening of an untrodden path to the more exact investigation of truth. Although we have so little to offer, there is something that sharpens our respectful gratitude. This indeed is the pleasure with which we bear our poverty, while everything overflows in the greater abundance of glory for Your Highness, who, having yourself done whatever new, good, and great things will ever be found in the wealth of the sciences, have enfeebled in others all power to equal you. So much and no more are we in a position to offer Your Highness. Filled with reverence and homage and begging your continued protection, we pray God to grant you the highest prosperity and greatness.

Your Highness' very humble servants,

THE ACADEMICIANS OF THE CIMENTO

Florence, 14 July 1667
The Secretary for the *Saggi*[1]

---

[1] *Il Saggiato segretario.* "Il Saggiato" is really one of the fanciful names assumed by Italian academicians under a long-standing and, as Targioni Tozzetti says (*TT* 1, 412), boring and ridiculous custom.

PREFACE[2]

Without any doubt the first born of all the creatures of the Divine wisdom was the idea of Truth, to which the eternal Architect held so closely in his plan for the building of the universe that nothing whatever was made with even the slightest alloy of falsehood. But when later on, in the contemplation of such a high and perfect structure, man acquired too immoderate a desire to understand the marvellous power of it and to take the measure and proportions of such a beautiful harmony, then, wishing to enter too completely into the truth, he came to create an indefinite number of falsehoods. This had no other cause but the desire for those wings that Nature did not want to give him, perhaps fearing that she might at some time be surprised in the preparation of her most stupendous works. On such wings he began to soar, though oppressed by the weight of his material body, gaining the strength to rise higher than the gamut of perceptible things, and so tried to fix himself in a light such as is no longer light when it enters the eyes, but, fading, darkens and changes color.

This was how the first seeds of false opinions arose from human rashness. It is not that the brightness of God's beautiful creations is at all obscured by this, or that they are in any way spoilt by their commerce with it; for these defects all lie in human ignorance, where they have their root. Meanwhile, man, improperly fitting causes to effects, takes their true essence neither from one nor the other, but putting them together forms a false science within his own mind. But it is not that the sovereign beneficence of God, when creating our souls, does not let them glance for an instant, so to speak, at the immense treasure of His eternal wisdom, adorning them with the first gleams of truth as with precious gems. We see that this is true from the notions, preserved in our spirits, that we cannot have learned here, but must inevitably have gathered from another place.

[2] Notes for this preface appear in G. 267, 6r–v (see DIS, 274) and others on fol. 3r–v, 7r–v, and 8r–v (DIS, 274–77 passim), and in G. 264, 4r–8r (DIS, 277–78).

But it is indeed our misfortune that these most noble jewels, badly held in the setting of a soul still too tender, just as it comes down into its earthly house and wraps itself in this common clay, at once fall out of their fastenings and are so befouled as to be worthless, until at length by assiduous and eager study they are brought back to their places. Now this is exactly what the soul is trying to do in the investigation of natural things. Here it must be confessed that nothing is better for this than geometry, which at once opens the way to truth and frees us in a moment from every other more uncertain and fatiguing investigation. The fact is that geometry leads us a little way along the road of philosophical speculation, but then abandons us when we least expect it. This is not because it does not cover infinite spaces and traverse all the universal works of nature in the sense that they all obey the mathematical laws by which the eternal Understanding governs and directs them; but because we ourselves have not as yet taken more than a few strides on this long and spacious road. Now here, where we are no longer permitted to step forward, there is nothing better to turn to than our faith in experiment. As one may take a heap of loose and unset jewels and seek to put them back one after another into their setting, so experiment, fitting effects to causes and causes to effects—though it may not succeed at the first throw, like Geometry—performs enough so that by trial and error[3] it sometimes succeeds in hitting the target.

But we must proceed with great caution lest too much faith in experiment should deceive us, since before it shows us manifest truth, after lifting the first veils of more evident falsehood, it always makes visible certain misleading appearances that seem to be true. These are those indistinct outlines that show through the last veils that cover the lovely image of Truth more closely, through the fineness of which she sometimes appears so vividly pictured that anyone might say that all was discovered. Here, then, it is our business to have a masterly understanding of the ways of truth and falsehood and to use our own judgment as shrewdly as we can, so as to see clearly whether it is or is not. There is no doubt that to be able to do better we must at some time have seen Truth unveiled, an advantage possessed only by those who have acquired some taste for the study of geometry.

Yet besides trying new experiments, it is not less useful to search among those already made, in case any might be found that might in any way have counterfeited the pure face of Truth. So it has been the aim of our Academy, besides doing whatever experiments occurred to us, to make them, either from fruitful curiosity or for comparison, on

---

[3] *Provando e riprovando,* the motto of the Academy. See p. 52.

those things that have been done or written about by others, as unfortunately we see errors being established by things that go by the name of experiments. This was precisely what first moved the most perspicaceous and indefatigable mind of His Highness Prince Leopold of Tuscany, who, as a recreation after the diligent management of affairs and the ceaseless anxieties that arise from his exalted state, began to exercise his intellect upon the steep road of the most noble kinds of knowledge. Therefore it was quite easy for His Highness' sublime understanding to see how the trust in great authors is most often inimical to talented men. Either from excessive trust or from reverence for a name, such people dare not consider doubtful what is conjectured on such great authority. His Highness decided that it should be the task of his great spirit to verify the value of their assertions by wiser and more exact experiments; and when the proof or the refutation had been obtained, to present it as a gift, so precious and so desirable, to whomsoever is anxious to discover Truth.

These prudent precepts of His Highness our protector, embraced by the Academy with the veneration and esteem that they deserved, have not had it as their aim to make them indiscreet censors of other laborious scholars, or presumptuous dispensers of truth or error; but have mainly been intended to encourage other people to repeat the same experiments with the greatest rigor, just as we have sometimes dared to do to those of others, even though in publishing these first examples we have for the most part abstained from this, the better to confirm, with proper regards to all our readers, the sincerity of our dispassionate and respectful opinions. In fact, to give full scope to such a noble and useful enterprise we should wish for nothing else but a free communication from the various Societies, scattered as they are today throughout the most illustrious and notable region of Europe. These, opening such a profitable mutual correspondence with the same purpose of reaching such important ends, would all go on searching with the same liberty, according to their means, and participating in Truth. For our part, we shall contribute to this work with the greatest sincerity and ingenuousness.

Wherever we have reported the experiments of others, we have always cited the authors, as far as they were known to us, and have often freely confessed being much helped in these trials, which we did not after all succeed in bringing to as happy an end. But as the clearest proof of the open sincerity of our procedure, all may see the liberality with which we have always shared them with anyone passing through these parts who showed a desire to enjoy some account of them, whether as an act of courtesy, or because he esteemed learning, or from the incentive of noble curiosity. This we have done since the first days of our Academy, founded in the year 1657, during which early days were discovered, if not all, then the greater part of the things of which these examples are now being printed.

If, after all, someone perceives that among the things that we give out as ours some are to be found that were first imagined and published by others, this will never be our fault; for as we can neither know all nor see all, nobody should marvel at the correspondence between our intellects and those of others, just as we really are not surprised at the correspondence between theirs and ours.

We should certainly not wish anyone to persuade himself that we have the presumption to bring out a finished work, or even a perfect pattern of a great experimental history, knowing well that more time and labor are going to be required for such a purpose. Everyone may perceive this from the very title we have given this book, merely of examples. We should never even have published this without the powerful stimulus that we have had from honorable people whose kind entreaties persuaded us to suffer the shame of printing such imperfect beginnings.[4]

Finally, before everything else we protest that we would never wish to pick a quarrel with anyone, entering into subtle disputes or vain contradictions; and if sometimes in passing from one experiment to another, or for any other reason whatever, some slight hint of speculation is given, this is always to be taken as the opinion or private sentiment of the academicians, never that of the Academy, whose only task is to make experiments and to tell about them. For such was our first intention and also the purpose of that exalted Person who with his particular protection and great knowledge led us along the way, and whose wise and prudent counsel we have always exactly and regularly obeyed.[5]

[I]

EXPLANATION OF SOME INSTRUMENTS FOR DISCOVERING THE ALTERATIONS OF THE AIR THAT RESULT FROM HEAT AND COLD

It is very useful, indeed necessary in making experiments on nature, to have an exact knowledge of the changes that take place in the air. For, since the air takes all things into its bosom and weighs on them with the towering height of its realm, everything groans under its

---

[4] See p. 80 above.

[5] In the surviving copy of the diary of the Academy (G. 262) there is a much briefer and less pompous preface (fol. 1r–v), apparently intended for publication (DIS, 274).

pressure, and as this is stronger or weaker, so all things become more oppressed, or breathe more freely. Thus according to the various states of the air the quicksilver rises or falls in the tube that contains a vacuum. This, some think, is when the air becomes rarer or denser, lighter or heavier, according to the varied warmth that it gets from sun or [II] shade, heat or cold, as well as from being clear and free, or encumbered with clouds, or heavy with fog. Thus, pressing with a varying force on the mercury beneath, it constrains it to rise more or less in the tube that is immersed in this. Therefore it is needful, both for this experiment, which will first be fully dealt with, and for others that will be recorded later in this book, to have instruments of a sort that will give us assurance that they are telling us the truth, not only about the greatest alterations of the air but, if this be possible, also about the smallest differences. We shall therefore speak of those in- struments that have served us, although some of them may by this time have gone into various parts of Europe,[6] so that they will not now be news to many people; but in any case there may be someone who may wish a more detailed description, if not of their use, which is easily understood, at least of the manner and craft of their construction.

The first instrument is that shown in figure I. This, like the others, serves to find out the changes in the heat and cold of the air. It is commonly called a thermometer. It is entirely of fine crystal glass,[7] made by those workmen who, using their own cheeks as bellows, blow their breath into a glass tube at the flame of a lamp, either with one wick, or several small ones. Gradually blowing, as the necessities of their work require, they succeed in forming the most delicate and marvelous works in glass. We call such a craftsman the glassblower.[8] It is his task to shape the bulb of the instrument of such a size and capacity, and to attach a tube of such a bore, that when it is filled with spirit of wine up to a certain mark [IV] on its neck, the cold of ice or

[6] The Grand Duke saw to it that the thermometer of his invention was sent far and wide. They were in Poland in 1657, France by May 1658, and the first one reached England in 1661. See also chapter 2, p. 21.

[7] Throughout the *Saggi* we read of *cristallo* and of *vetro*. *Cristallo* is lead glass, used for fine tableware and for the best instruments of the Academy. Unless the two terms appear in the same sentence I have translated either by "glass."

[8] The best of the surviving glassware, some of which unquestionably dates from before 1657 and was made for Ferdinand II, is almost certainly the work of the glassblower Antonio Alamanni, called "Il gonfia," who died during the times of the Academy and was succeeded by another glassblower, perhaps Iacopo Mariani, somewhat less able than Alamanni. See Florence, Museo di Storia della Scienza, *Catalogo degli strumenti* (Florence, 1954), p. 27. Vincenzio Antinori, in his introduction to the third Florentine edition (1841) of the *Saggi*, p. 41, calls this man Giuseppe Moriani.

III

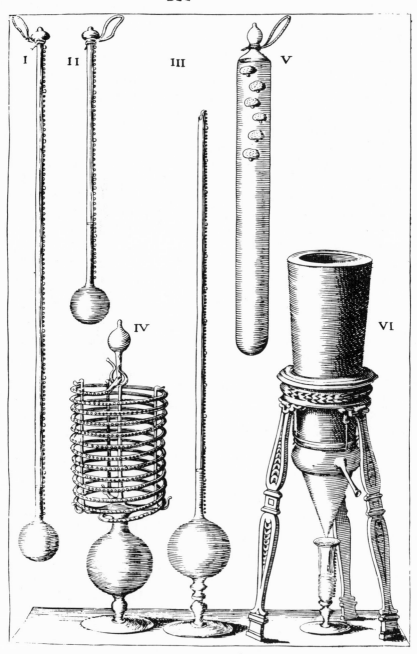

snow alone may not suffice to condense the spirit below the 20th degree on the little tube,[9] while on the contrary the greatest activity of the solar rays, even in the height of summer, may not have the strength to expand it beyond 80 degrees.

The way of filling the instrument is to heat the bulb red-hot and at once plunge the open end of the tube into the spirit of wine, so that this will be gradually sucked up. But because it is difficult, if not entirely impossible, to get out all the air by rarefying it, and because the bulb remains partly empty if a little is left in, the filling can be finished with a glass funnel that has had its neck drawn out extremely fine, which can be done when the ball of glass is red-hot, for then it can be drawn into a very fine thread, hollow and empty inside, as is clear to anyone who knows glass-working. With such a funnel, then, the filling of the thermometer can be finished, introducing the narrow neck of the funnel into the tube and driving the liquid in with the force of the breath, or sucking some out if there should be too much.

It must also be mentioned that the degrees on the tube should be accurately marked, and so they all need to be divided carefully with compasses into ten equal parts, marking each division with a little bead of white enamel. Then the other degrees between are marked with black glass or enamel beads. This division can be done by eye, as practice, industry, and study of the art teaches a man to regulate the spaces in this way and adjust the divisions very well. Whoever has had the training usually makes but small errors. When these things are done the amount of spirit of wine is adjusted by experiment in the sun, and with ice; and then the end of the tube is closed with the seal commonly called hermetic, that is, made with the flame; and the thermometer is finished.

The reason for the custom of using spirit in these instruments, [V] rather than ordinary water, is first of all that it is more sensitive and so feels the smallest alterations of heat and cold earlier and, taking them into itself more quickly, moves immediately because of its great lightness. In the second place, ordinary water, however fine and pure it may be, always leaves some residue in the course of time, or a deposit of sediment that gradually stains the glass and obscures its clarity; while the lightest spirit of wine, or *acquarzente*[10] as we like to call it, always remains beautifully clear, and never comes to lose

---

[9] This "fixed point" is confirmed by an explicit passage in G. 264, 84r (*DIS*, 461); but there are many experiments in the diaries in which temperatures below 20° on the 100-degree thermometer clearly refer to media at temperatures above the ice point, often well above it.

[10] *Il sottilissimo spirito del vino, o acquarzente.* On the basis of this I have everywhere translated *arquarzente* as "spirit of wine."

that choice limpidity characteristic of the glass that contains it.[11] Indeed, because it is so clear and crystalline, so that at first glance one may not succeed in discerning the boundary between the spirit and the empty neck of the instrument, it has sometimes been the practice to color it with an infusion of Kermes, or of those drops commonly called dragon's blood. But as it has been observed that no matter how light and delicate the tint the glass picks up a little of it and after some time becomes soiled and makes the confusion greater, the custom of coloring the spirit has now been given up.[12] It requires only that one should look intently in order to use it clear and limpid.

Much might be said about many other operations and refinements of glassblowing; but these things are too difficult to explain on paper, so that it is really impossible to teach the subject in writing. Therefore one must have a moderately well instructed glassblower, as the art is improved by long practice.

The second instrument (figure II) is only a smaller copy of the first. There is no difference between them except that when they are put in the same surroundings the first moves considerably more than the second. One is divided into 100 degrees, the other into 50. [VII] In the most piercing cold of winter the former goes down to 17 or 16 degrees, the latter usually to 12 or 11; and one year in the most excessive cold it reached 8, and another year 6.[13] Then at the other extreme, in the greatest and most oppressive heat of our summer, exposed to the sun at midday, while the first does not pass 80 degrees, the second either does not rise above 40, or else just a little. Now the rule for making them so that they correspond in this way is only acquired with practice, which teaches one how to proportion the bulb to the tube, and the tube to the bulb, and how to adjust the amount of spirit in order that they may operate without irregular variations.[14]

The third (figure III) is also a copy of the first, but made larger. On

[11] There are, of course, much better reasons for rejecting water: its higher freezing point and, above all, the fact that it comes to a minimum volume at about 4°C.

[12] This change had been made by 1658. See my *A History of the Thermometer* (Baltimore, 1966), pp. 35–36.

[13] It was not until November 23, 1661 that a determination of the ice point of one of these 50-degree thermometers was reported. It was found to be $13\frac{1}{2}$ degrees (G. 262, 120v).

[14] When a case containing a great deal of the glassware of the Academy was discovered in Florence in 1829, G. Libri was allowed to calibrate a number of the 50-degree thermometers. (*Ann. Chim. Phys.* 45 (1830): 354–61.) He found the ice points very uniform, $13\frac{1}{2}$ degrees, and that their zero corresponded to $-15°R$ ($-18.75°C$), their 50-degree point to $44°R$ ($55°C$). A table in *DIS*, 75, seems to be simply a linear conversion, neglecting the lowest of the three points.

that account it becomes a good four times faster and more sensitive,[15] although divided into 300 degrees. Its construction is the same as that of the other two; but as has been noted, the craft of making it cannot be taught by rote, needing long practical experience, reducing and increasing by trial and error either the volume of the bulb, the bore of the tube, or the quantity of spirit, until it is just right. A renowned craftsman in this trade,[16] who used to serve the Grand Duke, was accustomed to say that if 50-degree thermometers were desired he was quite able to make two or three or any number that would always agree when in the same surroundings, but certainly not 100-degree ones, much less those with 300 degrees, seeing that inequalities can more easily occur in the larger bulb and the longer tube, and since every little error made in working them is able to produce very great disturbances and alter the equal proportion that there has to be between them.[17]

The fourth thermometer (figure IV), with the helical tube, is also [IX] made in the same way as the others. It is true that it is not comparable with them, for the very long tube cannot possibly be kept the same size and bore over its whole length, because in bending it one must of necessity pass it again and again over the flame, and when the glass is thus reheated one cannot help it being squeezed and reduced in some places and in others relaxed and enlarged. Therefore, let the bulb be blown with a large capacity and let the long neck be bent in easy turns, closely spaced and rising smoothly, so that it may have as small a height as possible and be less subject to vibration and to the risk of breakage. At the top it should have another empty bulb, sealed with the fire, to be a receptacle for the air in the tube, where it may find refuge from the force that the spirit exerts on it in rising; for otherwise it will have to oppose its advance, having too little room, and the vessel will break.

A thermometer made in this way will be so touchy, and, one might say, of such exquisite sensitivity, that a small candle flame just blown in its direction will be able to put to flight the spirit of wine that it contains. This effect is greater the more ample the bulb, so that it may be made of any size desired, and without observing any other rule, as this instrument is rather made for a whim, and through curiosity to

[15] *Più geloso e veloce.* It had a much more extended scale, but would be more sluggish because of its larger bulb.

[16] Antonio Alamanni, who is named in an early draft of this passage (*G.* 264, 86r), where he is paid the posthumous compliment of saying that now that he is dead they have no hope of finding his equal.

[17] Thermometers with other numbers of degrees are occasionally mentioned; for example 150 (*G.* 262, 83v; *TT* 2, 580).

see the liquid run through tens of degrees when it is simply breathed on, than for finding out with it the just and unfailing proportions of heat and cold.

The fifth instrument (figure V) is also a thermometer, but lazier and more slothful than the others. For the others at once begin to change with every little change in the temperature of the air, but this one is not so quick [X] and requires more than minute and insensible differences to move it. Nevertheless, since some of these thermometers have also gone to various places in Italy and abroad, their construction will be dealt with briefly here.

To make such an instrument, a glass vessel will be taken, filled with the most refined spirit of wine, and strongly cooled with ice. In it will be immersed a 100-degree thermometer. In the same liquid will also be put a number of little balls of blown glass, hollow, but all perfectly sealed with the flame. These should float on the liquid because of the air inside them, and if by chance one that is slightly heavier sinks to the bottom, take it out, and with fine emery on a leaden plate grind it off at the stem until it becomes lighter and floats. Next the vessel will be taken out of the ice and brought into a room the air of which has been considerably heated by fires, so that the very cold liquid receives heat equally in every part. Thus, as it gradually warms up and becomes lighter by its rarefaction, those little balls that had great difficulty in floating when the liquid was coldest will be the first to move toward the bottom, and at the same time the liquid in the thermometer will be seen to rise.[18]

The ball, then, that begins to descend when the thermometer is at 20° is marked no. 1 as being the heaviest, since it sinks when the liquid is still very cold, and not warmed, or hardly at all. The one that sinks when the spirit in the thermometer is at 30° will be no. 2; at 40°, no. 3; at 50°, no. 4; at 60°, no. 5; and at 70°, no. 6, which will be the last and lightest. [XI] Thus six balls will have been taken in a scale of equal differences, each of ten degrees. And this is how this thermometer comes to be more coarsely divided than the others, since each of these balls that rises or falls means 10 degrees of the 100-degree

---

[18] The origin of this sort of thermometer may well be in a passage in Galileo's *Discorsi . . . intorno a due nuove scienze* (Leyden, 1638; *G.OP.*, VIII, 114) in which he notes that a ball of wax, weighted to be in equilibrium with water, would rise or sink with the least change in temperature. It may date from about 1641, for on June 20, 1657 it is recorded that a "thermometer made 16 years ago was cut open with a diamond" in order to find out whether the spirit would still burn (G. 262, 3v). Such thermometers were used as clinical thermometers, bound to the arm of a fever patient (*DIS*, 464).

thermometer, about 4 degrees of the 50-degree one, and more than 40 degrees of the 300-degree instrument.

When the six balls (which should preferably be of colored glass or crystal, the better to see them in the liquid) have been chosen, they may be enclosed in a tube of crystal glass containing spirit of wine, hermetically sealed, taking care that it is not quite full, so that there will still be room for the spirit to expand when the advent of summer heat forces it to do so.

If the warmth of the room should not be enough to make the thermometer rise to 60°, it will be assisted by putting the glass vessel into a bath of tepid water, adding boiling water as needed, so that the spirit of wine contained in it may not be heated more in one part than in another but may warm up gently, as we said, and as regularly as possible.[19]

[XII]

EXPLANATION OF ANOTHER INSTRUMENT, THAT SERVES TO DETERMINE THE DIFFERENCES IN THE MOISTURE OF THE AIR

Now that we have seen some instruments that serve to determine the changes that the air receives from heat and cold, we must next consider another that can show us the changes that come simply from moisture. And although those that have already been imagined by divers ingenious men are many and various, we shall add to them only one. This, even if it has lately been described by others,[20] was nevertheless born in this Court, of the highest Royal intelligence,[21] and we shall say something about its invention and use, so that it may redound to the credit of our country.

It is a truncated cone (figure VI) made of cork, hollow within, and sealed with pitch, and covered with tin on the outside. At the narrowest part this is inserted into a glass vessel like a lamp chimney,

[19] The idea of stirring a liquid to promote uniformity of temperature seems not to have occurred to them.

[20] First in *Trattato della sfera di Galileo Galilei, con alcune prattiche intorno a quella . . . di Buonardo Savi* [pseudonym of Urbano Daviso] (Rome, 1656), pp. 196–97. The dedication of this book to Cardinal Giovanni Carlo de' Medici is dated March 20, 1656 (probably *stile commune*). On page 196, Daviso says that the Cardinal had been sent one of the Grand Duke's hygrometers, which arrived while he was writing. This would probably be late in 1655.

[21] Without any question Ferdinand was the inventor and the date 1655, in the summer, for experiments with it on August 27, 1655 are recorded (G. 259, 14v); and its invention is ascribed to the observation that "when something iced was put in a glass, the surrounding air seemed to change into water" (G. 259, 25r).

extended in the shape of a cone with a very sharp and narrow point. When the instrument has been so prepared and placed on its stand it is first filled with snow or finely crushed ice, the water from which will drain away through a tube leading from the upper part of the glass [XIV], as shown in the figure. The subtle moisture that is in the air is gradually ensnared by the cold of the glass, first covering the glass with a thin veil, and then as new moisture arrives it collects into larger drops and flows down the slippery surface of the glass and so distils little by little. So let there be a tall cylindrical vessel, divided into degrees, to receive the water dropping from the instrument.

Now it is quite evident that according as the air contains more or less moisture the power of cold will distil more or less water, which, dropping rapidly or slowly, will more or less fill the cylinder. So if we wish to compare one air with another, we observe, in that which we would try first, how much of the glass is filled in a given time. Then, throwing away the water, we transport the instrument into the air that we wish compared with the first and observe similarly to what mark the glass is filled in the same length of time. Thus, having found the difference in the moisture condensed into water on the two occasions, we shall have approximately that between the humidity of the air in the two places being compared.

By exposing this instrument to the air when different winds are blowing, we can also get to know which of them are more pregnant with moisture and which are drier and more thirsty. In this way we have found that when southerly winds prevail the glass sweats most copiously, because the air is excessively moist, perhaps because most of the sea is to the south of us; and peradventure because the sun works very strongly there and draws vapor out from these seas, which mixes [XV] with the winds. In a strong south-westerly gale we have collected as much as 35 or even 50 drops per minute. On one occasion, when the north wind and the south-west wind were struggling together, the weather cloudy, with the clouds touching the mountains, 84 were counted in the same time; but the north wind prevailing, the sweating gradually ceased, and in little more than half an hour the glass was dry,[22] although there was much snow inside it. It stayed this way all night and all the following day while the same winds continued to blow. The vessel has also been observed to remain very dry when west winds are blowing.

It is true that no certain rule can be given in these matters, for they

[22] Clearly this phenomenon was associated with the passage of a cold front.

can vary because of many circumstances, not only with the season and the wind but also with the region and the country itself, so that the properties of the above-mentioned winds sometimes change. And we know that in certain cities and regions the south winds are cooler than with us, for the reason that they have mountains to the south that are covered with snow, so that the winds turn colder in passing over them. For all that, our instrument will not fail to act faithfully wherever it is used, and its operation will correspond very exactly with the ordinary indications of the nature of these winds.

[XVI]

EXPLANATION OF SOME OTHER INSTRUMENTS, USED
FOR MEASURING TIME

To go no farther in seeking experiments in which we must have an exact measurement of time, such as those with projectiles and with sound, the nearest to hand is the preceding one, the testing of the humidity of the air and of the winds. The test here is to see the different amounts of moisture that are distilled from various airs by the ice-cold glass in a constant length of time. This difference is always so small and imperceptible that it is beyond the reach of the accuracy of the finest clocks. This is because we must either take the times from stroke to stroke,[23] in which the ear may easily be deceived, or from the spaces passed by the hand, in which the eyes may err more than ever. Thus we are obliged to have recourse to an instrument that will divide time more finely than the sound of the four strokes of the clock or the minutes marked by the hand; because in such things the judgment of the senses is in great danger of error. For (leaving aside the errors that there may be in the division of the dial or in other material parts of the instrument) it is difficult to judge whether the hand is or is not exactly on a mark; and with sound we must finally say [XVIII] that at the moment the clock strikes, the time denoted by the sound has passed. We have decided that the most exact instrument may be the pendulum, or *dondolo* as we like to call it, the going and returning of which is counted as one entire vibration. We did not believe that whenever in counting many vibrations one is missed (which rarely happens to anyone after a little practice) this small error would ever amount to as much as that made in the above-mentioned procedures.

But the ordinary pendulum with one string, in its freedom to wander insensibly, comes to leave its original path (whatever the cause of

---

[23] I.e., of a striking clock.

this may be), and towards the end, as it approaches rest, its motion is no longer in a vertical arc, but appears to be an oval spiral in which the vibrations can no longer be distinguished or counted.[24] Therefore, solely to make it keep the same path to the end, we thought of hanging the ball from a double thread, the ends being tied to a metal bracket, a small distance apart, as shown in figure VII. The ball, thus attached to the thread by its hook, pulls the thread, and with the proper weight stretches it into an isosceles triangle; for should it form a scalene one in its first vibration the ball, being free upon the thread, will at once run to the lowest point possible, by virtue of its weight, and remain there. So the movement of the pendulum comes to be regulated by this triangle, since (if we may be allowed this simile) the threads that form its sides serve as reins to the ball, so that it may not deviate more on one side than on the other, but will always move directly in the same arc.

It is true that not all the experiments to which [XX] the pendulum is applied require the same division of time, a very coarse one, such as can be had with the longest vibrations, sufficing for some of them. Others need so minute a division, made by vibrations so fast and crowded one upon another that the eye is fatigued in counting them. Therefore, in order to be able to shorten and lengthen the triangle easily, as needed, without having to untie and re-tie the upper ends of the thread, the lower bracket was added, also of metal, attached by a square clamp to the straight shaft of the instrument, so that it can run up and down on this and be made fast by a screw where desired. This second bracket is sawed apart lengthwise and divided by a cut, then put back or rather re-joined by means of two other screws, clamping the threads of the larger triangle in the middle, leaving its upper part or quadrilateral fixed between it and the bracket above. In this way the smaller triangle, which starts from the very narrow junction of the two separated parts and has that for its base, has free play for its vibrations. These will become more frequent the shorter the suspension of the ball or the less the height of the triangle.[25]

This seems the place to say that experience has shown us that not all the vibrations of the pendulum occur in precisely equal times[26] (as was also noticed by Galileo, after the observation that he made in

[24] They observed this on November 28, 1661 (G. 260, 172r; G. 262, 123r-v; TT 2, 669).

[25] In G. 269, 48r–51r, there are calculations, apparently by Viviani, for the variable pendulum; also a dimensioned sketch (fol. 51r). It was not bifilar at the time this drawing was made.

[26] They tried this on November 19, 1661 (G. 262, 118v; TT 2, 390).

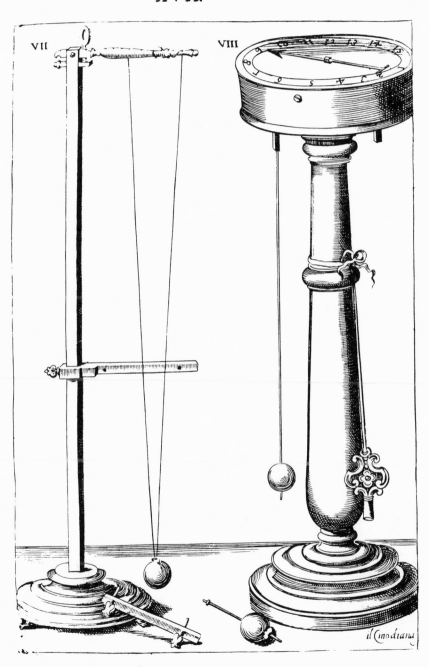

about 1583, before anyone else, of their approximate equality).[27] As they gradually approach a state of rest, they are performed in a shorter time than they were at first, as will be shown in its place.[28] Therefore, in those experiments that require [XXII] a greater exactness and that need such a long period of observation that the least inequality in such a great number of vibrations finally becomes noticeable, it was thought well to apply the pendulum to a clock, in imitation of the one that Galileo first invented and that his son Vincenzio Galilei put into practice in 1649.[29] In this (figure VIII) the pendulum is obliged by the force of the spring or the weight to fall always from the same height. Thus, with mutual benefit, not only are the times of the vibrations made perfectly equal but the defects of the other mechanisms of the clock are also, in a way, corrected.

To make such an instrument serve for different experiments that need time more or less finely divided, we have made various metal balls, strung on very thin steel wires of different lengths, all so that they can be inserted into the same female screw thread, as needed. The shortest of these completes a whole vibration in half a second, which is the smallest division that we have succeeded in making, as shorter pendulums turn out to be so quick that the eye cannot follow them.[30]

Enough has been said here about these instruments, which will come into use more often in the experiments that follow.[31]

[27] See appendix E.

[28] This promise does not seem to have been kept.

[29] About 1636, Galileo had envisaged the application of the pendulum to the clock, and had offered the idea to the States-General of Holland as part of a method of measuring the longitude. His confident description shows that he had not reduced the idea to practice: "I have such an instrument for measuring time that if 4 or 6 of them were made and allowed to run, we should find (in confirmation of their accuracy) that the time measured and shown by them, not only from hour to hour, but from day to day, and from month to month, would not differ among them by even a second, so uniformly do they run." (Galileo to the States-General, August 15, 1636; printed in Huygens, *Oeuvres*, III, 496.) This is the most breathtaking example I know of the great man's self-confidence. At the time of his death in 1649, Galileo's son Vincenzio seems to have been making a model after his father's design. See also Silvio A. Bedini, "Galileo and the measure of time," in *Saggi su Galileo Galilei* (Florence, 1967).

[30] The mechanism of the clock shown in our figure can only be guessed at; but it was clearly spring-driven. In his translation, Waller omitted this figure. This clock may have been made by Philip Treffler of Augsburg, who was in the Grand Duke's service, and made him several clocks. See *G.OP.* XIX, 658–59. A clock without a pendulum is figured in G. 261, 24r, together with a pendulum and a sand glass, to illustrate an experiment.

[31] A very small clock with an alarm bell (*un orivuolo carico con la sveglia*) was enclosed in a submerged flask on July 5, 1657 (G. 262, 14r).

꧁꧂

[XXIII]

EXPERIMENTS PERTAINING TO THE
NATURAL PRESSURE OF THE AIR

That famous experiment with the quicksilver that in 1643[32] presented itself before the great intellect of Torricelli is now known in
every part of Europe,[33] as is also the high and wonderful idea that he
formed about it when he began to speculate upon the reason for it.[34]
This, he would say, is the air, which, weighing upon all things below
it, forces them out of their places whenever there is an empty space
for them to go to; particularly [XXIV] liquids, because of their great
inclination to move. When we attempt to move solid bodies, on the
contrary—such as gravel, sand, and the like, or heaps of larger stones—
they interfere with each other and pack together, thanks to the rough-
ness and irregularity of their parts, in such a way that they hold and
support each other so as to resist more strongly the force that is trying
to remove them. Liquids, on the other hand—perhaps because of the
slipperiness or the roundness of their very small corpuscles or from
some other shape that may favor motion[35]—though standing in equilibrium, yield in every direction and spread out as soon as they are
pressed. Thus we see water disturbed by any tiny shaving that falls on
it and drawing back on every hand take wing, so to speak, in the most
orderly circles. Who knows whether the looseness of its parts may not
result in its being seldom or never still, even in the most suitable
vessels, even though it may always appear stagnant? Whence it comes
that every lightest zephyr disturbs and ripples it. And in lakes, even
those that seem most sheltered, water is so mobile because of its nature
that, though we may not perceive this, it is very obedient to invisible
undulations of the air above it and peradventure never still.

This is not more true of water than of other liquids, in all of which,
some think, the pressure of the air is marvellously evident, particularly
when they are held in places where over one part of their surface they
have an empty or almost empty space into which they can retire. Since

---

[32] In my *The History of the Barometer* (Baltimore, 1964), pp. 29–30, I
have given reasons for preferring the date 1644. See also W. E. K. Middleton, *Physis* 10 (1968): 241–42.

[33] Regarding its spread, see *ibid.*, chap. 3.

[34] *quand'ei ne prese a specular la ragione.* The 1841 "corrected" edition
has *specificar*, which seems quite arbitrary, and is contradicted by no less
than three drafts, all in Magalotti's handwriting, in G. 264 (fol. 148r, 144r,
143r, in chronological order).

[35] This idea is most probably derived from Descartes, especially his *Les
météores* (Leyden, 1637).

the surrounding air, itself pressed by so many miles of air heaped up over it, presses from one direction, and nothing weighs [XXV] on the other where the liquid is not retained or confined by the vacuum, which weighs nothing at all, the pressure will cause the liquid to rise until the weight of the raised liquid comes to equal that of the air pressing on the other side. With various liquids this equilibrium is reached at different heights, according as their greater or lesser specific gravity makes them able to resist, with lesser or greater heights, the force and power of the air. As is the common practice, and as was also first done by Torricelli, we have used quicksilver, which is so marvellously heavy and thus lends itself to a convenient operation for making a vacuum in a smaller space than can be done with any other fluid whatever. The following experiments will show what we have succeeded in finding out on this subject.

[XXVI]

AN EXPERIMENT FROM WHICH IT OCCURRED TO TORRICELLI, ITS FIRST INVENTOR, THAT THE SUSTAINING OF QUICKSILVER OR ANY OTHER FLUID AT A DEFINITE HEIGHT IN THE VACUUM COULD RESULT FROM THE NATURAL EXTERNAL PRESSURE OF THE AIR[36]

Let the glass tube ABC (figure I) be about two ells[37] [117 cm] long and open only at C. Fill it through this opening with quicksilver and then, closing it either with a finger or with a piece of bladder somewhat moistened and firmly tied on, turn it over and plunge it lightly into the mercury[38] in the vase DE. Then open it. The mercury will at once descend in the tube through the whole distance AF, where, having arrived at its level, it will remain after a few oscillations. The cylinder of mercury FB that is held up in the vertical tube and remains above the surface DE of the mercury in the vase, will be about an ell and a quarter [ca. 73 cm] high.[39] Though this height has been observed to vary a little by the external circumstances of heat and cold, and somewhat more in the different seasons and various states of the air, as is

---

[36] The original draft of this is in G. 266, 30r–v, 33r–v, transcribed in *DIS*, 281–82. It was comparatively little altered.

[37] *Braccia* (plural of *braccio*). *Braccio* has been variously translated as ell, cubit, and even yard. Between "ell" and "cubit" I have somewhat arbitrarily chosen the first. See also p. xiii.

[38] *Argento*. This (which cannot reasonably be rendered by "silver") is used interchangeably with *argentovivo* in the *Saggi* as dictated by the rhythm of the sentence. In general I have used "quicksilver" for *argentovivo* and "mercury" for *argento*.

[39] An ell and a quarter is taken everywhere in the *Saggi* as the mean height of the barometric column. They first described the experiment on August 29, 1657 (G. 262, 29r; *TT* 2, 393).

XXVII.

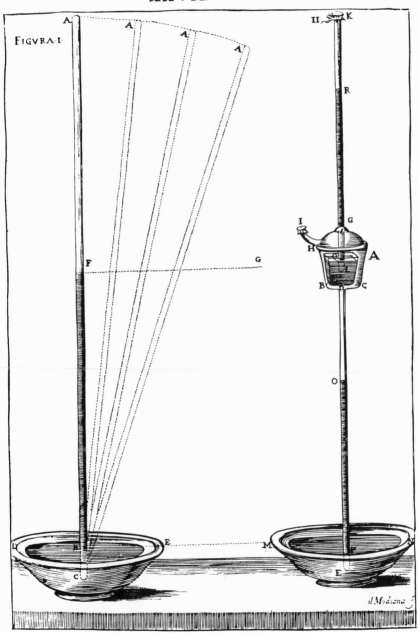

clear from a very long series of our observations,[40] yet as such variations are very small, it will hereafter always be indicated by the same measure of an ell and a quarter as the nearest approximation.[41]

The space AF will remain empty of air, and this will be manifest, inasmuch as if the whole tube AC is inclined, moving it round the point C as center, the internal level F will be seen to move correspondingly towards A, without ever rising above, but always touching, the horizontal line FG through the [XXVIII] point F, the first position of the mercury when the tube was erect. If the tube is inclined so that its extremity A touches FG, it will remain full of quicksilver, except for some very small part at A into which is reduced either some air with which the mercury may have been impregnated or other invisible exhalations that come out of it and rise above the level of the mercury.[42] This is seen most clearly whenever a little water is introduced into the tube. In making the vacuum this water, rising above the mercury, discloses in the passage that they make for themselves those very fine bubbles that rise towards the vacuum, as will appear in another place.

The same absence of air may be demonstrated by pouring water over the mercury DE, and then lifting out the mouth C in such a way that it stays in the water.[43] The quicksilver will at once fall, the water rising to the top and filling the entire tube, provided that its height does not exceed about $17\frac{1}{2}$ ells [about 10.2 meters], at which elevation (as we shall see later)[44] water is sustained, perhaps[45] by the same power that sustains the quicksilver at an ell and a quarter. But not even in this case does there appear at the summit of the tube any considerable

---

[40] These meteorological observations, made in Florence, are in G. 305. Others, from Vallombrosa, make up several volumes of the ms. (G. 296 to G. 304). They seem to have been started on the orders of the Grand Duke.

[41] It is worth noting that in the original description of the Torricellian experiment in his letter of June 11, 1644 to Michelangelo Ricci (Torricelli, *Opere* [Faenza, 1919], III, 186–88), he gives the height as "an ell and a quarter and a finger more." This would be about 75 cm, a more likely value at the elevation of about 50 or 70 meters above m.s.l. at which the experiment was probably made.

[42] July 30 and 31, 1660 (G. 262, 99r, 99v; *TT* 2, 393–94). They concluded that a perfect technique for introducing the mercury would get rid of the little bubble. The experiment with the inclined tube had been performed in Rome in 1645 and described by Emmanuel Maignan in his *Cursus philosophicus* (Toulouse, 1653), a work that was known to the Academy.

[43] Torricelli had described this very ingenious demonstration.

[44] Not in the *Saggi*; but see p. 268 below.

[45] *Forse dell'istessa potenza.* It would seem that the Academy managed to be more skeptical than even the rigorous Pascal, who had had no doubt about this particular matter as early as 1647.

amount of air, although here alone are condensed into a nearly invisible space those very tenuous exhalations that, as we said, rise from the quicksilver, or other subtle kinds of matter that have somehow been able to get in.

For these reasons and for greater brevity we shall hereafter call the space *AF*, and any other space that may be left in similar vessels by the descent of the quicksilver, an empty place or space; that is to say devoid of air, at least of that which, not at all changed from its natural state, surrounds the tube and stands free [XXX] in its region.[46] We do not presume to exclude from the space fire or light or aether or other very tenuous substances, either finely distributed with very small empty spaces between or filling the whole of the space that is called empty, as some would have it. Nevertheless, it has been our intention only to discuss the space filled with mercury and to understand the true cause of the wonderful balancing of its weight, intending never to pick quarrels with those who oppose the vacuum.[47]

We have made many experiments[48] to this end, some of which have indeed been described by others, as there are also those invented by our Academicians. The results of all these will be seen here, faithfully described, always observing our custom of historical narration and of never defrauding the inventors of their discoveries and their credit.

[XXXI]

ROBERVAL'S EXPERIMENT[49] IN FAVOR OF THE PRESSURE OF
THE AIR ON BODIES BENEATH IT, VERIFIED IN OUR ACADEMY

In figure II, *A* is a glass vessel to whose bottom *BC*, pierced at *D*, is attached the tube *DE*, two ells [117 cm] long. After the square beaker *F* has been placed over the opening, the vessel *A* is closed with

---

[46] This sounds scholastic.

[47] The peripatetic philosophers, who were in the great majority in 1667. Even one of the Academicians, Alessandro Marsili, was inclined that way (Cf. *TT* 1, 525).

[48] Two of the most interesting (Dec. 29, 1661 and Jan. 3, 1662; G. 262, 128r; *TT* 2, 394) are not mentioned in the *Saggi*; they involved weighing the barometer tube, at first hanging in its cistern, when it weighed as much as the tube and the mercury in it together; next with a tube closed with bladders. In this they could not detect a difference in weight between the air and the vacuum. The first experiment had been made by Marin Mersenne. See *Novarum observationum physico-mathematicarum*, Tomus III (Paris, 1647), first preface.

[49] There is some doubt whether this experiment was due to Gilles Personne de Roberval or to Adrien Auzout. Roberval referred to it in a letter written in 1648 (Paris, *BN*, ms. f. lat. 11197, fol. 26 sqq.). Its first published description was in the *Gravitas comparata* of Étienne Noël (Paris, 1648), pp. 77–80. Noël merely says that Roberval showed him the experi-

the cover *GH*, also of glass. This has an open spout *HI*, and at *G* there is a hole through which passes the tube *KL*, open at top and bottom and also two ells high, or not less than an ell and a quarter [73 cm]. This tube enters the beaker but does not touch its bottom and is held in place by mastic or other hot cement in the hole *G* through the cover. If this cement is made of brick, reduced to an impalpable powder by long grinding and incorporated with turpentine and Greek pitch, it will be excellent for cementing glassware so that the outside air may be excluded.[50] Similarly, the above-mentioned cover may be fastened all round with this cement, where it fits in the vessel *A*.

The lower end *E* being closed with a piece of bladder, we begin to pour in quicksilver through *K* until, overflowing the beaker *F*, it spills on to the bottom *BC* and then descends through the hole *D* to fill the tube *ED* and finally the entire vessel *A*, the air having a means of escape through the open spout *HI*. This is carefully closed with a piece of bladder at *I* as soon as the mercury starts to overflow. The whole tube is then filled up to *K*, and here also the quicksilver is allowed to overflow a little so that when the end of the tube is closed no air at all may remain in it.

After we have closed this end, the other bladder that closes the end *E* is pierced while it is under the level *MN* of the stagnant quicksilver [XXXIII] in which the tube is immersed. This will empty the upper tube *KL* and the vessel *A*, only the beaker *F* and the part *OP* of the tube *DE*, which will be an ell and a quarter above the level *MN*, remaining full.

Next let the air enter *A* by removing or puncturing the bladder *I*. This will at once cause the cylinder of mercury *OP* to fall into the lower vessel, and another, *QR*, will rise in the tube *LK* out of the mercury in the beaker *F*. This will be equal in height to the first, *OP*, an ell and a quarter, and will not descend until the tube is opened at the top, *K*, and the air comes down upon it through the tube *KL*.[51]

---

ment. In draft "A" (G. 266, 34r; *DIS*, 282) it was called the "experiment referred to by Mr. Pacquet" [*sic*]. The apparatus, essentially as in figure 10, was described in 1651 by Jean Pecquet, *Experimenta nova anatomica*, etc. (Paris, 1651) pp. 56–58, where it is ascribed to Auzout. For further discussion see my *The History of the Barometer*, pp. 48–50, 56. See also R. M. McKeon in *Actes XIe Congrès Int. Hist. Sci.* (Warsaw-Cracow, 1965), III, 355–63.

[50] Recipes for hot and cold cements will be found in G. 243, 4r.

[51] Made on August 2 and 3, 1657 (G. 262, 22v–23v; *TT* 2, 395–96). On the second occasion they let water in through *I*, but it did not send the mercury up, except that a certain amount of air came out of the water and produced a small effect.

If a little bladder, carefully removed from the intestines of a fish, is left in the vessel *A* after the air that is naturally found inside it has been carefully pressed out so that very little remains in its folds and the orifice has been closely tied, then, at once, when the descent of the quicksilver has left it in the vacuum, the little air that remains in it will inflate. It will then only collapse when the vessel is opened at *K*, so that the air from outside can weigh upon it.[52]

We have also more clearly observed the dilatation of air in a vacuum in another vessel such as *ADB* (figure III), enclosing in this a lamb's bladder that has been twisted and almost entirely deflated. This experiment was made as follows: the vessel was filled with quicksilver through the opening *D*, with the lower opening *E* tightly closed with a finger. *D* was then sealed with a piece of bladder, and then after the tube had been immersed in the quicksilver of the vessel *FG* the mercury was allowed to run out freely. The bladder *C* would then be inflated in the empty vessel *ADB* and remain in that state until the mouth *D* was opened and the external [XXXIV] air came down, at the same time letting the cylinder of mercury fall into the lower vessel *FG*.[53]

Similarly, if when the mouth *D* is sealed a small quantity of foam made from white of egg or soap beaten up with water is left on the mercury, then as the vessel *AB* goes on emptying, the air imprisoned in those tiny bubbles will so expand them that it finally breaks their extremely thin walls and will be liberated and entirely separated from the water, which will rain down on the mercury, free from the fine scattering of air that made it into froth.[54]

EXPERIMENTS ADDUCED BY SOME PEOPLE AGAINST THE PRESSURE OF THE AIR, AND THE REPLY TO THESE

There were two experiments on which some of our Academicians[55] thought they could base a considerable argument against the pressure

---

[52] This experiment is referred to by Étienne Noël, *Gravitas comparata* (Paris, 1648), p. 57. There are notes on this work by Viviani in *G.* 269, 5r.

[53] August 9, 1657 (*G.* 262, 24v; *TT* 2, 396–97).

[54] August 11, 1657 (*G.* 262, 25r; *TT* 2, 397). The original draft "A" (*G.* 266, 34–35v, printed in *DIS*, 282–86) differs in no important respect.

[55] Not identified. Marsili is a likely one. When it became difficult to deny the absence of air in the Torricellian tube, the peripatetics fell back on the allegation that the apparently empty space was full of "exhalations" from the mercury. An experiment designed to detect these was suggested by Marsili and performed on August 13, 1660 (*G.* 262, 104v–5r; *TT* 2, 403). It was repeated, with no certain result, on August 21 (*G.* 262, 106v–7r; *TT* 2, 404). See also p. 265 below. The concept of vapor pressure was far in the future.

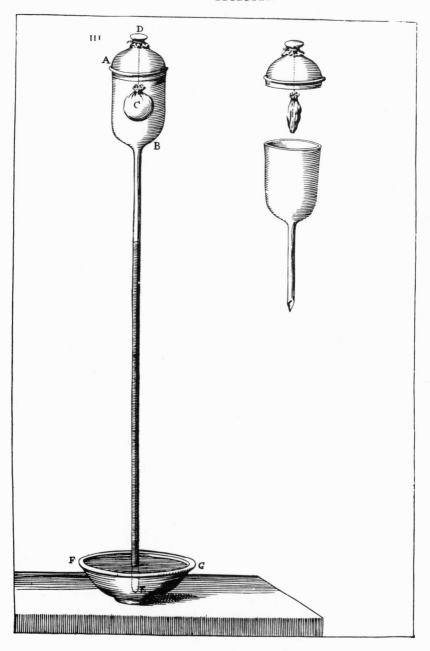

XXXII.

of the air on bodies below and so take away the power of sustaining fluids attributed to it by others.

One was by covering the vessel *A* and its tube with a large glass bell jar *BCD* (figure IV) cemented on a table all round. They persuaded themselves, then, that if it were true that it was the weight of all the region of air above that drove the quicksilver up the tube, and if it was in equilibrium with this, then, if the stagnant quicksilver were protected by a glass wall from such great pressure, the imperceptible weight of what little air is included under the bell jar ought to remain unable to keep the mercury at the same height as that to which the weight of such a vast region of air had pushed it. But in spite of this it was seen [XXXVI] not to fall in the least from its usual height *EG*.[56]

The second proof was similar to the first, in fact exactly the same, though perhaps more greatly refined.

A small vase *AB* (figure V)—but at first without the spout *CD*—is filled with quicksilver, and the tube *EF*, already filled, plunged into it, and the vacuum made in this in the usual way. Then a very little mercury is poured out of the vase *AB*, so that the amount of air in the space *AH*, pressing on the stagnant surface *GH*, will be very small. This is then protected from the pressure of the external air by carefully filling with hot cement the circular space *A* between the mouth of the vessel and the tube. And not even in such a case, when the amount of air pressing on the mercury was reduced to almost nothing, does there appear an appreciable lowering of the cylinder of mercury *IF* below its usual level.[57]

But those who adhere to the doctrine of the pressure of the air answered these experiments by saying that the phenomena just recounted, far from contradicting their opinion, favored it wonderfully. For, according to them, the immediate reason for the mercury being violently pushed to a height of an ell and a quarter and held there is not really the weight of the superincumbent air, which is taken away by the bell jar in the first experiment and by the cement in the second, but, on the contrary, the compression that had been produced by this weight in the air *BCD* of one figure and *AH* of the other. It is therefore not to be wondered at that as the same state of compression is maintained

[56] July 24, 1657 (G. 262, 19v; *TT* 2, 397); repeated on August 4, 1657 (G. 262, 23v–24r). Viviani claimed this and the next two experiments (G. 269, 259r).

[57] August 6, 1657 (G. 262, 24r; *TT* 2, 398). They were not aware at that time that Torricelli had used a similar, though imaginary, experiment to convince Ricci of the pressure of the air (Torricelli to Ricci, June 28, 1644, in *Opere*, III, 198–201).

XXXV.

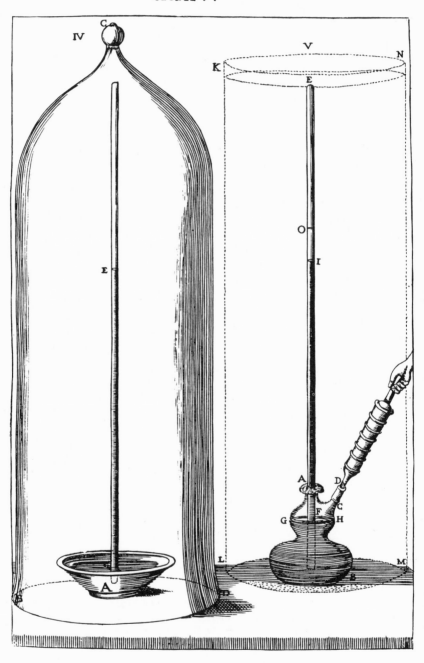

(as is imposed upon it by the resistance of the cement or the glass, in place of the entire height of the air) the elevation of the quicksilver does not fall below its usual measure.[58]

Although it was also believed by some people that the supposed spring of the air was entirely [XXXVIII] responsible for this effect, so that without it the phenomenon could not happen at all, there were others who tried to suggest the contrary with the following experiment: Taking the same vessel *AB* with its tube *EF* (figure V) and, before pouring out any mercury and cementing it at *A*, submerging it in a large vessel of water *KLMN*, the quicksilver was seen to fall visibly from *A* to *GH* and, on the contrary, to rise in the tube from *I* to *O*, this rise being about one-fourteenth the height of the water *EF*. The mouth *A* was then closed, so that only the mass of water *AGH* pressed on the mercury, but nevertheless this lost none of the height above its first level *I* that it had newly acquired from the weight of the superincumbent water *EF*. In this case the water in the space *AGH* ceded to it by the quicksilver in rising from *I* to *O* continues to hold it there by force and to resist its return, not by its springiness (they said), which perhaps it does not have, but by being already pushed down by the weight of all the height *EF*. Exactly the same, they said, happens in the air.

Finally, others wished to see what effect was produced by a greater or lesser dilatation of the air enclosed in the space *AGH* and made the following test: They added to the same vase *AB* the spout *CD*, in which was fastened a metal mounting, threaded inside. They applied to the latter the end of a syringe with its corresponding screw thread. Then whenever they drew out some of the air from *AGH* with this syringe, attenuating what remained, the level *I* was seen to fall, and if, on the contrary, it was compressed by the introduction of new air, this level rose.[59]

---

[58] This was precisely Torricelli's explanation, unpublished until 1663. See [Carlo Dati], *Lettere a Filaleti di Timauro Antiate, Della vera storia della cicloide, e della famosissima esperienza dell'argento vivo, 24 Gennaio 1662* (Florence, 1663). The Academicians almost certainly had not heard of this in 1657, for on August 3, 1658 Borelli wrote to Prince Leopold from Rome that Ricci had shown him a letter of Torricelli, "in which he notes the same explanation that I found, because when he stopped up the mouth of the vessel of quicksilver below and cut off the pressure of the whole region of air, the quicksilver nevertheless kept the same height of about an ell and a quarter; and to my greater discomfiture he adopts the same example of wool" (G. 275, 104r). Borelli's surprise is evident. This is a very clear indication of the secrecy surrounding the Torricellian experiment (see my *History of the Barometer*, pp. 30–32).

[59] September 13, 1657 (G. 262, 33r–v; *TT* 2, 398–99).

The same thing likewise happened in the vicinity of fire [XXXX] or of ice, because whenever the mouth C had been closed and heat was applied externally to the air AGH, the mercury rose, and fell when the outside of the vessel was rubbed with ice.[60] In nearly the same way as happened in the two opposite operations of the syringe, air might be condensed by the heat and dilated by the ice [sic].[61]

From all these experiments it seemed to them that they could believe with greater probability that this suspension of fluids derives, not absolutely from the weight of the air, but rather from the compression caused in a very small portion of it by that weight.

AN EXPERIMENT TO FIND OUT WHETHER THE AIR NEAR THE EARTH'S SURFACE IS COMPRESSED BY THE WEIGHT OF THE AIR ABOVE, AND WHETHER IF SET FREE IN A VACUUM, EVEN IF NOT ALTERED BY A NEW DEGREE OF HEAT, IT WOULD EXPAND INTO A LARGER VOLUME, AND TO WHAT EXTENT

The ingenious observation made by Roberval[62] of the little bladder that distends in the vacuum caused some of us[63] to think that they should determine to what extent air, set at liberty, has the power of

[60] September 15, 1657 (G. 262, 34r; TT 2, 399).

[61] The application of heat and cold is not included in draft "A" (G. 266, 37r–v, 39r; DIS, 286–87), which indeed was altered a great deal before the book appeared.

[62] Described in a letter to Des Noyers written in May and June 1648 and printed in the Gravitas comparata of Étienne Noël (Paris, 1648), p. 57; also in Jean Pecquet's Experimenta nova anatomica (Paris, 1651), pp. 50–52.

[63] Borelli, De motionibus naturalibus (1670), Cap. V, prop. CV, claims this; the account begins: "Ex Robervalii pulcherrima observatione illius vesicae cyprinae, quae in vacuo fistulae dilatatur, ego conjeci reperiri facile posse . . . maximum amplitudinem, ad quem aër non compressus à vi externa . . . dilatari queat . . . ." He refers to the apparatus that he thought out for the Academy, and then describes a simpler variant. Except for the absence of the bulb A, Paolo del Buono described this experiment quite clearly in a letter of which the original has been lost, but which was copied by Ismael Boulliau from a copy sent him by a Frenchman at Warsaw, Des Noyers (BN, fonds fr. 13039, fol. 126r–40v). I have published this document in Archive for History of the Exact Sciences 6 (1969): 1–28. On internal evidence his letter was written in September or October 1657, about five years before the experiment to be described was performed on August 4, 1662 (G. 262, 134r–35r; TT 2, 400–1). It was repeated on September 6 and 7, 1662 (G. 262, 153v–54r). It is almost certain that Paolo del Buono's letter never completed the long journey from Warsaw to Florence. In any event, the experiment is clearly ascribed to Borelli and described, with an excellent figure like that in the Saggi, in G. 268, 6v–7r. In the margin this is dated "A dì 31 Lugl. 1662."

expanding. Hence it seemed very probable to them that in a given vessel a space might be assigned that would suffice for the entire expansion of a certain volume of air; whence any other greater volumes, which might need a more ample space for their expansion, ought to depress the cylinder of quicksilver below its ordinary height of an ell and a quarter [73 cm], and on the contrary all lesser volumes, being (we said) too well provided for, would have to let its mercurial boundary rise as usual.[64] The experiment is as follows: [XXXXIII][65]

Let *ABC* (figure VI) be a glass vessel with a tube *BC* two ells [117 cm] long open at *C*. Let there be also a long glass *DEF* which, full of quicksilver, may be a vessel for the immersion of the tube *BC*; but such a vessel that the tube can not only be immersed in it, as in others, but can also be received wholly or partly within it, as in a sheath. Let *GHI* (figure VII) be another vessel similar in all respects to *ABC* and as far as possible equal in size.

In the latter, after the vacuum is made, is observed the height *KL* at which the mercury stays in equilibrium on this particular day. Then the vessel *ABC* is filled with quicksilver through the opening *C* as far as *M*, and the remaining space *MC* is left full of air. It is clear that when *C* is closed with the finger and the vessel inverted, the small amount of air left in *MC* will rise through the mercury and take its place in *A*. The mouth *C* is then plunged below the level *DE*, the finger is removed and the vacuum is made. The mercury falls to the height *PQ*. If this is measured and found equal to the height *KL* in the vessel *GHI*, where no air has been left that could alter it, this will be a sign that the cylinder of mercury *PQ* is not pushed down at all by the small amount of air *MC*, since the space left empty in *A* down to *P* must be larger than is needful for its entire dilatation and total expansion.

The tube *BC* is now pushed down little by little under the mercury *DF*, so that as the level *P* gradually rises,[66] for example to *R*, the space *PBA*, left free for the air, decreases. This immersion is continued until the height *RQ* is seen to begin to be less than *KL*. Note that the point *R* is a fixed and immutable limit of all the heights of the mercury cylinder equal to *KL*, because all the subsequent [XXXXIV] ones to-

[64] The twentieth-century reader scarcely needs to be told that this experiment would not measure the quantity that they sought, which in fact does not exist. What they really measured, though very awkwardly, was the precision of their barometric observations. It would be hard to find a neater illustration of the way in which experiment is conditioned by hypotheses.

[65] It is quite differently done in Draft "A" (*G.* 266, 40r–v; *DIS*, 277–78).

[66] In relation to the vessel *ABC*.

XXXXIII.

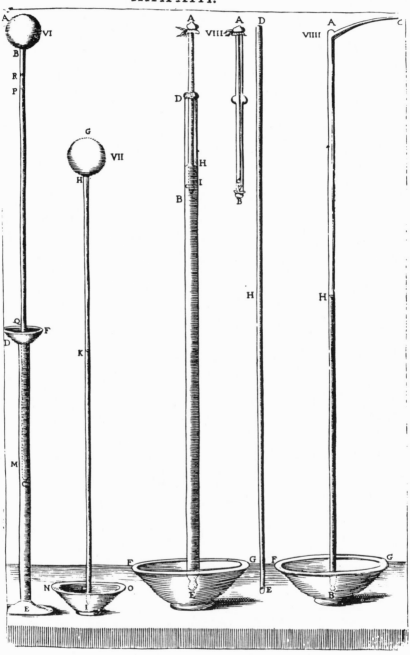

wards *B,* depending on the deeper immersion of the tube, are found to diminish gradually. From this it appears that we may probably believe that the remaining space *RBA* in the vessel is all occupied by dilated air, since from *R* upwards it is clear that the cylinder of quicksilver below it suffers some force, an evident sign (as it appears to some of us) that the volume of air *MC* needs a volume not smaller than *ABR* in order to have its full breathing space. The measurement of this volume, and therefore of the dilatation of the air *MC,* can be done as follows:

Let us imagine that these things have happened in the vessel *ABC,* where the air *MC* has received its entire natural dilatation in the space *AR.* We wish to know how large is the space *MC* occupied by naturally compressed air compared to the space *AR* occupied by the same amount of air when it is dilated. This will be found by the simple operation of weighing the water that can be contained in *MC* and in *AR.* It is found, for example, that the former is to the latter as 1 to 174. The same will be true of the air, and so this in its dilatation takes up 173 spaces besides the one that it occupied in its natural state of compression.

It should be mentioned that we have repeated this experiment several times in various kinds of weather, and it did not always yield us the same proportion. At first we made it with a vessel of another design, although the operation was the same as this, and the proportion turned out to be 1 to 209.[67] Then, using the present instrument, we obtained 1 to 182; and, finally, the third time, which also appeared to be more exact than the others, it was 1 [XXXXVI] to 174, as we set it down above in our description. But this diversity caused us no surprise at all, considering that the experiment was made with a different sort of air each time, more or less compressed as the season was warmer or colder, as well as according to the sites, higher and lower. It is impossible that it should always dilate in the same way so as to maintain the same fixed proportion.[68]

Note that the bulb *GH* was added to the simple tube *HI,* so that the air that is disseminated in invisible portions through the quicksilver and makes the latter bubble in its descent by rising into the vacuum, might have plenty of room to expand in such a large space,

[67] This result is the only one given in Draft "A."

[68] This attracted the admiration of Robert Boyle (see *Works,* ed. Birch, [1772], III, 497) who said in 1670 that this was better than the results published in his *New Experiments,* expt. VI (*Works,* I, 22). But he had later discovered some old notes that showed him that on June 2, 1662, he had expanded a bubble of air more than 10,000 times (*Works,* III, 499).

without having to alter with its pressure the natural height *KL* at which, by its nature, the mercury must be balanced.[69]

In figure VIII, *AB* is a narrow tube of glass or crystal[70] less than an ell and a quarter [73 cm] long. The lower end *B* is closed with a piece of bladder. The tube is filled with quicksilver, and through *A* a small lance *AC* is immersed in this. The lance bears lightly against the bladder at the bottom, its other extremity reaching the mouth *A*, which is also closed with a piece of bladder. At the same time another tube *DE*, more than an ell and a quarter long, is made in such a way that the end *E* can easily be closed with a finger, while the other end *D* is large enough to receive the tube *AB*. The latter, filled with quicksilver as described, is inserted, taking care to put it far enough into the bore of the tube so that its end *B* is less than [XXXXVIII] an ell and a quarter from the stagnant level *FG* of the mercury, measured towards *D*. The small tube is then fastened at *D* with mastic or hot cement in such a way that any tiny hole through which the outside air might leak in is perfectly closed. The tube *ED* is then completely filled with mercury and the end *E* stopped with the finger and immersed in the mercury *FG*. The vacuum is made in the part *DH*, the end *B* of the small tube *BA* remaining completely immersed in the mercury *HI*. The end *E* is again closed with the finger, without taking it from beneath the level *FG*, and thus, communication with *FG* being cut off, the tube *DE* becomes the vessel in which the tube *AB* is immersed.[71]

The little lance *AC* is then pressed down from outside at *A*, breaking through the bladder at *B*. As soon as this is done, the tube *AB*, although less than an ell and a quarter long, will be seen to empty

[69] There exists another suggestion for measuring the maximum degree to which air could be compressed as well as dilated (*G.* 268, 119r–21r), which the editors of *DIS* (p. 421) think may be due to Rinaldini. This makes use of the principle of the hydraulic press and could yield data of the sort required to deduce Boyle's law. It involves two separate instruments, one for compression, one for dilatation. There is no evidence that it was ever tried.

[70] Lead glass.

[71] October 13, 1657 (*G.* 262, 38r–39r; *TT* 2, 404–5). Draft "A" was briefer (*G.* 266, 46r–v; *DIS*, 288–89). The experiment is also described in Viviani's handwriting in *G.* 269, 63r and 64r, and in *G.* 269, 259r he claims it as his own.

itself of its mercury, which is the opposite of what would have happened if the space *DH* had been full of air, as will be evident from the following experiment.

AN EXPERIMENT PROPOSED FOR DETERMINING
WHETHER SUSTAINED FLUIDS WILL FALL AGAIN
WHEN THE PRESSURE OF THE AIR IS TAKEN AWAY
AND RETURN TO THEIR FORMER HEIGHT WHEN IT
IS RESTORED

Let *AB* (figure VIIII) be a glass tube about two ells [117 cm] long. Near the upper end *A*, which is hermetically sealed, let the spout *AC* be drawn out so fine that it can easily be opened by breaking the point off with the fingers and as easily closed again with a candle flame. Let the tube be filled with quicksilver through the mouth *B*, which (like the mouths of all the other similar tubes and vessels that are used in making a vacuum) should be made [L] with a ground or flattened edge that can be securely closed with the finger. Let there be also a tube *DE*, of exactly the same length as the tube *AB*, and also closed at *D*, and with its opening at *E* not circular, or cut straight across, but with rather a long cut. This tube, filled with quicksilver, is to be put like a sword in its scabbard into the tube *AB*, which is large enough for *DE* to move up and down inside it. Let the mouth *B* then be closed with the finger and the two tubes inverted and immersed in the vase *FG* as usual, so that the vacuum may be formed. This will happen in the same way in both tubes, the level of the quicksilver in each being at *H*.

Next let the mouth *B* of the outer tube be closed again with the finger beneath the level *FG*, so that the mercury *BH* may no longer be in communication with that in the vessel, but so that the tube *AB*, closed in this way, may serve (as in the preceding experiment) as a vessel for the inner tube *DE*, whose mouth *E* remains open thanks to its oblique cut. The point may now be broken off the spout *AC*, so that the air may come down through it onto the mercury *H* surrounding the internal tube *DE*, which air, pressing on this, will at once make the latter tube fill completely, since in the tube AB there is enough mercury to fill it, and the empty space, as we said, is not longer than an ell and a quarter. This experiment is very easily made, and can be repeated several times with great rapidity.[72]

[72] This seems not to be in G. 262; nor did Targioni Tozzetti discover it in the diaries. He did find in some loose papers a rather decorative drawing that seems to correspond to figure VIIII (*TT* 2, 405 & fig. 96). Draft "A" (*G.* 266, 48r–v; *DIS*, 289–90) has the essentials, but was largely rewritten.

AN EXPERIMENT FOR THE SAME PURPOSE OF FINDING OUT WHETHER THE AIR ACTS IN THE SUPPORT OF FLUIDS

Let there be a small glass phial such as *ABC* in figure X with such a narrow opening at *C* that when it is filled with any [LII] desired liquid this will not pour out even if the phial is open and its opening pointed downwards. This is filled [73] with quicksilver by way of a very fine glass funnel and, after the opening *C* has been sealed with mastic or sealing wax, is put into a glass vessel such as *DE* in such a way that the opening touches this. The cover *F* is carefully sealed with the usual cement all round the joint. Let the whole vessel *DE* be then filled with mercury through the opening *G* and let the vacuum be made, and when this has been done let a lighted candle be brought to the outside of the phial near the mouth *C*, and held there until the sealing wax melts and unseals this. As soon as it is opened the little phial will be seen to begin to empty itself; but if the air is introduced into the vase *DE* this will immediately cease. [74]

If in place of quicksilver the phial is filled with oil, wine, or some other liquid, the result will be the same. [75]

AN EXPERIMENT TO SHOW THAT IN VESSELS TALLER THAN AN ELL AND A QUARTER, FULL OF QUICKSILVER AND INVERTED IN THE AIR, THE VACUUM WILL BE MADE IN ALL THE SPACE ABOVE THE HEIGHT OF AN ELL AND A QUARTER, PROVIDED THAT THEIR MOUTHS ARE VERY NARROW

In figure XII, let *AB* be a glass tube of any length and thickness, provided that it is not shorter than an ell and a quarter [73 cm], closed at *A* and with a very small opening at *B*. Let it be filled with quicksilver and hung up vertically in the air with the opening downwards. The mercury will at once be seen to spurt out, not in drops but as a continuous jet, until it is reduced to its usual height *C* of an ell and a quarter, when it will stop pouring out. [76]

---

[73] Partly.

[74] October 13, 1657 (*G.* 262, 39r–v; *TT* 2, 406). Draft "A" (*G.* 266, 49r; *DIS*, 290) has the interesting additional information that the added mercury does not change the height of the mercury in the tube above that in the cistern.

[75] It was repeated with water on August 9, 1662 (*G.* 262, 137v; *TT* 2, 407).

[76] August 7 and 8, 1662 (*G.* 262, 135v and 136v; *TT* 2, 406). Apparently not in Draft "A." In *G.* 266, 113v, there is a note expressing the belief that water, oil, etc. would behave in a similar way, if they were in a tube that was longer than the height at which they would be balanced by the air pressure.

LI.

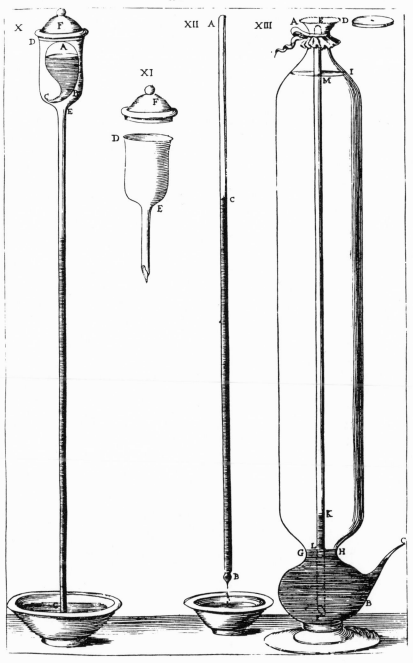

[LIV]

AN EXPERIMENT PROPOSED TO SHOW MORE CLEARLY
THAT THE SUPPORT OF FLUIDS IN ANY LENGTH OF TUBE
BECOMES LESS WHERE THE PRESSURE OF THE AIR BECOMES
WEAKER, AND THAT IF THE PRESSURE RETURNS THE
FLUIDS RISE AGAIN

Let *AB* in figure XIII be a glass vessel about two-thirds of an ell
[39 cm] high, having a very narrow spout, open at *C*. Let all the bulb
GFB be filled with quicksilver through the mouth *AD* so that as the
mercury in the spout levels itself with that in the bulb it pushes out
the air before it. When the mercury arrives at *C* the spout is closed
by means of a flame. Provide also a narrow tube *EF*, closed at *E* and
cut on an angle at *F*, and somewhat shorter than the internal height
of the vessel *AB*. Because of its thin bore and because it is less than an
ell and a quarter long, this can be lowered, full of quicksilver, through
the air in the vessel *AB* without emptying itself, until its mouth
plunges into the mercury *GH*. When it is in this position, let the
vessel *AB* be filled to overflowing with boiling water, and then let its
mouth *AD* be sealed with a glass disk cut to fit it and having a small
hole drilled in the center, this being covered with a piece of bladder
tightly tied on.

The water begins to cool gradually and, in cooling, to condense, so
that part of the vessel, such as *AI*, will remain empty; and at the same
time the tube *EF* will empty itself down to a certain point such as *K*.
When the mercury arrives there it stops and goes down no farther.
Then let the bladder be pierced where it is seen to be depressed in
the hole in the glass, and at once, on the entry [LVI] of the air, the
mercury will be seen to rise again with the greatest fury and fill the
whole tube *EF*. Even if this tube had been longer it would still have
been filled again, provided that it was not longer than an ell and a
quarter.

Note that the height *KL* would have to be about one-fourteenth of
that of the water *ML*, as will be shown later.[77] But even if it exceeds
that height, as most often happens, this may occur for two reasons.
One is that the water with which the vessel is filled was not put in hot
enough to make the space that it leaves in condensing large enough to
receive all the mercury that would have to come out of the tube *EF*;
and so, as every bit that comes out pushes the water up, the vessel is
full before the tube can empty itself as much as it ought to do. The
other reason is that even if this same empty space is sufficient for the
mercury in the tube, it may not be enough for the air that rises out of

[77] See the next experiment.

the mercury in the bulb or the water in the vessel, and this air, need-ing more room to expand in the empty space *AI*, may sometimes exert a certain force on the water and consequently push up and sustain the mercury inside the tube rather higher than it would be held up by the simple weight and pressure of the water.[78]

[LVIII]

AN EXPERIMENT IN WHICH THE PRESSURE OF ANOTHER FLUID, ADDED TO THE PRESSURE OF THE AIR, ACTS ON THE CYLINDER OF QUICKSILVER

Let us assume that the vacuum has been made in the narrow tube *ABC* (figure XIV), in which the quicksilver rises, on account of the pressure of the air alone, to *D*, usually to the height of an ell and a quarter. Let water then be put in above the stagnant level *EB* and made to rise to *A*. The level *D* will be seen to have risen to *F*, and *DF* will be about one-fourteenth the depth *AB* of the water. This is be-cause the weight of the cylinder of mercury *DF* is equal to that of a cylinder of water having the same base and with the height *AB*. And if instead of water the same space *AB* is filled with oil, the mercury will rise only to *G*; if with spirit of wine, to *H*. Whence from the pro-portion between the height *AB* of the fluid around the tube to the increased height of the cylinder of quicksilver over its original height of an ell and a quarter, produced by the same fluid, we shall be able to obtain the ratio between the specific gravity of the mercury and that of each of the fluids. Thus the relative specific gravities of these fluids themselves can very easily be deduced.[79]

The same thing can be done without any vacuum, using a simple cylindrical beaker *AB* (figure XV), into which a little quicksilver has been put, with a thin tube *CD*, open at top and bottom, immersed in it. Then pouring in various fluids above the level *EF*, all to the same height, from the various amounts of elevation produced in the mer-cury in the tube by their weights [LX] we shall have not only the pro-

---

[78] This cannot be found in G. 262, nor does Targioni Tozzetti give a reference to the diaries, but it is described in nearly this form by Viviani in a memorandum attached to his draft of a letter to Leopold dated January 5, 1658 (G. 275, 52r). It is also described in G. 261, 63v. A much rewritten version appears in Draft "A" (G. 266, 53r–54r; *DIS*, 290–91), but this lacks the last paragraph. There is also a variant draft in G. 263, 2r–4v; *DIS*, 385–87, by Viviani, in which the siphon barometer is described quite incidentally as an invention of Borelli. Actually it was invented by Blaise Pascal no later than 1650 and described in his *Traictez de l'équilibre des liqueurs et de la pesanteur de la masse de l'air*, etc. (Paris, 1663), p. 100.

[79] August 12, 1657 (G. 262, 25v; *TT* 2, 408–9). The same experiment was made with spirit of wine (*ibid.*), and somehow they obtained a ratio of 42.

LIX.

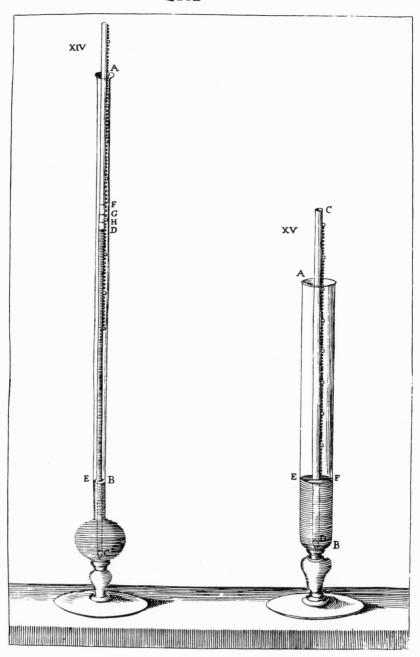

portions of their specific gravities to that of mercury but also the specific gravities of these liquids in relation to each other.[80]

Note that in this and other similar experiments, where it happens that the internal or external levels of quicksilver change their height because of the pressure of some fluid or for some other reason, the letters in the figure demonstrating such operations must always be understood to move as needed and accompany the levels to where, from time to time, they are found.

AN EXPERIMENT BY WHICH IT IS SHOWN THAT
WHERE THERE IS NO AIR PRESSURE, A VACUUM
MAY STILL BE MADE, NOT ONLY WITH MERCURY BUT
ALSO WITH WATER, TO ANY HEIGHT IN A TUBE,
ALTHOUGH LESS THAN THAT AT WHICH THE LIQUID
IS USUALLY SUSTAINED

In figure XVI, let *AB* be a glass vase holding about six pounds [2.0 kg] of water,[81] and with its mouth *A* wide enough to pass the tube *CD*, one ell [58 cm] high, closed at *C* and cut off obliquely at *D*. At about *E*, where it begins to emerge from the vase *AB*, let the tube have two little glass rings a very short distance apart, so that the piece of bladder *FEG*, pierced at *E*, may be very tightly tied between them (figure XVII).

The vase *AB* is filled with water as hot as it can bear, and the tube *CD* with cold water. A little plate of glass, suitable for closing the mouth *A* of the vase, is threaded on to the tube from the end *D*. The tube is plunged into the vase, and the bladder turned down and folded around the neck [LXII] of the vase and tied closely after the air has been expelled from the folds. The water in cooling then gradually empties part of the neck *AI*, and at the same time (as in the preceding experiment)[82] empties the tube for some such distance as *CK*. When the water has arrived there it will stop and move no more, if some new external accident of heat or cold does not change it. Then when the bladder is pierced, so that the air may return to press on the surface

---

[80] This was not found in G. 262, but is in Draft "A" in somewhat different terms (G. 266, 70r–v; *DIS*, 298) where the first instrument is called "υγροστάθμιον o vero stadera dei liquidi" (balance for liquids), and (in a passage crossed out) is ascribed to Candido del Buono. Vincenzio Viviani is given credit for the second, simpler version in another cancelled phrase. He claimed it in G. 269, 259r.

[81] In draft "A" (G. 266, 57r; *DIS*, 291) this is "a vase with about the capacity of a flask" (*fiasco*, i.e., a two-liter wine flask). The draft differs only in detail from the final version; but the last paragraph of the latter is missing.

[82] Actually the last but one.

LXI.

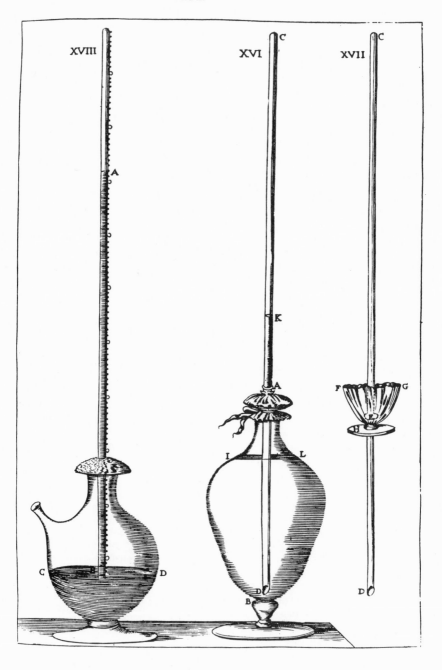

of the water at *IL*, the tube again becomes full, as it was in the beginning.[83]

It was believed by some that the reason why the water in the tube did not at first sink to the level of that in the vase when the vacuum was made was similar to that mentioned in the previous experiment,[82] namely because of the air that escapes from the water and rises into the empty space, which may be too small for its entire expansion. So they felt that if this experiment were made with wine, oil, spirit, or other liquids, it might be determined, from the larger or smaller vacuum that would remain in the tube, which of the liquids has more air distributed between its particles.[84]

AN EXPERIMENT FIRST MADE IN FRANCE AND LATER
VERIFIED IN OUR ACADEMY, FROM WHICH, IT APPEARS,
A STRONGER ARGUMENT FOR THE PRESSURE OF THE AIR
MAY BE DERIVED

In his book about new anatomical experiments, Pecquet writes that many people have observed that the height of the quicksilver in the vessels in which the vacuum is produced varies according to where the experiment is made, so that in more elevated places it is less, and greater in lower ones, provided that such an elevation is [LXIV] very considerable, as it is indeed in the highest mountains of Auvergne, on whose summits the mercury ought not to hold at a fraction of its ordinary measure.[85] This has been said to happen because the higher air that is found on the lofty ranges, that has so much less weight above, presses more weakly; nor has it the strength needed to sustain the mercury at the same height as the deeper air of the valleys and the lower plains. Whatever may be the truth of this reasoning, a discussion of which is not now our intention, we have also observed this same effect at the top of one of the highest towers in Florence,[86] which is 142 ells [83 m] high, and also on some of the various hills that sur-

---

[83] This has not been identified in the diaries; but a way of making a vacuum over water, due to Viviani, was tried on August 20, 1660 (*G*. 262, 105v–6v; *TT* 2, 442 and fig. 143).

[84] Again, the concept of vapor pressure is entirely lacking.

[85] Jean Pecquet, *Nova experimenta anatomica*, etc. (Paris, 1651) p. 55. It is interesting that the Academicians read about the famous Puy-de-Dôme experiment in this work, and not in Blaise Pascal's famous *Recit de la grande expérience de l'equilibre des liqueurs*, etc. (Paris, 1648). It is quite possible that the latter, which was published in a very small edition, did not reach Florence, though his earlier *Expériences nouvelles touchant le vuide* (Paris, 1647) did; notes on it by Viviani have been preserved (*G*. 259 fol. 5r).

[86] That of the Palazzo Vecchio.

round the city.[87] Now it was clearly seen that the height of the mercury was different in different places on the tower or the hill, becoming lower the higher we went up and rising the farther we descended, until it was balanced at its usual height when brought back to the bottom. Nor is a greater altitude than fifty ells [29 m] necessary, in order to render this effect very clearly visible.[88]

Observations made in this way[89] put it into the minds of some to make such an instrument serve as a very exact meter of the state of compression of the air, believing that the various heights of the cylinder of mercury AB (figure XVIII) ought to show without fail the changing pressure that it has on the stagnant surface CD, thanks to the differing heights[90] that it has in its region. But this idea was rendered doubtful by the many variations and the irregular movements that appeared in a long series of observations. Whereas, when this instrument was left fixed and immovable in the same place, it showed those variations [LXV] that occurred only because of the changes between hot and cold weather to be very small, and rarely more than two or three degrees,[91] yet on the contrary there were very notable ones, sometimes of more than twelve degrees, resulting from other causes unknown to us and not apparent. Nevertheless, in order to have the same information in a more certain way, the construction of the instruments to be described next was thought of.[92] In these, although the external occurrences of heat and cold may do much to impair their

[87] One was the Monte d'Artimino.

[88] This should change the height of the mercury by about 3 mm, a figure that gives an indication of the precision of the observations of atmospheric pressure at the time.

[89] Chiefly in the latter part of September 1657, when the Court was at Artimino (G. 262, 34v–35r). The Grand Duke had ordered Borelli to verify this important experiment (Cf. *TT* 1, 206).

[90] The idea that an increase in the barometric height could be the result of the accumulation of air overhead by continued strong winds, a concept far ahead of its time, seems to have been due to Prince Leopold himself, and to date from March 1660. I have dealt with this episode in *A History of the Theories of Rain* (London, 1965; New York, 1966), pp. 67–69.

[91] Degrees: The Florentine barometers seem to have been rather variously divided. One that had been sent to Borelli as official observer at Pisa stood normally at about 460 degrees (Borelli to Prince Leopold, March 16, 1660, G. 276, 12r–15r), but another document (G. 269, 83v; *DIS*, 434) refers to an instrument in which the mercury fell 3 degrees when it was carried from the bottom to the top of the bell-tower, "and the said 3 degrees is the 140th part of the height at the bottom."

[92] In draft "A" (G. 266, 604–v; *DIS*, 292–93), "immaginò il sig. Vincenzio Viviani a" was first written, crossed out, and "fu pensato alla" substituted, thus preserving the obligatory anonymity.

true and correct operation, yet they are not so inevitable that they cannot easily be avoided by the sagacity of the diligent observer.[93]

⁕⁕⁕

[LXVI]
DESCRIPTION OF THE INSTRUMENTS FOR SHOWING THE VARIOUS CHANGES THAT OCCUR IN THE NATURAL STATE OF COMPRESSION OF THE AIR

The First Instrument

Let a narrow glass tube be chosen, as uniform as can be found and somewhat wider than an ordinary quill pen, and let it be bent as shown at *ABCD* in figure XVIIII, so that it has two parallel branches *AB, CD*, about as long as they are represented in the figure.[94] These are divided into degrees with care and exactness, in such a way that the ends of the 10-degree divisions of both come out on a level. So that this may be done better than is possible with the usual enamel beads, a strip of parchment, finely divided into uniform degrees, may be stuck to the outside of each tube.[95] These are to be looked at [LXVIII] through the glass and are visible to the observer because of its transparency. The branch *CD* is to be expanded into a trumpet mouth at *D*, and the branch *AB* communicates with one or more hollow bulbs, also of glass, such as *E, F*, to hold a quantity of air. The last of these terminates in a very long spout such as *GH*, for sealing with the flame, and therefore drawn out extremely fine. A good deal of quicksilver is put in through the mouth *D*, and this will stand exactly at the same level at *I* and *K*, because the vessel is open and the branches *AB, CD* are of equal bore.

The instrument, thus prepared, is to be carried to the foot of a tower, where it is allowed to remain long enough for the air inside it to take up the temperature of the surrounding air. Then, quickly bringing a small flame to *H*, the spout is sealed with great rapidity, so that the air in the bulbs may not be altered by the heat coming from the flame. When this is done, let there be someone on the tower to

[93] In his comments on Draft "A," Borelli refused to believe in the utility of such instruments (*DIS*, 330; G. 267, 18r). Nor was Rinaldini impressed, for he wrote to Viviani on January 9, 1658 (G. 283, 33r) that in using one he had had difficulties with heat and cold.

[94] In this book the figures are 33 per cent smaller than in the original.

[95] This is the only indication that I have found that the graduation of tubes with enamel beads, so characteristic of the Academy's glassware, was ever felt to be a disadvantage. The design of their instruments undoubtedly owed much to the artistic feeling of the Duke's glassblowers.

LXVII.

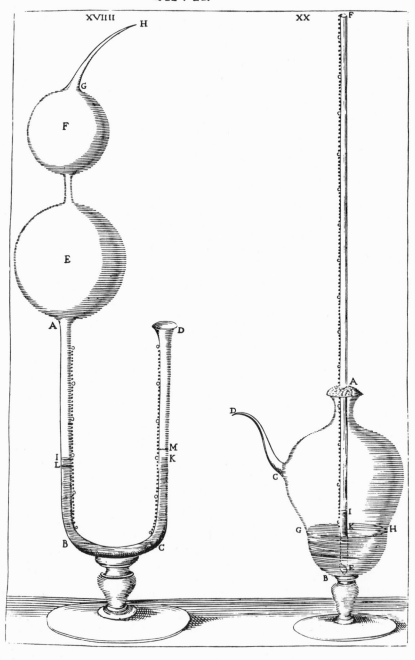

draw the instrument up by a cord that has already been tied to it so that nobody has to remain near it after the spout has been closed; and when it has been taken to the highest point of the tower, it is to be set level, as it stood at the base. Then when the temperature of the air aloft has been examined with a delicate thermometer and found equal to that of the air at the bottom,[96] it may be observed that while at the foot of the tower the mercury levelled off at *I, K*, at the top the level *I* becomes visibly depressed, say to *L*, and the level *K* rises by an equal amount, to *M*, thanks (they say) to the stronger and more violent pressure exerted on *I* by the lower air carried up inside the bulbs *E, F*, in comparison with that of the air above; whence the surface *K* is more gently pressed.

It may be remembered that any very small difference of heat or cold that there may be between the upper and the lower air is [LXX] able to produce a change in the levels in the two branches *AB, CD*, and sometimes show the opposite of what would have to follow if we assumed that only the different pressure of the air operates, inasmuch as this instrument is a kind of air thermometer, and these are for the most part very sensitive. On that account, when it is desired to make this experiment we select the early dawn,[97] or else an overcast day, so as to have the air above and below as nearly as possible at the same temperature. Care should also be taken not to leave much time between the first observation made at the foot of the tower and the second made at the summit; and to avoid approaching the instrument, except at the moment when the scales are to be observed, which must be done expeditiously, being careful not to breathe on it lest the bulbs be warmed. The thicker the glass in these, the better they will defend the air they contain from external impressions.[98]

All these precautions are also to be taken in the use of the three following instruments, which are themselves no less sensitive and are subject to the same errors as this first one.[99]

[96] What if it were not? This passage underlines the staunch empiricism of the Academy. On January 2, 1667/8 (O.S.)—while a presentation copy of the *Saggi* was on its way to the Royal Society—Robert Hooke described his "marine barometer," (so many times re-invented!) on exactly the principle of these instruments, but with the addition of a spirit thermometer and a way of making corrections for temperature. See my *History of the Barometer*, p. 375.

[97] There might, of course, be a very large temperature inversion at dawn in clear weather.

[98] Down to here, this is substantially as in Draft "A" (*G*. 266, 62r–63v; *DIS*, 293–95), which lacks the following paragraph.

[99] I have not found this instrument, nor the three that follow, described in *G*. 262; and Targioni Tozzetti gives no references to the diaries.

The Second Instrument

In figure XX, let *AB* be a glass vase that can hold about four pounds[100] [1.36 kg], with an open spout CD. Let enough quicksilver be put into this to cover the mouth *E* of the thin tube *EF*, half an ell [29 cm] high, and open at top and bottom but cut off on a slope at *E*, and straight across at *F*. This, which is divided into degrees, is to be immersed in the mercury *GH*, and the space left around the mouth of the vase sealed with mastic or other cement that will make it airtight. [LXXII] Prepared in this way, let it be carried to the foot of the tower, and when the air inside it has been allowed to reach the temperature of that outside, the spout will be sealed and the instrument drawn up by means of a cord onto the summit of the tower. When it has been set level there, the mercury will be seen to have risen inside the tube a few degrees, as to *I*.

This rise is also said to occur from the cause that we have stated in the description of the previous instrument, that is to say, because the lower air enclosed in the space *ACGH* exerts a greater force on the annular surface of mercury around the tube than does the upper air pressing on the surface *I* through the opening *F*. Therefore the raising of the little cylinder *IK* restores the equilibrium between these two pressures.

The Third Instrument

Let *A* in figure XXI be a glass bulb one-third of an ell [19.5 cm] in diameter, having a neck *BC*, about two-thirds of an ell long, finely divided into degrees and somewhat wider than it appears in the figure. Into the bulb let there be put as much water as half the neck *CD* will hold, and, closing the end *C* with the finger, plunge it into the water contained in the bladder *EF*. This is to be prevented from taking on its greatest spherical dilatation by a suitable weight attached at *F*. Then the folds of the bladder are to be taken and tied closely around the neck *BC* at *E*, being careful to pour in water during the process, making it overflow so as to be sure not to include any air, which, later changing in any way, might interfere with the correct operation of the instrument. When all is arranged at the foot of the tower, the bulb is to be attached at *G* to the cord [LXXIV] let down from the summit. After observing the level of the water, draw the instrument up, where it will be observed that it is depressed through some degrees, as at *H*, more or less, according to the present state of the air and the greater or lesser height of the tower.

---

[100] Of water. In Draft "A" (G. 266, 63v–64v; *DIS*, 295–96) it is simply "a vase of large capacity." The final version is much shorter than the draft.

LXXIII.

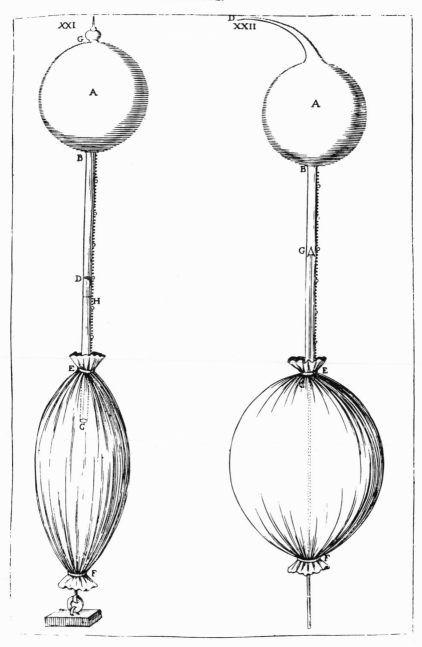

In the same way, this is said to happen because the bladder *EF* is there surrounded by the upper air, whence, not being furnished externally with sufficient resistance to bear the force of dilatation due to the lower air preserved in *GD*, it is obliged to yield and enlarge its internal volume, which is filled up again by the descent of the small volume of water *HD*.[101]

The Fourth Instrument

In figure XXII let the glass bulb *A*, with its neck *BC*, be exactly similar to that of the third instrument, except that it is open by way of the very thin spout at *D*. Let the bladder *EF* be closely tied around the mouth *C* of the neck *CB*. This bladder will have been closed by tying the lower end around a very thin glass rod or copper wire which, passing inside the bladder, reaches into the neck *BC* of the bulb *A*, where it serves to indicate the degrees into which this is finely divided. When this instrument has been carried to the foot of the tower it is sealed at *D* like the others, and the degree indicated by the pointer *G* is observed. Then when it has been raised to the top it is to be observed again, and the pointer will be found to have risen several degrees.[102]

To explain such an effect, consider this vessel, full of the lower air. Because a part of the vessel is less solid than glass, in fact yielding and marvellously extensible [LXXVI] as is the bladder *EF*, when it is raised this air at once feels the shackles of the air around it loosening and makes an effort to recover and expand. It succeeds, making the bladder distend somewhat. Now during its inflation this approaches a spherical shape more and more nearly, and its diameter *EF* becomes less as the bottom *F* gradually rises. Therefore the index *FG*, fastened to this, partakes of its motion and runs farther into the neck *BC*, finally touching a graduation higher than *G*.[103]

[LXXVII]
Various Experiments Made in the Vacuum

Torricelli's concept of the pressure of the air on bodies beneath it now seemed to be well enough established by the series of experiments already described. Although it may be presumptuous and full

---

[101] Greatly rewritten from Draft "A" (G. 266, 64v–65r; *DIS*, 296).

[102] Entirely rewritten from Draft "A" (G. 266, 65v; *DIS*, 296–97).

[103] Instead of this paragraph, the Draft "A" (G. 266, 66r–v; *DIS*, 297–98) has a "Fifth instrument" illustrated in the annexed figure 4.1, from a sketch by Viviani in G. 270, 101v. The description begins: "Fifth instrument.

of danger to make assertions about those things on which no lamp of Geometry shines to help our eyes, yet the presumption is never so excusable, nor the danger more certain to be avoided, than at the moment when, purely by way of many experiments all concordant, our intellect journeys to the attainment of its desire. Although it sometimes fails to reach this, yet in approaching it as closely as we can we are satisfied. So, as it appeared that we had gained from the above-mentioned results some reasonable probability that the pressure is produced in that way, we decided that it would be no loss of labor to go on with various experiments in the vacuum, seeing whether their results would turn out oppositely or in some respects differently from those that they show when made in air.

To learn how great is the expansion of the compressed air collected at the foot of the tower, and measure it by reducing it to a regular and known volume, enclose the slightly inflated bladder *C* in the glass bulb *AB*, and hold it to the bottom *B* by means of a copper wire. Heating the bulb externally, bring in through the spout *D* enough spirit of wine not only to cover the bladder *C* but also to fill part of the tube *DA*."

The method of operation will be obvious to the twentieth-century reader.

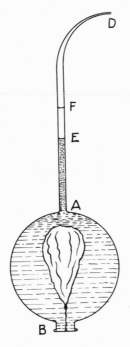

FIGURE 4.1   The fifth "air barometer" of Viviani.

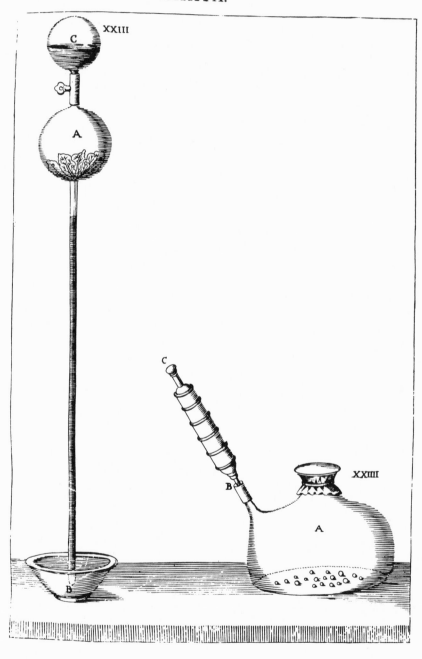

XXIII

C

A

C

B

XXIIII

A

B

[LXXVIII]

AN EXPERIMENT TO FIND OUT WHETHER DROPS OF LIQUID, FREED FROM THE PRESSURE OF THE SURROUNDING AIR, LOSE THE SPHERICAL SHAPE TO WHICH THEY NATURALLY CONFORM

Some people used to attribute to the pressure of the air the phenomenon commonly observed in drops of quicksilver and every other fluid,[104] which always become round when they are sprayed or rain down into the air or settle on a dry substance. On that account we wished to see them in the vacuum, imagining that it could easily happen that some notable difference might be observed. But this experiment made it clear that the cause of this phenomenon was something other than pressure. For when the vacuum had been made in the vessel *AB* (figure XXIII), and the tap turned to open the bulb *C*, the water or mercury kept in the latter, falling in drops onto some cabbage leaves put in with the dew on them as they were gathered, remained in drops as round as if they had been on the plant.[105]

Similarly, if the air in the vessel *A* (figure XXIIII) was compressed or rarefied by means of the syringe *BC*, the drops of water or quicksilver that had been sprayed over the bottom of the vessel did not change from their customary shape.[106]

[LXXX]

AN EXPERIMENT ON THE EFFECT OF HEAT AND COLD APPLIED EXTERNALLY TO EMPTY SPACES

Let a bladder *ABC* be tied beneath the bulb *D* (figure XXV), and when the vacuum has been made let the bladder be turned upwards so that it surrounds the bulb. Then with a glass rod, or something similar that will not bend, take the exact height of the cylinder of mercury *GH* from the stagnant level *EF*. When this has been done, fill the bladder with hot water. Then measuring again after a short time, the cylinder will be found somewhat depressed below its original height. When this observation has been made, let the hot water run out, and after the mercury has been allowed to return to its first height at *H*, put cold water in, mixed with crushed ice and salt. Soon after-

---

[104] Galileo attributed it to an "antagonism between air and water," in his *Discorsi e dimostrazioni matematiche intorno à due nuove scienze,* etc. (Leyden, 1638). See *Opere*, ed. naz., VIII, 115; or the English translation by H. Crew and A. de Salvio (New York, 1914), p. 71.

[105] July 16, 1660 (G. 262, 92v–93r; *TT* 2, 417).

[106] July 21, 1660 (G. 262, 94r; *TT* 2, 417). These are in Draft "A" (G. 266, 74r–v; *DIS*, 301).

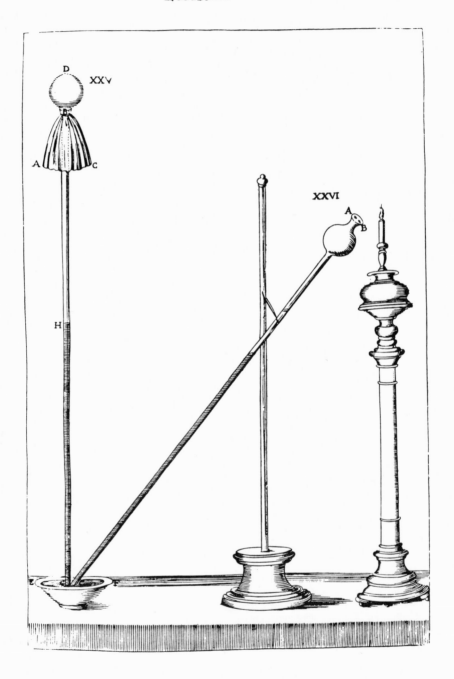

wards, measuring again in the same way, the cylinder will be found notably higher.[107]

We shall not omit to say here that the hot water used by us in this experiment took the 50-degree thermometer to 48 degrees [52°C], such a heat lowering the mercury 1/140 of its height; and it was raised 1/58 by the cold, in which the same thermometer came to 11½ degrees [ca. −2°C].

If a little air is then introduced into the bulb *D*, although this becomes extremely rarefied by the expansion that it suffers in the vacuum, still, taking up the heat and cold very rapidly, by its expansion and contraction it makes the changes of the mercury in rising and falling more rapid and more easily discernible.[108]

[LXXXII]

AN EXPERIMENT TO MAKE IT CLEAR WHETHER IT IS THE AIR WHICH, ACTING AS A LAMINA ON THE REAR SURFACE OF A GLASS LENS, REFLECTS THE SECOND INVERTED IMAGE THAT APPEARS THERE, FAINTER AND MORE OBSCURE, OF A LAMP OR OTHER OBJECT REFLECTED IN IT, AS KEPLER BELIEVED (ASTRON. OTT.)[109]

Let a glass lens such as *AB* in figure XXVI be fastened with hot cement to the mouth of the vessel *AC*. The mouth is to have its edge somewhat turned down outside and flattened, so that the lens may easily be cemented to it all round. Then when the vessel has been filled with quicksilver, the vacuum made, and the room darkened, let a lighted candle be brought near the lens, and observe that the usual two images will be seen there in the same way. The smaller but very bright one is always erect and is that which comes from the outer surface. The other, larger but always more subdued and weaker and most frequently inverted, is not lost, even though because of the vacuum

---

[107] Musschenbroek (1731, p. 64; see Appendix A, no. 15) could not believe this and tried it himself, coming to the conclusion that air must have been present.

[108] July 6, 1660 (*G.* 262, 87r-v; *TT* 2, 399). Repeated August 9, 1662 (*G.* 262, 138r) and January 10, 1667 (*G.* 262, 169r-v).

[109] Kepler, *Ad Vitellionem paralipomena, quibus astronomiae pars optica traditur,* etc. (Frankfurt, 1604), pp. 13–14. Kepler's idea seems to be that if light is to bounce, it must have something to bounce off. He certainly would not have considered an artificial vacuum. But elsewhere (*ibid.,* p. 142), he explains abnormal terrestrial refraction (looming) by reflection "from the upper surface of our atmosphere" (*à superiore superficie aeris nos ambientis*). It was then thought to have a definite boundary. Targioni Tozzetti (*TT,* 420–21) quotes from a manuscript of Viviani's in the *Real Segretario Vecchio* describing this experiment in a slightly different form.

LXXXV.

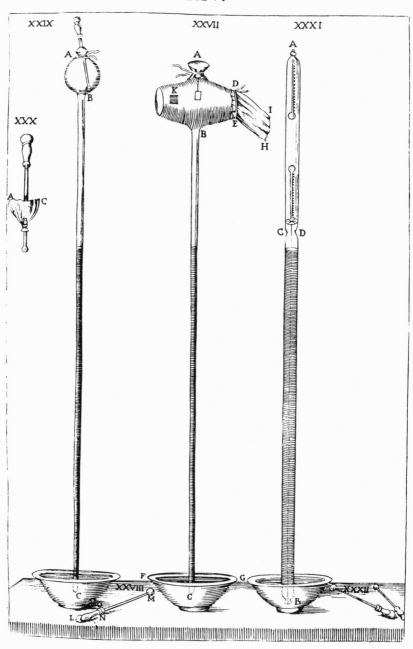

the internal concave surface of the lens lacks its imagined lamina of air.

In making this experiment it has always[110] been our practice to finish filling the tube with three or four fingers of spirit of wine.[111] When this rises through the quicksilver as the vessel is inverted to make the vacuum, it washes the lens and clears it marvellously well of every bit of tarnish that the mercury could leave on it, which might otherwise be said to be able to serve as a foil to it in place of the air. But notwithstanding this, as we have said, the appearance of the two images is the same, and when the space is again filled with air it makes not the slightest difference.[112]

[LXXXIV]

AN EXPERIMENT TO DETERMINE WHETHER AMBER AND OTHER ELECTRICAL SUBSTANCES NEED TO HAVE AIR AROUND THEM SO THAT THEY MAY ATTRACT

Provide a large vessel of thick glass, such as *ABC* in figure XXVII, large enough in its upper part *AB* for a hand to move and work inside it. Let it have three openings *A*, *C*, and *DE*. The mouth *A* is to be left open and *C* to be closed with a piece of bladder and set upon a pad of cotton or other soft cushion floating on the mercury in the bowl *FG*, so that the great weight of mercury that it must bear may not break off the rim to which the ligature is tied, or break the tube. The opening *DE*, made large enough to receive a hand, should have all around it an edge or projection to which is tied and strongly bound a large bladder open at both ends, such as *DEHI*. Through this the hand is introduced into the vessel, grasping a piece of yellow amber of the best quality, after a very light pendulum of paper or straw has been arranged in the vessel in a place where the amber may easily be presented to it after having been stroked and warmed[113] on a little strip of cloth *K* glued inside the glass. Then the bladder is tied at its end *HI* somewhat above the wrist joint, so that the hand may remain

[110] Only two dates are mentioned in G. 262: June 8, 1660 (73r–v) and June 9 (74r–v).

[111] In G. 262 vinegar is mentioned, not spirit of wine; and they left the apparatus set up over night so that the film of vinegar might have time to evaporate from the lens. "The idea was proposed by Sig. Vincenzio Viviani" (73v). Draft "A" (G. 266, 78v–78r; *DIS*, 304–5) has it spirit of wine.

[112] Quite apart from using a vacuum they tried excluding the air from the back surface with wax, pitch, etc., but always saw more than one reflection (G. 265, 122r; *DIS*, 355).

[113] William Gilbert, *De Magnete* (London, 1600), chapter 2, says that amber must be warmed in order that it may attract, but is careful to note that the warming must be done by friction, not before a fire.

free to move in the vessel. The place where the ligature has to be made should be armed with a leather bracelet very strongly bound to the flesh. On this bracelet, besides the very firm binding, the edge of the bladder may be stuck to the arm all round.

When this has been done, the whole vessel is filled with quicksilver through the aperture *A*, being careful in filling it that the folds and wrinkles [LXXXVI] of the bladder are entirely full, so that the air may escape as much as ever it can. When it is full the opening *A* is closed with a piece of bladder in the same way and the lower ligature undone beneath the level *FG*, letting the mercury out to make the vacuum. Now take the amber in the fingers and, when it has been strongly rubbed on the cloth *K*, present it to the suspended paper or straw and see whether it still attracts it as it does in air.

We have had little success with this experiment; for whenever we have made it[114] the air has come in so quickly to fill the vacuum that it has never been possible to see what the amber would have done in it. So, reflecting on what vent or aperture it could be that admitted such a large amount of air in so short a time, we decided that it could be nothing else but the binding of the arm. But as the arm could not be more tightly bound without receiving considerable harm, especially to the veins and the blood, a small stick such as *LM* in figure XXVIII was used instead, with a little ball of amber at one end. Then when the same end *HI* of the bladder had been tied between the two projections *N*, *O* on the stick, the vessel was again filled with quicksilver and the vacuum remade.

In truth nothing was got by this new way of doing the experiment, inasmuch as although the air came in more slowly—it will take every means of entry—the pressure of the external air still pushed the bladder inwards to such a degree that it carried the amber so far past the piece of cloth that it could not be warmed on it; while it was impossible to withdraw the stick and to move it back and forward as would have been necessary[115] until, the vessel again becoming full, the air inside balanced the external air.

However, as we desired to obtain some results [LXXXVIII] from this experiment, we thought of another vessel such as *ABC* in figure XXIX, persuading ourselves that with it we could more easily avoid both the leaking of the air and the difficulty of moving the stick forwards and backwards. Then this vessel was filled with quicksilver through the opening *A*, after the other end *C* had first been closed and

[114] Four dates are known: August 5 and 6, 1660, August 31, 1662, and September 2, 1662 (*G.* 262, 100v–101v, 151r, 151v; *TT* 2, 423, but incorrectly dated in *TT* 2).

[115] The force would have been of the order of 100 kg.

supported on a cushion as the previous experiment taught us to do. Then, when the bladder *ABC* (figure XXX) had been tied round the stick, this was plunged into the opening *A* (see figure XXIX) so that the amber would come to rest at *B* on a little piece of cloth attached to the glass as before. Next, after several bits of straw, chopped extremely small, had been placed on the mercury, the bladder was tied down directly below the rim of the mouth *A*. The vacuum was made, and we commenced to warm the amber on the cloth by moving the handle of the stick to and fro outside and to present it, when it was believed to be warm enough, to one or another of the bits of straw that were left scattered about the bulb by the falling mercury. But not one of them was ever seen to be attracted.

We would nevertheless point out that we cannot base any firm opinion on this experiment, nor attribute such a result unconditionally to the absence of air, more or less of which always got into the vessel, for we were never able to tighten the binding so that it would not leak in by some hidden path. Perhaps this occurred because of the motion that had to be made in this experiment to warm the amber; it being practically impossible that the binding should not loosen and yield, at least enough for the very subtle air to get in, when this was being done.[116]

It was also observed that after the vessel had again been filled with air the amber would not attract, even when it had been rubbed on the cloth *B* with great force. This at first led us to suspect [LXXXIX] that some kind of deposit had been left on the cloth by the quicksilver, so that in rubbing it the amber would receive a slight tarnish that might block up the invisible mouths of those passages through which its virtue comes out.[117] This suspicion grew, inasmuch as we already knew that amber, and all other jewels endowed with the same virtue, refuse to attract when they have been bathed in some liquids. But after it had been seen that the same amber, rubbed on another cloth that had been washed and washed again with mercury, nevertheless attracted with great force, it was thought that the cloth in the vessel might perhaps be harmed by the humidity absorbed from the gum used to attach it. Therefore a strip of chamois stuck on with sealing wax was substituted for the cloth in order to avoid the absorption of moisture; but this diligence was without result, since whether the vessel was empty or full of air the amber never attracted. This is as much as we can truthfully say about an experiment tried in so many ways, but to no effect.

[116] Down to here this is in Draft "A" (*G.* 266, 77r–78r; *DIS*, 303–4), but somewhat more concisely.

[117] See Gilbert, *De magnete*, tr. P. F. Mottelay (London, 1893), p. 96.

AN EXPERIMENT TO FIND OUT HOW THE INVISIBLE
EXHALATIONS OF FIRE MOVE IN THE VACUUM[118]

As it had become clear to us by way of other experiments[119] that the heat of fire does not move in every direction equally, but diffuses upwards incomparably more than in any other direction, one of us[120] thought that in an empty space, on the contrary, some difference might be found, from which might be drawn very firm conjectures about the principles and also about the natural movement of fire.[121] This may be tried by means of an instrument like the following.

Let there be a tube such as *AB* in figure XXXI, two ells [117 cm] long, into which, while it is still open at *A*, is lowered a 50-degree thermometer, upside down. [LXXXXI] The sealed end of this is made in such a way that it can rest on the internal swelling above the neck *CD*, worked in the tube expressly for that purpose. And so that in putting in the quicksilver the thermometer may not have to move, and, striking the bulb of the one that must be placed above, break both, it is to be secured to a thread that is brought out of the end *B* and that can serve to hold the thermometer up when the tube is inverted for filling. The first one having been installed, the other is put in, as similar as possible to the former, and in sealing the end *A* hermetically, this is fixed with the same piece of red-hot glass.

The instrument being thus prepared, the quicksilver is put in and the vacuum made, taking care to keep the neck *CD* higher than an ell and one-quarter [74 cm] so that the thermometer placed upon it may not remain buried in the mercury, but may stay accessible to the observer with all its graduations. With the tube held steady a great deal of heat is communicated to the empty space with two red-hot iron balls (figure XXXII) held at an equal distance from the tube, but at an unequal distance from the bulbs of the two thermometers. They should be held somewhat closer to the lower one, so that the heat, which always rises in air, may thus be distributed equally.[122]

After having repeated this experiment a great number of times,[123] we can say only that the upper thermometer indeed feels more heat than the lower one. It is true that the difference is very small in com-

[118] In Draft "A" (*G.* 266, 80r; *DIS,* 306) the title is: "An instrument used to try to find out if there is positive gravity in fire." It was a peripatetic dogma that heat is a quality, not a substance.

[119] See p. 341 below.

[120] Vincenzio Viviani (see *G.* 262, 88v; *TT* 2, 425).

[121] "Natural movement" sounds like a relic of the school philosophy, as it undoubtedly is. Cf. p. 341.

[122] The experiment was made in a vacuum in order to reduce the effect of convection, an effect that was clearly understood by Carlo Rinaldini.

[123] On five occasions in July and August 1660 (*G.* 262, 88v–90r, 91r, 92v, 103r–v, 103v–104r; *TT* 2, 426).

parison with that observed when the tube is full of air, since in the latter case it sometimes reached five degrees, but in the vacuum it did not exceed two. Nor did it seem to anyone that it ought to be otherwise, since the air around the bulbs, being heated more in its upper part, heated the thermometer nearest it more strongly.[124]

[LXXXXIII]
AN EXPERIMENT ON THE MOTION OF SMOKE IN THE VACUUM

In the bulb of the vessel *AB* (figure XXXIII), fasten a pastille of bitumen, black or of some other dark color, which can easily catch fire. Then when a vacuum has been made, ignite it in the brilliant rays of the sun with the burning mirror. Smoke will at once be seen to emerge, and instead of rising as usual immediately on leaving the pastille, it descends, forming a parabola like the jet of a fountain. If the air is let in and the smoke again produced, it at once rises towards the top of the bulb.[125]

Now as in this apparatus there were made many experiments that did not require the construction of a special vessel, as did most of those described up to this point, it will be well, in order to avoid proxility, after a very brief description of the vessel and its dimensions—the size of the paper not being sufficient for a full-scale figure of it, such as we have had made, for greater clarity, of some other things that belong to this vessel—to speak in detail about the means we have adopted for its easy and convenient use. In this way others who also desire to see our experiments and verify them with their own will be able to make use of it, at least until another apparatus, easier and safer, is available.[126]

---

[124] Draft "A" is much more discursive and contains a table of observations (*G.* 266, 81v and *DIS*, 307).

[125] June 12, 22, and 28, 1660, and July 16, 1660 (*G.* 262, 75v–76r, 81v, 83r, 93r; *TT* 2, 427–28). It is interesting that Descartes suggested precisely this experiment in a letter to Mersenne dated December 13, 1647 (Descartes, *Oeuvres*, eds. Chas. Adam & Paul Tannery, 12 vols. [Paris, 1897–1911], V, 98–100): "I also wish that you would try to kindle fire in your vacuum, and observe whether the smoke goes up or down, and what the shape of the flame is. This experiment can be made by hanging a little sulphur or camphor at the end of a thread in the vacuum, and setting fire to it through the glass with a lens or burning mirror." This letter could not possibly have been known to the members of the Academy. The experiment is ascribed to Borelli in a letter from Magalotti to Ricci, July 5, 1660 (*LIUI*, 2, 88–90). Boyle had tried unsuccessfully to make a similar experiment (*New Experiments Physico-mechanicall*, etc. [Oxford, 1660], expt. 15, pp. 102–5; *Works*, ed. Birch [1772], I, 32.) Borelli adduced it as one more argument against "positive levity," as is indeed stated in the unpublished preamble to a draft in *G.* 263, 78r, and another in *G.* 265, 139r.

[126] This is, of course, the air pump. See also p. 264 below.

LXXXXII.

XXX III

So *AB*, figure XXXIV, is the glass vessel, the mouth *AC* of which extends outwards with a flat rim. The opening is three fingers wide and the height of the neck *AD* is four fingers. The diameter of the bulb *DE* is one-third of an ell [19.5 cm][127] and the length of the tube *EB* about two ells [117 cm]. The lower end *B* is closed with a [LXXXXIV] piece of bladder, and when it has been placed on a leather cushion put to float on the mercury in a bowl (figure XXXV), the filling of the vessel is started. But because in pouring the quicksilver through the mouth *AC* it falls down violently through the tube and a great deal of air remains trapped between the wall of this and the mercury itself, the very thin funnel, *ABC* in figure XXXVI, is used. This is also of glass and as tall as the whole vessel. Care is taken to keep its body *AB* always full, so that the neck *BC* may never be able to fill with air, and thus the mercury gradually fills the vessel, driving out the air little by little with the quiet rise of its level. The filling completed, the mouth *AC* is covered with a plate of glass, somewhat crowned as in figure XXXVII, and this is tied on with a piece of bladder bound with waxed string beneath the rim of the opening. Then, applying the palms of the hands under the bulb on both sides, it is lifted enough to expose the end *B* to the quicksilver when the cushion has been removed. Then the knot of the binding is loosened and the mercury itself exerts its weight to open the end and freely flows out, producing the vacuum.

Then when we have to put into the bulb something that may not be covered with mercury, or something that is not to be spread over it, such as liquids, which are put in the little vessel *A* (figure XXXVIII), or something, such as animals, that would drown there, we are accustomed to leave enough air in the neck *AD* to serve for the small vessel or for the animal that we wish to enclose. When the vacuum has been made this air, expanding into the volume of such a large bulb, becomes so rare that it may be said not to be there at all, as, thanks to its extreme lightness, it does not in the least impede any of the effects that it is desired to observe. [LXXXXVI]

Now when we wish to put in fishes no air is left in, nor is the whole bulb filled with mercury, but enough water is put in so that when the vacuum has been made it stands above the sustained cylinder and finally fills about half the bulb, so that the fishes can move in it and even dart about. At other times, when we have desired to put in small animals, such as lizards, leeches, and such things, we put in with them a little ball of solid glass (figure XXXIX) formed with gores. This floated on the mercury when the vacuum was being made, finally

[127] The bulb alone would hold 53 kilograms of mercury!

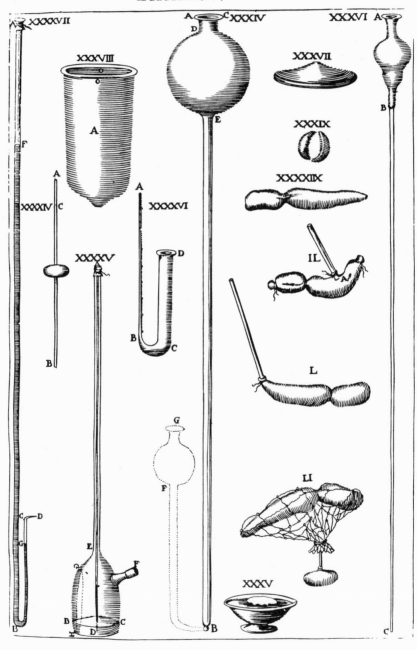

closing the mouth *E* of the tube, so that the animal would have to remain within the bulb for more convenient observation.

To some people all this information will perhaps seem superfluous; but those who have been trained in experimenting by long practice and know by experience the difficulties encountered in making an experiment because of the impediments so often produced merely by the use of physical instruments, will appreciate these minutiae instead of despising them. It is unbelievable how fruitful they are and how considerable is the loss of time that is avoided by their use.

AN EXPERIMENT ON SOUND IN THE VACUUM

After a bell had been suspended from the same thread in place of the pastille[128] and the vacuum made, we began to shake the bulb vigorously, and the bell made itself heard with the same tone as if there had been air in the bulb; or if there was any difference, it was certainly not observable. It is true that in this experiment the sonorous instrument ought not to communicate with the vessel in any way—an impossibility—[LXXXXVII] for otherwise we cannot say for certain whether the sound is formed from the very rarefied air and the exhalations evaporated from the quicksilver in the vacuum, or from the shaking that the vessel, and consequently the external air that surrounds it, receives from the strokes of the metal by way of the thread.[129]

Consequently we thought of making this experiment with a wind instrument, as being one that creates the vibration, not with percussion like the bell, but from the impetus of the air that comes out of it. And because it would have been too difficult, if not entirely impossible to put such an instrument in the vacuum that could be made with quicksilver, we resolved to enclose it in a vessel from which the air could be drawn out, in the way latterly practiced with wonderful success by Boyle for the purpose of his very beautiful and noble experiments, among which this one also occurred to him, though he did not then put it into practice for the lack of a craftsman able to make the device.[130]

[128] See p. 147.

[129] This had been emphasized by Emmanuel Maignan (Magnanus) in his *Cursus philosophicus*, etc., 4 vols. paged as one (Toulouse, 1653), pp. 1933–34, and this book was one of a list suggested to Prince Leopold by Rinaldini in a letter dated November 6, 1656 (G. 275, 44r). The list is copied in *DIS*, 51–52. In Draft "A" (G. 266, 76v; *DIS*, 302) there appears the interesting suggestion that "if possible" the bell should be supported by "a magnetic force applied externally."

[130] Boyle, *New Experiments Physico-mechanicall*, expt. 27, p. 212. (*Works*, I, 64).

Even if we cannot succeed in emptying vessels in this way as perfectly as with the quicksilver, in any case the method is able to make the air thin enough so that from the variation that is clearly seen to appear in these phenomena, which really depend on its ordinary pressure, it then becomes very easy to form an opinion on what they would do in a perfect vacuum.[131] We shall relate what we have succeeded in observing, declaring that we refer to it more with the intention of sharing the way in which we thought of making this experiment, than because we have succeeded in getting sure and infallible results. We can say that we have outlined the experiment rather than made it.

So a little organ was built, *ABCD* in figure XXXX, with only one pipe and a bellows on a stand, communicating with its wind chest hollowed out in the thickness of [LXXXXVIII] the base *BC* itself. We enclosed this in a small copper box *F* (figure XXXXI) and introduced through the opening *G* the handle *HI* (see figure XXXX), hinging it at *K* on the column or support *KL*, after inserting it in the ring *M*, soldered to a small piece of wire. This wire passes through holes in the cheeks of the bellows on each side, embracing them between sharp bends at its extremities. Then by moving the handle to and fro, one or the other side of the bellows opens and shuts,[132] thus sending wind to the pipe. We then took a circle of leather with a hole in the middle and passing the mouth *G* through the hole we tied it all around and, taking the outer edge of the leather, brought it around the handle and bound it tightly. Thus we managed to keep the passage closed to the air and, by the softness of the yielding leather, to allow the motion necessary for pushing the handle to and fro.

When everything had been adjusted, and the joint of the cover *E* carefully fastened on with hot cement, we began to exhaust the air from the box with a pump (figure XXXXII) screwed in to the upper opening *N*. After each upward stroke the tap *O* was closed, so that the air that had been sucked out was expelled through the valve *P* by pushing the plunger down again and could not re-enter the box and render the exertions of the pumper useless. After many strokes, when the remaining air had become so rare that the leather of the mouth *G* was all sucked inside the opening, and the force of an exceedingly strong man no longer sufficed to draw the plunger up, we began to move the handle backwards and forwards so as to send the very thin

[131] In the seventeenth century such experiments almost never succeeded, partly because no one realized that a very good vacuum would be required. Boyle succeeded in preventing the ticking of a watch being heard (*Works*, I, 64).

[132] It had a fixed central board.

LXXXXIX.

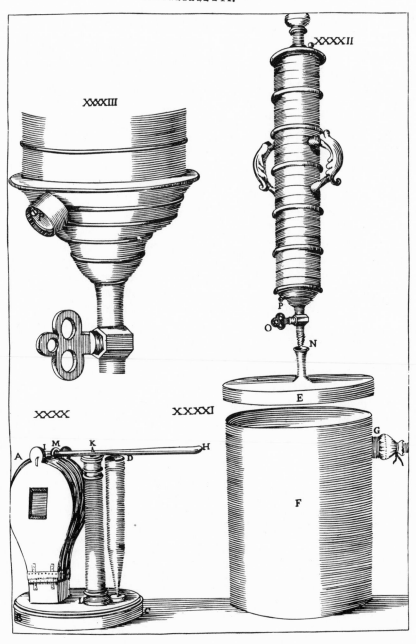

air from the bellows into the pipe, and listened to its sound. But the truth is that it did not seem to differ at all, not only from that produced in the same box full of air in its natural state, but not even from that produced [C] after we had driven and crammed in a very large quantity of air with the same pump. So, some of us said in jest, either the air has nothing to do with sound, or it can produce it in any state whatever.[133]

Figure XXXXIII shows on a larger scale the valve *P* that was made to give an exit to the air that was gradually removed from the box.[134]

AN EXPERIMENT ON THE ACTION OF THE MAGNET IN THE VACUUM

When a needle was attached to the thread that had supported the bell, and the magnet brought up outside the bulb, it was attracted from the same distance as when the bulb was filled with air.[135]

AN EXPERIMENT ON THE RISE OF LIQUIDS IN THE
BORES OF VERY NARROW TUBES IN THE VACUUM

Among the other results of the pressure of the air some people have also included the rise of almost all liquids in very narrow tubes that are dipped into them.[136] They suspect that the very thin cylinder of air that presses down through the tube, on water for example, exerts its pressure more weakly because of the resistance offered to its descent by its large area of contact with the internal surface of the exceedingly narrow vessel. On the contrary, in their opinion, the air that presses freely on the ample surface of water round the tube and is allowed to come down on it with all its force, raises the water inside the tube until the pressure of the lifted water [CII] combines with the very weak internal pressure to make a pressure equal to that of the external air.[137] In order to throw some light on the truth of this matter, we sought to see what would happen to this phenomenon in a vacuum.

[133] This experiment has not been found in the diary.
[134] This apparatus was as near to an air pump as the Academy got.
[135] August 6, 1660 (G. 262, 102r). *TT* 2, 431, ascribes it erroneously to August 5. Draft "A" has it in slightly different words (G. 266, 77r; *DIS*, 303). Boyle (see *Works*, I, 32) had done a similar experiment (*New Experiments*, pp. 105–6, expt. 16); but in 1647 Marin Mersenne knew by experiment —probably made by Roberval—that the magnet would attract through the vacuum (*Novarum observationum Tomus III* [Paris, 1647], 2nd preface).
[136] I do not know who first wrote about this phenomenon. At some time in the 1630's Niccolo Aggiunti was concerned about its cause (G. 128, 60r).
[137] It is remarkable that this view could be held even by people who were quite well aware that mercury is depressed in a capillary tube, rather than rising in it. Robert Hooke, for example, in a pamphlet entitled *An Attempt for the Explication of the Phenomena Observable in an Experiment Published by the Honourable Robert Boyle, Esq.*, etc. (London, 1661), after

Therefore the usual bulb was prepared as we have described it for putting in fishes, that is to say with the upper half filled with water. In this was immersed the very thin tube *AB* (figure XXXXIV), open at both ends and threaded through a little hollow glass bead cemented on around it, and counterpoised in such a way that it would stand upright in the water. The mouth *AC* [see figure XXXIV] was closed in the way described, the vacuum was made, and the water stopped at about the middle of the bulb. The little tube remained erect, above the water surface from the bead up, and the water inside rose as far as *C*. When the lower end of the vessel was closed with a finger, so that the arrival of the air should not empty it, the mouth *AC* was opened in order to see whether, when the air came down upon the water, this would make any sort of variation from its former level *C*, because of such a major and violent impulse. But in fact it did not move.[138]

However, it was suspected after this experiment that the wetting of the whole internal surface of the little tube, when it was all immersed in the water before the vacuum was made, might have served as a cement for the very narrow cylinder *CB*, so that it was held up by adhesion rather than by the external pressure. It was therefore resolved that the air in the vessel where we wished this experiment to be made should be expanded and rarefied beforehand, so that the first immersion would be made with the air already dilated and with the tube dry. In this way the only water that would have to rise would be that which the weak pressure of the very thin air might be thought able

---

developing a qualitative explanation of capillary phenomena that would pass muster today, and even attempting to account for the different behavior of mercury, finally forgets all about mercury and in the end adopts the hypothesis of differential pressure examined here. The attraction of this idea was probably its delightful simplicity. The Academy noted the opposite behavior of mercury as early as August 11, 1657 (G. 262, 25r–v; *TT* 2, 637–38).

[138] There is an unusually long account of the progress of this experiment in the diary for June 14, 15, and 16, 1660 (G. 262, 76r–79r; *TT* 2, 435–37). They tried a cork float, but too much air came out of this, and a hollow wax float, but it blew up. They were bothered by the violent bubbling in the water. This always happened with water in a vacuum and had been a mystery ever since Berti's experiment, made about 1640 (see Middleton, *History of the Barometer*, pp. 10–15). On June 16 Vincenzio Viviani explained that these bubbles were "either exhalations, or sublimations of volatile particles in the mercury. They are produced in every vacuum, and only now noticeable, for they must of necessity rise in a medium heavier than themselves, and when it is transparent, like water, they are discovered" (G. 262, 79r; *TT* 2, 437). Viviani believed that a better vacuum could be made over water than over mercury because of these "exhalations" (see G. 262, 105v; *TT* 2, 442).

to lift up. Therefore, when the air was brought back to its natural state, [CIV] or indeed artificially compressed as well, it was thought that we should see what movement the water would make in the tube.

Therefore a thick glass vessel was taken, such as *ABC* in figure XXXXV. In it was put the narrow tube *AD*, and when the mouth *A* had been closed with a piece of bladder the vessel was laid down so that its neck *AE* was horizontal, as also was the tube *AD*. With the vessel in this position red wine (the better to see the level in the tube) was put in through the mouth *F* until it came to the level *GH*, using such care in this that the end *D* of the tube did not become wet. Then the end of a pump was screwed into a metal nut cemented to the opening *F*. When several vigorous strokes had been made with it the vessel was again stood upright, so that the wine that formerly came to the level *GH* now had its surface at *BC*, leaving the end *D* of the tube immersed.

Then the wine at once rose to some point *E*, and this rise was equal to that which would have taken place in air that is naturally compressed, since not only when the air was allowed to resume its natural state by opening the mouth *F* but also when air was forcibly driven in with the pump so that the bladder *A* became very hard, the wine was not seen to rise as much as the thickness of a hair above the elevation to which the very rarefied air in the vessel had driven it in the first part of the experiment.[139]

Another experiment was also made, as follows: In the usual bulb[140] was put a U-tube such as *ABCD* in figure XXXXVI, suspended in such a way that after the vacuum was made it would remain upright in the middle of the bulb and full of quicksilver. Then, observing the graduation at which the mercury remained in the narrow branch *AB*, it was not seen to move away as the air was again let in. This experiment was repeated a great number of times, and always with the same result.[141] [CVI]

Finally, those who held the sustaining of the liquid at a definite height to be a most certain effect of the pressure of the air still wished to see whether the air that presses on a stagnant surface, if it is constrained to pass through the bore of a very narrow tube and has to pass through this in order to exert pressure, is weakened enough to enable us to observe a visible fall in the height of the liquid on which

[139] September 9, 1662 (*G.* 262, 154r–v; *TT* 2, 433). The account in the *Saggi* lacks a few sentences of explanation that occur in the diary. That in Draft "A" (*G.* 266, 95r–96r; *DIS*, 300–1) is much less extended. Viviani claimed to have designed these experiments (*G.* 269, 259r).

[140] Figure XXXIV, plate LXXXXV.

[141] August 29, 1662 (*G.* 262, 149r–v; *TT* 2, 433, where it is erroneously ascribed to August 19). Repeated August 30, 1662 (*G.* 262, 149v–50r; *TT* 2, 434).

it presses in this way. This would probably have to happen, according to them, because if one of the forces were to be lost or weakened, it is a necessary consequence that the other would have to preponderate, altering the original equilibrium.

So a tube was taken, like *ABCD* in figure XXXXVII, whose height *AB* was two ells [117 cm] and the upturned part *BD* half an ell [29 cm], drawn out to a fineness more extreme than is shown in the figure. This tube was open at *A* and *D*, and we began to fill it with quicksilver through the end *A*, until the mercury in the part *BCD* arrived at *D*, when the spout *CD* was sealed with a flame. Then when the tube had been filled right up to *A*, it was closed with a piece of bladder as usual. The point was broken off the spout at *D*, and the mercury began to come out with great difficulty, in contrast to what it does when the air presses on it from the other end. In the present tube, instead of air, there was nothing but the vacuum gradually being formed at *A*, so the mercury was pressed out by no force other than that of its own height above the ell and one-quarter, taken from *C* towards *A*. When it had arrived at *F*, the same height above the level *C* as the mercury was observed to stand on the same day in another tube immersed in a very ample vessel, it at once ceased to come out. Then, holding the tube perpendicular to the horizon, the mercury in it was set in motion by raising and lowering it gently, [CVIII] so that vibrating with reciprocal rises and falls in the two branches, a little of it in the branch *BCD* finally came out of the spout *D*. When the tube was held still and the mercury had come to rest, a part of the tube such as *GCD* remained free from it. So therefore the air pressing on *G*, even though strained through the very narrow channel *DCG*, did not lose enough of its force for us to be able to perceive any shortening of the cylinder *FG*.[142]

Thus from all these experiments, and some others of the same kind, of which there is not now time to tell, it appeared to some of us that we could confirm that the weaker pressure of the air through very narrow spaces, taken generally in this way, is not alone sufficient to explain these phenomena and other similar ones. We thought that some other causes would have to act together with it, at least.

AN EXPERIMENT ON WATER IN THE VACUUM

The fine observation, made by Boyle,[143] of the ebullition of lukewarm water in a vacuum, not only rendered us immoderately curious to see such a beautiful and marvellous effect but also opened our minds and made us wish to make the same experiment with natural

---

[142] August 29 and 30, 1662 (G. 262, 149r–v, 150r; *TT* 2, 434–35).
[143] *New Experiments*, expt. 43. (See *Works*, I, 114).

water and with water reduced with ice to the greatest coldness that it could take without congealing.

Therefore into the little vessel *A*, represented in figure XXXVIII, plate LXXXXV, we put natural water, not changed from its ordinary temperature. When the vacuum had been made, there appeared in this a rain of very minute bubbles which, although they were in great plenty, [CX] nevertheless became very sparse, and did not cause the water to lose its transparency. They moved upwards, until, the rain becoming slower and slower, the water remained as quiet as it was at first.

Lukewarm water, as soon as the vacuum had been made, began to boil furiously up to the top of the vessel, bubbling like a cauldron when the heat rises higher. When the bulb was opened and the little vessel taken out it did not seem that its heat had been increased by such boiling.

The cold water made four or five very tiny bubbles, and then stopped, with no other visible change.

It should be noted that on the entry of the external air, both the shower of little bubbles in the water at ordinary temperature and the boiling of the tepid water immediately stopped.[144]

AN EXPERIMENT WITH SNOW IN THE VACUUM

The first time, a very small piece of snow was put in, and when the mercury had gone down we could scarcely see anything of it except the water. Such rapid melting appeared strange to us, so to gain better information the experiment was repeated with another, larger piece, roughly shaped into a cylinder, and of the largest length and thickness that could go into the bulb. Into this, then (when it was full of quicksilver) we tried to put the cylinder of snow, pushing it forcibly under the mercury. But in some way or other it slipped from the hand of the man who was immersing it and floated again, and we saw that in the mere act of immersion the mercury had eaten away a large part of it, the water from which was seen to float [CXI] upon the mercury. In this way we realized that what had melted the small piece of snow so rapidly had been the mercury and not the vacuum as it had seemed at first sight. The above-mentioned cylinder was put in again, the vessel closed, and the vacuum made, and what little remained was seen to liquefy just as slowly as it usually does in the air.[145]

This experiment was done in summertime, so that the snow was not

[144] August 8, 1662 (*G.* 262, 136r–v; *TT* 2, 440). On the "boiling" of water in a vacuum, see also p. 268 below.

[145] August 9, 1662 (*G.* 262, 137v; *TT* 2, 440), but not on August 22, as also stated by Targioni Tozzetti; on the latter date *G.* 262 has the freezing of melted snow, not snow *in vacuo*.

soft[146] (as we say at Florence of snow when it falls in flakes and before it freezes), but was from that which had been trodden down and compressed in storage.[147]

AN EXPERIMENT ON THE SOLUTION OF PEARLS
AND CORAL IN THE VACUUM

We have learned this experiment from Boyle[148] also, and it is made as follows:

Pearls and coral, as everyone knows, dissolve in distilled vinegar. In air, however, this operation takes place very slowly and consists of a very delicate separation of the most minute bubbles that gradually rise from the pearls and the coral. But these do not become abundant enough to alter the transparency of the vinegar, particularly with coral, which dissolves with more difficulty unless it is very finely powdered. Pearls are more tender, so that with these the bubbles are more plentiful. We wished to see both separately in the vacuum, and from both we saw such a dense shower of bubbles come out, that the vinegar was all raised in foam and overflowed the little vessel, [CXII] which thus appeared to be full of milk or of the whitest snow. We let in the air, upon which the foam at once dispersed and the vinegar recovered its natural transparency and began to act as before.[149]

We shall not omit to mention here an effect incidentally observed in this solution. This is that pearls that are dissolving emit one or more little bubbles of air, which must naturally rise, and carry the pearls up, attached to them. But as soon as such bubbles emerge from the vinegar and strike the air they burst and their covering disperses in a very fine spray. Then the pearls sink again, while at the same time others that are being joined to new bubbles are rising. And so, all the time that they are being dissolved, we see them go up and down through the vinegar in a continuous ebb and flow.[150]

[CXIII]

AN ACCOUNT OF WHAT HAPPENED TO VARIOUS ANIMALS
PLACED IN THE VACUUM

From the time that Torricelli thought of the first experiment with the quicksilver, he also thought of enclosing various animals in the

---

[146] *solla.*

[147] The Tuscan nobility had snow and ice brought from the mountains in winter and stored in large *conserve.* Most of the experiments on artificial freezing (see below, pp. 166 ff.) were made in July and August.

[148] Boyle dissolved coral in his 42nd "New Experiment." (*Works*, I, 113.)

[149] August 7 and 8, 1662 (G. 262, 136r; *TT* 2, 441).

[150] This was observed on July 21, 1660, and was in an ordinary beaker, not in the vacuum. A sentence in the diary that makes this clear (G. 262, 95v; *TT* 2, 441) has been left out of the *Saggi.*

empty space, in order to observe their motion, flight, breathing and everything else that happened to them.[151] It is true that as at that time he had no instruments suitable for this test, he contented himself with making it as well as he could. However, the small and delicate animals, oppressed by the mercury through which they had to rise in order to reach the top of the vessel after it had been turned upside down and immersed, arrived there dead or expiring for the most part, so that it could not well be decided whether they received the greater injury from being smothered by the quicksilver or from being deprived of air. And this was because he either did not think of opening the ends of the vessel, or did not dare to, perhaps being diffident about the adequacy of the binding to close them so that they would hold the air, driven by its own [CXIV] weight. What is more, he was distracted, soon after the invention of this experiment, by other studies to which he was called, and had not time to set about refining it further, as he might perhaps have done had not his too early death prevented it forever.

Now when we had assured ourselves that the force of the air was not so violent that cements, plasters, and pieces of bladder firmly enough tied, would not resist it well enough, we used vessels open at both ends, as we have seen already, and as we have finally used in this experiment. We shall therefore tell what was observed[152] to happen to various animals enclosed in this vessel, as follows:

A leech that stayed there more than an hour remained alive and healthy, moving freely as if it had been in the air. A snail, one of the bare ones,[153] did the same; nor in these was the least thing observed from which it could be argued that the lack of air had any effect on them.[154]

Two crickets remained very lively in it for a quarter of an hour, moving all the time, but not jumping. At the entry of the air each gave a leap.

A butterfly, whether it had already been hurt in being handled when being put into the vessel or suffered afterwards from lack of air, certainly appeared deprived of movement as soon as the vacuum had been made. We could discern with great difficulty a weak trembling in its wings. At the entry of the air they indeed fluttered, but we could

[151] Dati, *Lettere a Filaleti* (1663), p. 20, says that Torricelli experimented with small fish, large flies, and butterflies.

[152] On no less than thirteen occasions between July 21, 1660 and August 29, 1662. Boyle did similar experiments (*New Experiments*, nos. 40, 41, pp. 326–83; *Works*, I, 97–112).

[153] I.e., a slug.

[154] Boyle, *New Experiments*, notes this about snails.

not tell whether the animal or the wind moved them. When it was removed from the vessel after a little while it was found to be dead.

There is a species of fly, larger than the others, commonly called bluebottles,[155] which when flying make a buzz in the air by the motion of their wings. One of these, which continued to buzz very loudly after the vessel was closed, [CXV] stopped this entirely immediately the vacuum was made and let itself go as if dead, and the whirring wings fell silent. When this was observed the air was at once restored to it, in which it recovered a little and made some motion. But the remedy was too late, for it died just as it was taken out.

A lizard at once appeared ill when it found itself in the vacuum, and shortly afterwards closed its eyes and seemed to be dead. We then noticed that it seemed to breathe from time to time, this appearing as an inflation beneath the front legs on each side of the thorax. This went on for about six minutes, after which, its respiration apparently lost, it once more seemed dead. Now it was given air, by which it was cured so well that when the vessel was opened soon afterwards the animal jumped out and ran away. Recaptured and enclosed once again, it again became ill, but revived once more when taken out after a short time. At last, put back for the third time, after a little while— it must have been about ten minutes—it writhed somewhat, as if it had taken poison, emptied its bowels and, collapsing altogether, fell dead in the glass.

Another lizard suffered the same writhings and convulsive motions in a shorter time. It sometimes took a little rest, and, as if it had thus recovered courage and vigor, tried several times to climb up the wall of the vessel. But soon the first symptoms returned, with horrid distortions of the mouth and swelling of the eyes, as if they would jump out of its head. Then it threw itself on its back, and in that position, after a few gasps, it died. It was later observed that it had discharged matter from its mouth and its lower parts, so that its belly had become soft and lean.

Another, which had begun to suffer the same symptoms, was quickly cured with the help of a rapid restoration of air. [CXVI]

Scarcely had the vacuum been made when a small bird at once began to gasp and breathlessly search for air, and to stagger, fluttering its wings and tail. When the air was restored to it after half a minute when it seemed near death, it appeared for an instant to revive, but in a few moments closed its eyes and died.

Although a goldfinch,[156] and later another, were very quickly re-

---

[155] *Mosconi.*

[156] *Calderugio.* The *Vocabolario della Crusca* identifies this bird as *Fringilla carduelis*, a goldfinch.

lieved with air, it was not in time, so rapidly do these gentle little animals receive an irremediable injury when they are deprived of it.

The almost sudden death of these birds might at first sight appear contrary to the experiment of Boyle,[157] who tells about a lark that lived in the exhausted receiver for ten minutes (although wounded in one wing), and a sparrow taken with bird-lime that survived for seven, at the end of which time it seemed dead, but recovered with the help of fresh air, and then when it was put in again and the exhausting of the vessel again started, died at the end of five minutes. But whoever will reflect on the different ways of making the vacuum in the two vessels will perceive that these two experiments, far from contradicting one-another, agree marvellously well, for whereas in one the air, by the successive strokes of the pump, is thinned out in very slow and scarcely perceptible stages, in the other it is suddenly reduced by the extremely rapid descent of the quicksilver to that ultimate degree of rarity and fineness which can no longer suffice for respiration when the air has reached this state. And perhaps if before making the vacuum we had first inclined our vessel in such a way that the mouth AC of the bulb (figure XXXIV, plate LXXXXV) had come below the height of an ell and a quarter [73 cm], taken on the perpendicular that falls from this mouth on to the plane of the stagnant level of the quicksilver, and if when it was in such a position we had opened the lower end B, then raised the vessel and [CXVII] brought it upright little by little, we should have observed the same effects as those described by Boyle. For as the air would have to pass through all degrees of rarity, successively greater and greater (much like what happened in the exhaustion of his receiver), it would not so quickly have become useless for the breathing of these animals.

A soft-shelled crab moved at first, then lost courage and very soon began to swoon. It remained much as if slothful or rather curled up, and as we did not see it make any further motion the air was let in. At this it roused itself, beginning to move slowly, but died a little while after it was taken out of the vessel.

A frog was at once stupefied and swelled considerably all over, but when the air came back it showed its recovery with sudden jumps.

On another occasion a hard-shell crab and a frog were put into the same vessel. As to the crab, it was seen to move until the end of the experiment, which must have been a good half hour, and changed not at all, unless it swelled a little. The frog, on the other hand, was seen to swell up all over in a nasty way by the time ten minutes had passed; for instance two very large bladders arose on each side of its snout.

---

[157] Boyle, *New Experiments*, pp. 328–29; *Works*, I, 97.

Vomiting a great deal of froth from its mouth, which stayed open and full of the tongue and other membranes and blisters all excessively swelled up, it remained, ever motionless, in this state. When the air was introduced it deflated in an instant, remaining meager and ugly with an extreme and dreadful leanness, so that it was judged to have been double the size when it was put into the vessel. When it was taken out it was dead. The crab was indeed alive, as noted above, but its life was prolonged only a few moments. [CXVIII]

Another frog also swelled up and became deformed, and after it had expelled some stuff from its mouth and made a great deal of foam, we found it dead when we came back after half an hour to see it. At the entrance of the air it also became spare and lean, as the other had done. When its thorax was opened by a careful anatomist, at first its lungs were not found, being so shrivelled up by being emptied of air. Yet, when he blew through a thin straw into that passage that it has under the tongue through which to breathe, they unfolded; whence it was seen that most of the air that was inside when the animal was enclosed had come out to enjoy the benefit of expanding into the empty space, without any lesion of these organs, since when they were inflated they did not leak.

Some very lively little fishes were enclosed, with sufficient water. As soon as the vacuum was made they seemed to swell visibly and floated with their bellies upwards. They made several efforts to right themselves, but without success, always returning to the supine position. Finally the entrance of the air sent them to the bottom, where they died without ever being able to recover. One was subsequently opened and compared with another that was cut open while alive and that had not been in the vacuum. Looking inside the former, we found the air bladder entirely deflated, while in the latter it was round and firm, as that of all fish should ordinarily be.

The eyes of a rather larger barbel swelled oddly and it turned on its back, stretching its fins as if they were stiff, opening its gill covers, and, its whole body swelling up, it came to the surface of the water. Several times it tried with various squirmings and greater efforts to return to its normal position, but could not. When the air was restored after six [CXIX] minutes the swelling of the eyes at once went down, and although the thorax returned to its proper size the fish was nevertheless obliged to go to the bottom, gasping constantly, without ever being able to regain the surface. Put in some other water it died soon afterwards. When it was opened its bladder was found quite shrunk, to the extent that that of another fish only a fifth as large, dissected while alive, seemed bigger and more turgid.

An eel stayed there a long time without fainting or losing any of its liveliness. But finally at the end of an hour it too died, and its bladder was found to be deflated like those of the other fishes.

Another barbel, placed in the vacuum and very quickly treated with air, by great good fortune came out alive. It occurred to us to put this fish in an aquarium where there were some other fishes. The water was more than an ell and a half [88 cm] deep. Then, either because it found it comfortable to do so, or because this was really imposed upon it by the deflation that its bladder had suffered, it is certain that in the whole time that it lived (which was about a month), however much it was chased by scaring it and agitating the water, it was never seen to rise as the other fishes did, but always to go along the bottom, swimming with its belly close to the ground. After its death its bladder appeared to the eye to be as much inflated as it ought naturally to be, but was much less difficult to compress than those of other fishes.

A bladder from another very big fish, closed while full of wind just as it was taken out, did not change at all when the vacuum was made. The vessel was therefore opened, for we thought that nothing could be deduced from this experiment except that the membrane that covered [CXXI] the bladder internally was of so strong a texture that the force of the air naturally within it was not sufficient to tear it. But the outside air had no sooner entered than the bladder deflated (figure XXXXIIX, plate LXXXXV) in exactly the same way as was found in the fishes killed in the vacuum. This was a clear sign that the greater part of the air in the bladder had escaped by opening or tearing the valve of some invisible passage, while what little air remained, expanding in the vacuum, served to keep the bladder inflated to the same extent as at first, just as was seen to happen in Roberval's experiment.[158]

Then, to see how the air got out of these bladders, whether by some passage made in them by nature or by one opened by the force of the air itself, another bladder was taken from another fish with every possible care, and its extremities bound tightly with silk threads, as we imagined that if there was a passage it might be in one of these. When this was put in the vacuum it remained inflated just as the other one had done, but the return of the outside air made it deflate in the same manner. Therefore, in order to find the path that the air inside had opened for itself so that it could get out, a little hole was made in the bladder, large enough to insert the end of a little glass tube. This was put in (figure IL)[159] and tied around the hole that had

[158] See p. 109 above.

[159] The engraver's Roman numerals were somewhat unusual, as the reader will have noticed.

been made, and, leaving the two extremities tied up, air was blown in through the little tube. This air, because it was plentiful, certainly inflated the bladder, but at the same time some came out of the little tear *A*, which must have been that made by the air inside in order to escape; and when a lighted candle was brought near the tear the flame was visibly moved. But looking [CXXIII] more closely at it while the bladder was being distended by vigorous inflation, it was not small enough to escape the eye of the observer.

We saw in this that the air had not leaked out through the ligatures we had made, and agreed that it had made a new rupture in order to escape. We wished to see whether it also comes out in the same way in the bodies of the fish that die in the vacuum, that is, by tearing the delicate membrane of the bladder, or blows out through some hidden passage. So after taking the bladder very carefully out of a carp that had died in the vacuum, we pierced it at the pointed end (figure L) and inserted a little tube in the same way as we had done with the other bladder. It was blown up very forcibly and held the air very well.

This was a good enough test to provide a clear indication that the air, without breaking out, is nevertheless able to find some passage invisible to our eyes. So we thought of letting the water itself reveal this, and taking another bladder from a living and healthy fish we enclosed this in a small piece of netting, loaded it with a suitable weight (figure LI), and put it in the water as usual.[160] When it was under water the vacuum was made and many bubbles of air were seen to come out of the pointed end; so that it appears that we may really believe that the natural passage through which it comes is in that place. When the vessel was opened the air deflated the bladder, like the others.[161]

Finally, desiring to see which way the air used to come out of the bodies of these fishes, that is to say through the gill slits or through the mouth, we enclosed a carp in the same net, so that being drawn to the bottom by the weight attached to it, the fish would be obliged to remain under water. Then the vacuum was made, and a great deal of air was seen to come out of the mouth in very large bubbles, in the same way as it had been seen to come from the submerged bladder. [CXXV]

This ought to have been the end of these experiments,[162] but while

---

[160] In the vessel of figure XXXIV, plate LXXXXV.

[161] Borelli suggested this. See *TT* 1, 422.

[162] I.e., experiments on the vacuum. The digression that follows is an illustration of the effect of the unconscionable time that it took to print the *Saggi*.

these pages were being printed there occurred to one of our Academicians[163] a way of making the operation of this last vessel (figure XXXIV) much easier, and we shall not omit to relate this, the more so because we have tried it and found it very convenient for the purpose of making the vacuum. The invention consists in adding to the tube *BE* of figure XXXIV the turned-up piece *BFG*; then, the quicksilver being put in through the mouth *AC* in the ordinary way, when it arrives at *G* the tube is closed there and then filled up to *AC*. After this has been closed as usual, it is sufficient to open the end *G*, when without any other immersion all the mercury above a height of an ell and a quarter [73 cm], taken from the level *G* towards *E*, will come out of it. It should be remarked that the bulb *GF* serves to retain the mercury in the reciprocal oscillations that it makes in the two branches of the tube before it becomes steady, because of the impetus engendered by its fall.

This is all for now about the natural pressure of the air and its various effects.

᠁

[CXXVII]
EXPERIMENTS ON ARTIFICIAL FREEZING

Among the stupendous operations of nature there is one that has always been held in great esteem:[164] that admirable process in which, taking its fluidity from water, she binds it and fastens it together, giving it solidity and hardness.[165] Although this work may go on every day before the eyes, yet like other more secret and uncommon operations it has given, in every age, ample material to the minds of men for the most subtle speculations. Consider that [CXXVIII] whereas fire, released in rushing flames, thrusting itself into the closest joints of stones and even of metals, opens these up and liquefies them and re-

---

[163] It was probably Borelli. See *De motionibus naturalibus*, p. 209 (Cap. V, Prop. CI), in which he refers to figure 34 of the "book of experiments of the Academia experimentalis Medicea."

[164] I have reproduced as Plate 3 (p. 71) evidence of the difficulty that Magalotti had in beginning the introduction to this chapter (G. 266, 151r).

[165] The seventeenth-century interest in freezing was more than idle curiosity and probably stemmed from a desire to examine one of the peripatetic arguments against the vacuum, that depended on the belief that water contracted on freezing, so that the rupture of closed vessels had to be due to the *horror vacui*. But the Academy's, or perhaps Leopold's interest in this phenomenon went far beyond this, for they froze various liquids artificially on more than fifty occasions between September 1657 and August 1662.

duces them to water;[166] cold on the other hand (and this is a more wonderful thing) makes the most fluid liquors hard and glassy, converting them into frozen snow and ice, which then with every warm wind that blows around them once more melt and become running waters. Indeed (and this causes more amazement) cold is seen to exert such violent force in the freezing of liquids that, penetrating not only into glass but through the hidden passages of metals—just as in deep subterranean mines the raging fire explodes with violence and furiously opens every path—so cold, in the act of congelation, breaks open closed vessels of the thickest crystal, stretches and thins and finally tears open those of the purest gold, and shatters those cast from rough bronze, and these of such a thickness that if they were to be broken by loading them with dead weight it would need perhaps—indeed, without any doubt—thousands and thousands of pounds.

Now as to the basis of this strange transition that water and almost all other liquids make in congealing, there are those who think that where the cold works down there in its mines with the most suitable materials, it is capable of putting the purest waters into a condition to receive so hard a temper that it forms them even into the hardest crystal rocks and into gems of various colors, according to the tints that the fumes of the nearby minerals may give them, and even into the invincible solidity of the diamond itself. And Plato was of the opinion that from the residues of water, whence he thought gold is created in the secret places of the earth, the diamond is engendered, and for that reason that divine philosopher in his *Timaeus* called the diamond the off-shoot of gold.[167]

Then concerning the cause of freezing, clever men have gone on through all ages making various speculations as to whether this really stems [CXXIX] from a real and characteristic substance (the "positive cold" of the schools) which, like light and heat in the mines of the sun, has its particular home in air, in water, or in ice; or makes its own reservoir and treasury in some other part of the world. It may perhaps appear that the words of the holy prophet in the Bible might

[166] This description of the action is very like that given by Galileo in the *Discorsi . . . intorno à due nuove scienze*. See *G.OP.*, VIII, 66; or the translation into English by H. Crew and A. de Salvio (New York, 1914), p. 19.

[167] The passage is in *Timaeus*, 59b, 5–7. I quote it in R. G. Bury's translation (Loeb Classical Library): "And the off-shoot of gold, which is very hard because of its density and black in colour, is called adamant." Bury conjectures in a footnote that *adamas* may be "perhaps haematite or platinum." Magalotti seems to have missed the word *melanthen* (black in color), which would seem to rule out the diamond, at least as known to him. Mr. Stephen K. Marshall has given me reasons for supposing that Plato was referring to native platinum.

be taken in this sense: "Hast thou entered into the treasures of the snow? or hast thou seen the treasures of the hail?"[168] Or else cold may be nothing but a total absence and expulsion of heat.

This and other curious observations to be made about the power that nature makes use of in her process of freezing: whether in doing so she contracts or expands water and other liquids; whether she transmutes them slowly, taking time, or really with instantaneous speed; these induced us to attempt such experiments by way of the artificial freezing caused by the extraneous power of ice and salts, in the belief that we should not in this way change, or vary in any way, the work usually done by nature, when with no other means than the simple and pure ice of the air[169] she leads the water to congeal.

In the following experiments we relate what we have had the good fortune to see up to now in this subject, so vast and needing so many and such uninterrupted observations.

EXPERIMENTS MADE TO FIND OUT WHETHER WATER EXPANDS IN FREEZING

[170] It was the opinion of Galileo that ice is water that is expanded rather than contracted; for contraction, he said, produces a decrease in volume and an increase in specific gravity, and [CXXXI] expansion a greater lightness and an increase in volume. But water increases in volume on freezing, and the ice thus made is lighter than water, since it floats, etc. [171]

Assuming this—and it is demonstrated clearly by experiment—we were curious to see what water could do when it found itself im-

---

[168] Job 38:22.

[169] *Col semplice, e puro ghiaccio dell'aria.* I am inclined to think that this cryptic phrase owes something to Gassendi, who held, for example, that the upper regions of the air are cold because of "nitrous corpuscles" or "seeds" of cold that abound there (*Opera omnia in sex tomos divisas* [Lyons, 1658–75], II, 70). Henry Guerlac has traced the origin of this theory back to at least 1604 (*Actes 7ᵉ Congrès Int. Hist. Sci.* [1953]: 332–49). The Academicians had studied Gassendi (see p. 237).

[170] Draft "A" begins here (G. 255, 33r; *DIS*, 308) as far as this chapter is concerned.

[171] Galileo, in his *Discorso . . . intorno alle cose, che stanno in su l'acqua,* etc. (Florence, 1612); (*G.OP.*, IV, 63–141. The reference is to p. 65). The reader who wishes a first-class example of the sort of polemic writing that such an apparently harmless statement could draw from the Aristotelians should consult this volume of the *Opere*.

It is worth noting that on August 20, 1635, Niccolò Aggiunti made a simple experiment to show that water expands in freezing (G. 128, 104v–5r). This reminds one of the ice calorimeter and was much easier to make than those of the Academy.

prisoned in a vessel where there was not the slightest room for expansion; for it is always seen, in conformity with the words of Galileo, that ice, whether formed into huge slabs or broken into the smallest pieces of whatever size and shape you will, always floats on water; an irrefutable argument that in the act of freezing, considering the whole volume that is being frozen, it acquires lightness, either by the interposition of minute empty spaces or by a very finely divided admixture of particles of air or other similar material which, like the little bubbles in crystal and in glass, can be discerned in ice, here more plentiful and there scarcer, when it is looked at in a good light, and which are seen to escape from it in large numbers if it is then broken into very small fragments under water.[172]

The First Experiment

Taking, therefore, a vessel of thin sheet silver with two screwed-on covers (figure I), such as are used in summer to freeze sherbet and other drinks, we filled it with water that had been cooled with ice and put it to freeze.[173] We were careful to cool the water before use, so that, being used in a state of minimum expansion, it would not contract when first cooled and in this way acquire room to expand in freezing.[174] [CXXXIII] Then when we thought that the ice outside it might have exerted its effect, we took the vessel out, and opening the first cover, which was quite full, we found the second burst open and all covered with a thin crust of ice made by the water that had come out, pushed by that which had expanded within the vessel in freezing. Nor can it be said that such bursting could result from anything but expansion, or that it came from the contraction of the water in freezing, since, being forced by the power of cold to contract itself into a smaller space for fear of leaving empty the space from which it went on gradually retreating, the water would have kept on squeezing the cover down until when this could stretch no farther it finally broke.[175] Such an argument is out of place, since in such an event we should have had to find the cover sunk inwards, whereas we found it forced outwards, and we saw that it had become notably convex above the

[172] Robert Boyle, *New Experiments and Observations Touching Cold*, etc. (London, 1665), pp. 222–36 (*Works*, 1772 ed., II, 536–40), described several experiments to show that water expands in freezing; but they are not very systematic.

[173] The diary (G. 262, 29v) says that it was left "in ice" overnight.

[174] They found out later that the volume of water passes through a minimum before it is cold enough to freeze (see p. 183), but at the time of this experiment they probably were not aware of this. But as this account has been considerably rewritten from the diary, the absence of a reference here is a little surprising.

[175] The "plenist" argument already referred to in note 165 above.

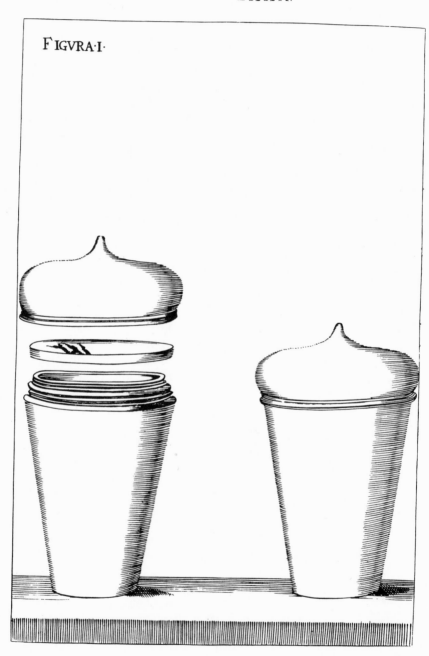

FIGVRA·I·

plane in which it lay. The surface of the ice within the vessel was also observed to be convex. Besides this, the edges of the opening were turned outwards, which shows how great must have been the force with which this was done. It would have been greater still if more water had been congealed than actually froze, for we found on breaking the first layer of ice that almost all the rest was liquid.[176]

The Second Experiment

Seeing that the force of freezing was greater than the resistance of this first vessel, we thought of making a ball of silver, but cast, and of the thickness of a piaster[177] and egg-shaped, made to be opened and closed by means of a screw and with [CXXXIV] another screw at the opening made at the end of the neck, as appears in figure II. The screw in the middle being closed and tightened vigorously in a vise and the vessel filled with water and carefully closed in the other place, we put it in ice strewn with salt, and when it was taken out after a little while we found it perfectly sound. Opening it in the middle we took out the core of ice, but this was rather tender and less transparent than ordinary ice, and perhaps somewhat denser and more compressed, since when it was put in water it did not seem to float quite as well as ice ought to do, everyone judging that it swam somewhat lower. In the middle it had a cavity as big as a large almond without its shell.[178] We repeated this experiment several times, always with the same result.[179]

The Third Experiment

Some of us marvelled at this unexpected result, which appeared at first sight not only to contradict the opinion of Galileo but what is more, to contradict itself. Even though the ice appeared denser and heavier than that made in the air with no power but that of natural cold, it still must be lighter than water, for it finally remained more or less afloat. Much less did they manage to stop worrying about the empty space they saw, that was always found in the center of the ball of frozen water. Thus it seemed necessary to say that all the water, which when liquid was enough to fill the ball, contracted when frozen by the volume of this empty space.

[176] September 1 and 2, 1657 (*G.* 262, 29r–v, 29v–30v; *TT* 2, 454).

[177] *Una piastra.* A coin, probably of Spanish origin, sometimes called a Spanish dollar. This would be about 2 or 3 mm thick.

[178] *G.* 262, 30v says a pigeon's egg.

[179] September 2, 1657, according to *G.* 262, 30r–v; but it is hard to believe that even the resources of the Pitti Palace could have supplied a special ball, cast (presumably by the lost-wax process) and machined, within a few hours of it being "thought of."

CXXXII.

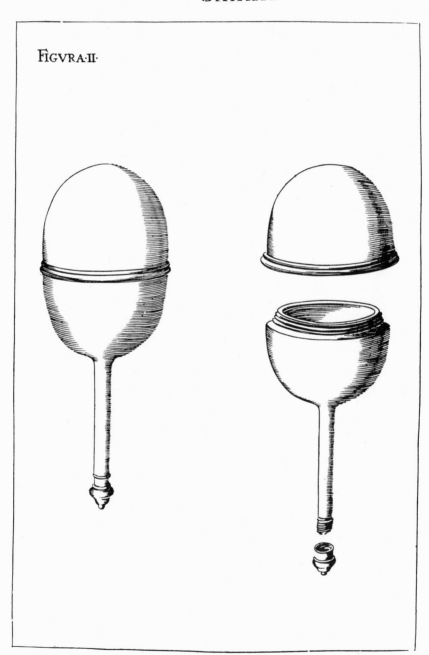

Therefore, made suspicious, by such a [CXXXV] manifest incongruity, of the presence of some fallacy, we began to observe with the very greatest care the whole progress of this freezing. So, raising the ball out of the ice frequently, and looking carefully at every part of it, we perceived a certain very slight bubbling that appeared from time to time around the center screw, a clear indication that the water (so great was the force of the expansion) was leaking out past the threads. Therefore these threads were waxed, and the bulb again filled and put into the ice.[180] Although it was taken out very frequently, the bubbling was no longer seen, nor was it heard to hiss as it had done before. Indeed, when the ball was taken out after the freezing was finished it was open, the energy of the cold in expanding it having torn apart the threads, as can be seen in figure III.

The experiment was repeated many times[181] and always showed the same effect. Made again in another ball of bronze[182] (figure IV), with a screw twice as long as in the silver one, it still had the same success.

The Fourth Experiment

To get over the difficulties that go with the screws, we had some glass balls made (figures V, VI), half a finger thick. When these had been filled with water and sealed with the flame[183] they were set to freeze. The result was not at all different from that with the first vessel made of sheet metal, as all of them burst in various ways, some breaking the neck cleanly off, some breaking open on one side because of an irregular shape or inequalities in the glass, and some developing small cracks all over. It was noted that the detachment of the neck occurred mainly when the entire ball was buried in the ice [CXXXVII] so that the water in the neck, having a lesser volume, was the first to solidify and perhaps to influence the glass to give away. For then in freezing, the rest of the water pressing in every direction, either because it found the neck already weakened or because the water frozen in it served as a wedge or cone against the internal bore of the neck, easily succeeded in detaching it. This did not happen when the upper part of the ball was allowed to remain exposed and quite out of the ice.

How great was the force of such expansion can be understood from the fact that when the necks were not put in underneath they flew

---

[180] They had worried for a month, for the threads were waxed on October 3, 1657 (G. 262, 35v; *TT* 2, 456).

[181] We do not know when. I think there is reason to doubt whether this statement, which often occurs, is always strictly true.

[182] On October 6 and 8, 1657 (G. 262, 36v, 37r).

[183] G. 262, 36v says that they were sealed with *stucco a fuoco*, hot cement.

CXXXVI.

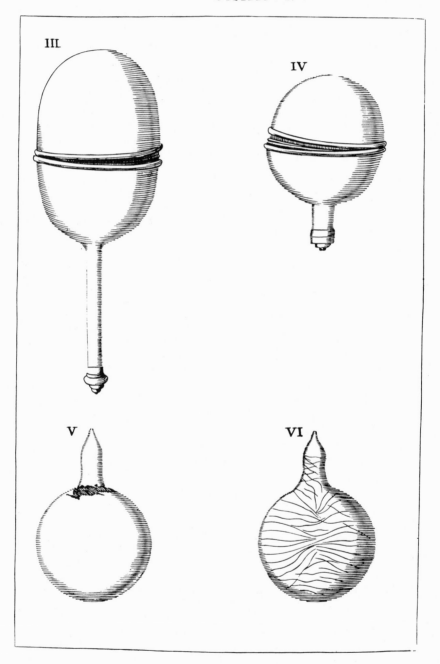

two or three ells [1.2 to 1.8 meters] into the air when they broke off, throwing around a great deal of the ice with which the ball had been covered.[184]

The Fifth Experiment

Finally we resolved to have a brass ball cast in one piece, about as thick as two piasters, with only one opening at its foot, as in figure VII, made so that it could be closed with a very sound and perfect screw. Then with the idea of being able to take the ball of ice out in one piece, a delicate scratch was made around it, so that the ball could be put back in the lathe immediately after the freezing, to be cut apart. But this gave a marvellous chance to the water, so that when it froze it split the ball there, making use of the insensible inequality that the very slight cut had produced in the thickness of the metal. Therefore, when we had made another ball, we put it in the ice without weakening it at all. Nevertheless this one burst like all the others— there were many—in whatever [CXXXIX] place the water might gradually find it easiest to crack it (figure VIII).[185]

The Sixth Experiment

At last we tried with a ball of the finest gold, the size of the drawing in figure IX.[186] This resisted many freezings without giving any sign of manifest rupture, at first causing no small astonishment, and some of us began to consider whether the space required for the expansion could by any chance be obtained from the thickness of the metal, which thanks to its softness might go on being sensibly compressed by the exertion of the water; as tin, silver, and gold itself are much compressed throughout when they are beaten. But it was then observed that whereas the ball would at first stand upright because it was somewhat flattened at the bottom, after these freezings it would no longer do so. We then all found it easy to understand where the space had been obtained.

And because the ball appeared very well reduced to a perfect spherical shape, for further assurance (if in making further freezings it should not break) as to whether it stayed the same size or went on growing a little, we had a circle or gauge made of brass, which would exactly take in its largest circumference. Examining it with this at

[184] October 5 and 8, 1657 (*G.* 262, 36v, 37r; *TT* 2, 457).

[185] October 12, 1657 (*G.* 262, 37v–38r; *TT* 2, 457). Nearly a decade later— though before the publication of the *Saggi*—we find Christiaan Huygens making similar experiments in Paris, using a heavy cannon barrel (see Acad. R. des Sci., Paris, *Registres de Physique*, vol. I, pp. 38–40 [ms.]).

[186] The illustrations in the present edition have been reduced by about 33 per cent in linear dimensions.

CXXXVIII.

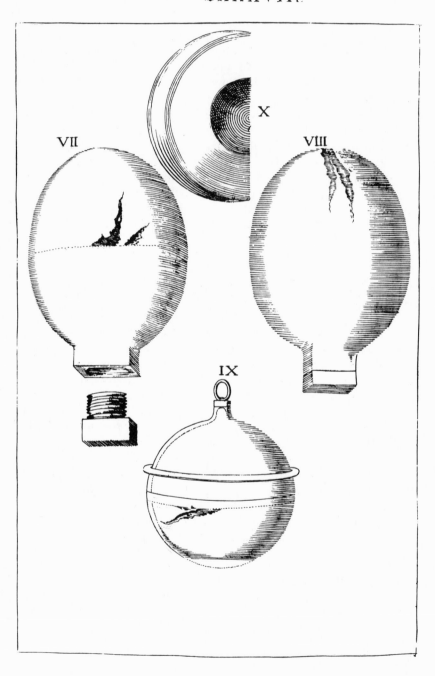

each freezing we found that it continually expanded, thanks to the softness and malleability of the very pure metal, which allowed it to expand more and more. And perhaps if the ball had been cast it would have gone even further; but being in two pieces, [CXXXXI] silver-soldered together, it finally ruptured, and the tear that began in the silver of the soldered joint extended even into the gold.[187]

AN EXPERIMENT TO MEASURE THE FORCE OF EXPANSION OF WATER THAT IS CONFINED WHILE FREEZING

To make this measurement we thought of having a metal ball made, like the others, but round, and thick enough so that the force of expansion would not, in our judgment, succeed in breaking it. This was to be filled with water, closed with its screw, and put to freeze in the usual way. So this was done, and at first we found that the water froze with no oozing out and without any visible rupture of the metal. The ball was therefore replaced in the lathe, and, endeavoring to maintain its shape as well as possible, we proceeded to take off a very thin layer, so to speak, uniformly all over. Then it was put in ice for the second time, with different water. Nor did this break open, although it was frozen, and it was returned again and again to be made thinner a little at a time, until a narrow crack was observed.[188]

The same experiment was repeated with three balls, the thickest of which was according to the section indicated in figure X, so that it seemed to us that we could say that that was the greatest thickness overcome by the expansion of water that is confined while freezing.

When we had got this far, we began to wish to reduce this force to that of a dead weight.[189] We considered that the way to go about this was to have cast, from metal of the same composition and hardness, a

---

[187] October 25, 1660 (*G.* 262, 110r–v; *TT* 2, 458). It is to be noted that the Academy did not publish any hypothesis about the mechanism of the expansion of water in freezing. With a letter to Prince Leopold from Rome, September 21, 1658, (*G.* 275, 119r–20r) Borelli had sent three possible explanations, all based on hypothetical shapes of the "atoms" of water and of "fire" (see *LIUI*, I, 100–11). This rather Cartesian exercise seems not to have found favor. See also *G.* 268, 84r–87v.

[188] I have not found this in *G.* 262, nor does it seem to be in Draft "A."

[189] In Borelli's remarks on Draft "A," this was suggested in some detail, but with a different technique for supporting the ring and applying the weights (*G.* 267, 21v; *DIS*, 334). But there is a marginal note in Viviani's writing, suggesting some degree of exasperation with Borelli: "This experiment has already been proposed by that fool Viviani [he uses a slightly obscene term for "fool"] and master Filippo [Treffler] took an order to make these and other conical rings and arbors, but nothing more was ever heard of it." Targioni Tozzetti says that this was on November 7, 1660 (*TT* 1, 426) but the original document has eluded me.

ring of the same thickness [CXXXXII] as the ball, and of a conical form, and to insert in this its own arbor of iron, made so that the outer surface of this arbor would mate perfectly with the inner surface of the ring. Above the ring the arbor would extend a distance equal to the thickness of the ring itself.

We thought that when this had been adjusted in this way, we would place it on a thick stone table with a hole in the middle a shade larger than the lower end of the hole in the ring. Then it was our idea to go on loading the arbor from above with dead weight, or else to weight it from below by hanging the same weight from a hook made on the axis of the arbor, so that the force of the weight, acting in its direction, would drive the arbor into the ring and strain it more uniformly. When the weight had reached a certain point, we would carefully add small pieces of lead until we found the least weight that would break the ring. Then, in order to assure ourselves that the resistance of the ring to breaking should not be strengthened by having its base touch the roughnesses of the stone, we had conceived the idea of cementing a sheet of polished steel around the hole in the table, and also of bevelling and polishing the lower surface of the ring in such a way as to reduce the contract to a mere circumference and to remove from that, by means of the slippery smoothness of the steel, every opportunity for the least resistance to expansion. But because an immense weight would have been required to overcome the resistance of such a thickness, it was thought that we might succeed by examining the resistance of much thinner rings, but at least of various thicknesses and heights and with much more manageable dead weights; for when we had found out by repeated experiments the various forces that would be needed [CXXXXIII] to break each of them we could, in the same way, find out what weight ought to have been enough to break the first ring, that was as thick as the ball. In this way we might have the approximate force of expansion of water that is confined while freezing.[190]

Such would have been our idea; but having observed in sawing apart the burst balls that there was nearly always found some defect resulting from the melt, or scale, or air bubbles, that would produce various inequalities in the resistance of the metal, we are not interested in proceeding further for the present in view of such uncertainties. But we do not for this reason wish to avoid saying frankly what our idea was, although we did not succeed in doing what we desired. At least it will serve as a warning to others not to set out on a road with no possibility of success, or perhaps to arouse their minds

[190] The theory of such an experiment would have been entirely beyond them, or anyone else at the time.

either to find a way around the above-mentioned difficulties or to set out more happily on another road.

EXPERIMENTS TO MEASURE THE GREATEST EXPANSION
THAT WATER RECEIVES IN FREEZING

The First Experiment

We have made this experiment in two ways: by measuring and by weighing. That made by measuring is as follows: We endeavored to select a small glass tube drawn as uniformly as it could possibly be, and closing it at one end we filled it with water up to the middle and thrust it into ice that had been pounded very fine and incorporated [CXXXXIV] with its salt, until it should freeze. Then comparing the height of the liquid cylinder and the frozen cylinder on the same base, we found the former to be to the latter as 8 to 9.[191]

The Second Experiment

We did not think that we should put our trust in this experiment alone, believing it to be little less than impossible to find a glass tube (which is finally drawn with no other control than the workman's breath) so perfectly cylindrical that it has no inequalities sufficient to falsify, at least a little, the proportion that we might claim to deduce from the heights of the cylinders of water contained in it.

So in order to have a more regular vessel we took a pistol barrel instead, and had it re-drawn inside, until it was given the most perfect cylindrical shape obtainable with existing tools. Then we closed it at the end that had the touch-hole—also closing this with a very perfect screw—with a flattened sheet of steel, and after putting in six fingers of water we thrust in a cylinder of boxwood turned to fit the bore of the tube exactly and well soaked in oil and tallow so that it might not chance to get wet. When it had gone in far enough to close the end, we turned the tube upside down so that the water would all run down onto the base of the cylinder, and opening the touch-hole we began to push the tube onto the cylinder until we saw the water squirt out of the touch-hole. Then we closed this again with its screw and turned the tube upright again. Before putting in the water we had already marked on the cylinder the place where the plane of the end of the tube [CXXXXV] intersected it when it was pushed down to the bottom. We now marked where it came with the water in and then put the tube into snow strongly reinforced with salt and sprinkled with spirit of wine, which, as everyone knows by now, strengthens the freezing-power of ice wonderfully well.

---

[191] August 7, 1662 (*G.* 262, 136r). This was due to Viviani (*TT* 2, 694). On August 3, 1662 (*G.* 262, 133v; *TT* 2, 460–61) there was another slightly different experiment for the same purpose.

After about 12 minutes, the mark touching the end began to appear raised by the thickness of a piaster, and in a very short time rose by the thickness of two more, after which it moved no farther, no matter how we endeavored to increase the cold by adding large amounts of snow and salt.

When the pistol was taken out after a full hour, we found it so cold that it could scarcely be carried in the hand, from which we supposed that it had ice inside, and a better proof of this was the observation that when the touch-hole had been opened, striking the cylinder of wood against the wall did not make it go in a hair's breadth farther; and, barring a few very small drops that came out of the touch-hole, not a drop was seen to come down between the tube and the cylinder. Finally, trying the touch-hole with a pin, the ice was felt to be solid.

In spite of all this we did not know what to say, as it was still possible that the water had not frozen throughout, and we could not make this clear because of the opacity of the tube. It could also be that water had leaked out at the touch-hole, lowering its height in the tube and leaving the base of the cylinder dry. And finally it may be that water indeed expands in such a great proportion when it has freedom to dilate, but when enclosed in a vessel, as here, even water does what it can, freezing with a much smaller expansion. We say that it was enclosed, because the cylinder was so firmly fixed in the [CXXXXVI] tube by the wetting it received from the water driven into the grain of the wood by such a very great force, notwithstanding the protection by the oil, that even after the ice had melted and the water had come out by the touch-hole it was quite impossible to extract the cylinder either with pliers or with a vise, so that we had to use fire to burn it out.[192]

The Third Experiment

Seeing the difficulties that were encountered in attempting to determine this proportion by way of the heights of cylinders above the same base with a metal tube, we turned to the other method, that of weighing. With a transparent glass tube, weighing the water put in to freeze and that required to fill all the space occupied by the same water after it had frozen, we found, with a balance that turned with one forty-eighth of a grain, that the weight of the former was to that of the latter as 25 to 28-1/19. This proportion is hardly any less than that found by measurement, 8 to 9, which is the same as 25 to 28-1/8.[193]

[192] September 5, 1662 (G. 262, 152v–53v; TT 2, 461–62).
[193] August 8, 1662 (G. 262, 136v–37v; TT 2, 463).

Then, when we saw such a close approximation between these pro-portions, in order not to delude ourselves we went back for curiosity's sake to repeat the experiment by measurement, and this turned out to give the same proportion, 8 to 9. We could be certain that the weight had not varied at all, because the tube was kept closed while the freezing was being done, and we found with our balance that the water always kept the same weight, whether it was frozen or melted.

[CXXXXVII]

EXPERIMENTS ON THE PROGRESS OF ARTIFICIAL
FREEZINGS AND THEIR WONDERFUL PHENOMENA

The first vessel that we used at the beginning of these experiments was a glass bulb (figure XI) about one-eighth of an ell [7.3 cm] in diameter with a thin neck about an ell and a half [88 cm] long, closely divided in degrees. In it we put natural water, making it come about one-sixth of the way up the neck. Then putting the bulb into the ice with its salt, as we usually do when we wish to freeze liquids, we began to observe with the most exact attention all the movements of the water, concentrating on its level.

We knew already, as everyone does, that cold at first acts on all liquids to shrink them and diminish their volume. We had not only the ordinary proof of this in the spirit of thermometers, but we had experimented with water, olive oil, quicksilver, and many other fluids. On the other hand we also knew that water, in passing from simple coldness to the removal of its fluidity and the receiving of resistance and hardness by freezing, not only returns to the volume that it had before being frozen but exceeds it by a large amount, so that we see it break vessels of metal and of glass with great force. But we did not yet know the rate at which the cold effected these various al-terations in it; and it was impossible to learn this by freezing it in opaque vessels, such as those of silver, brass, and gold that we had [CXXXXIX] used up to this time.

Therefore in order not to miss this observation, which seemed to be the essential point in all these experiments, we returned to crystal and to glass, hoping to have a quick assurance of how things were progressing, because of the transparency of these materials. At the same time, we could take the bulb out of the ice quickly at any move-ment that might appear in the water, and recognize in the bulb the corresponding alterations. But the truth is that we had to take a great deal more trouble than we had originally intended, before being able to discover anything certain about the time taken by these phenomena.

To relate more distinctly what happened it is necessary to know

CXXXXVIII.

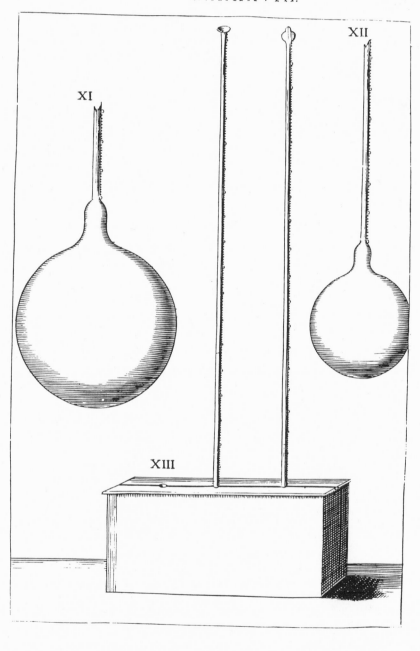

that on the first immersion of the bulb, as soon as it touched the icy water, there was observed a small but very rapid rise in the water in the neck, after which it steadily sank towards the bulb with a moderate speed until when it arrived at a certain level it continued down no farther, but stayed there for some time, motionless as far as the eyes could judge.[194] Then it gradually began to rise again, but with a very slow and apparently uniform motion. From this, with no proportionate acceleration, it made in an instant a very violent jump, and at this time it was impossible to follow it with the eye, for it covered many tens of degrees in what we may call an instant. And just as this violence commenced in a flash, so in the same way it ended, passing suddenly from that maximum speed to another rhythm of movement, also very fast but incomparably less so than the one before. Continuing to rise in this way, the water generally reached the top of the neck and ran over.[195]

During the whole [CXXXXX] time that these things were happening small bubbles of air, or it may be of some other subtle substance, were always seen rising through the water in greater or lesser number. This separation commenced only after the water had begun to take up the severe cold, as if the power of this cold had the faculty of discerning such substances and separating them from the water.

We now wished to see whether there was any kind of similarity between these changes and began to repeat the freezings. One sample of ice had scarcely melted when we put it to freeze all over again, and the water froze once more with the same series of alterations. Since these did not return every time to the same point or degree on the neck, we were beginning to believe that they had no firm and stable period, as a certain gleam of reason might appear to persuade us that they ought to have.

Meanwhile, in repeating these experiments it happened that on one occasion the water was by inattention allowed to freeze in the bulb near the neck. As was noted in the fourth experiment on freezing,[196] the bulb broke. We therefore made another, smaller one (figure XII),

---

[194] This was the point of maximum density (or minimum volume) peculiar to water. The academicians do not appear to have been particularly surprised to find this property. On the other hand Robert Hooke flatly refused to believe in it, interpreting the experiments that seemed to show it as being due to the properties of glass (see W. Derham, ed., *Philosophical Experiments and Observations of . . . Dr. Robert Hooke*, etc. [London, 1726], pp. 132–33). See also p. 169.

[195] These phenomena seem to have been first observed by Prince Leopold, as is noted in a description of them in G. 263, 50r–51r.

[196] P. 173 above.

so that the cold might insinuate itself into all the water more easily and quickly.[197] The neck was lengthened to two ells [117 cm] so that it would not have to overflow. The vessel was filled to 170 degrees and put into the ice.

Then, observing with careful attention, we found, firstly, that all the phenomena of falling, rising, resting, rising again, running up the tube, and slowing down always took place at the same points on the neck, that is to say when the surface of the water was at the same degrees. This was because in putting it in the ice we took care that the water was brought down to the same [CXXXXXII] point as when it was put in the ice the time before, which is the same as saying, at the same temperature,[198] for in such a case the whole vessel may be considered a thermometer, very sensitive because of the great capacity of the bulb and the extreme thinness of the neck.

When we had made sure of this, we began to try to find the precise time of freezing, and to this end we went on taking the bulb out of the ice at short intervals. But no matter how frequently we made such observations we never succeeded in detecting a slight veining with ice; it was always either all liquid or all frozen. From this it was very easy to conjecture that the operation of freezing must be extremely brief, and that whoever might have chanced to take out the bulb at the very moment when the water was taking its very rapid flight would have seen absolutely no visible change happen in the water.[199]

And because in putting the bulb into the ice and taking it out so many times it is disturbed all through its changes, we again let it return to its first state, put it into the ice, and waited for the point at which it usually began its impetuous motion. At half a degree before this should happen we took it out. Then we continued to look closely at the water in the bulb, which was still easily seen through the transparent glass to be clear and liquid. As the cold that it had acquired worked within it (although it had been taken from the ice), when it was at this point it rose in the neck with a great speed, too fast for the eye to follow or even for the mind to conceive and, in the

[197] The editors of *DIS* (p. 313n), by a comparison of sketches in *G*. 270, have deduced that the bulb of this was 110 mm in diameter, the wider part of the neck 13 mm in diameter and 25 mm long, and the tube about 3 mm in diameter and 560 mm long; but they could not determine the bore of the tube.

[198] *Tempera di calore e di freddo.* The concept of temperature had not become precise enough by this time to have acquired a word of its own. The best discussion is that of K. Meyer, *Die Entwicklung des Temperaturbegriffs im Laufe der Zeiten* (Braunschweig, 1913). No. 48 in the collection "Die Wissenschaft."

[199] In Draft "A" (*G*. 265, 38v; *DIS*, 313) this idea is ascribed to Prince Leopold and the ascription crossed out.

bulb, lost its transparency in an instant and, at once stopping its swift movement, froze.[200]

Nor was there room for doubt [CLIII] as to whether it had all frozen, or whether only a thin crust of ice had formed externally; for we observed quite well that in melting, the ball of ice gradually detached itself from the glass and became smaller, until when it was reduced to the size of a very small lentil it was lost to view at the end of the liquefaction.[201]

Trying the same experiment again and again, we were finally assured that this was the way the thing went, and that we had not made any mistake; so we were curious to see the process of congelation in various liquids. For greater brevity, we shall record these freezings in the tables that follow, in which:

"Natural State" signifies the graduation to which the water or other liquid came in the neck of the vessel before it was placed in the ice.

"Jump upon Immersion" means the first leap that the water is seen to make as soon as the bulb touches the ice. As will be made clear in a later experiment,[202] this does not proceed from any intrinsic change in the water, but from external causes relating to the vessel. Hence it varies a little from time to time, producing some variations in the other changes through which the liquid passes before it freezes. But as this entire jump is very small, so is its variation and so is its effect on the subsequent changes.

"Fall" denotes the graduation to which, after the above jump on immersion, the water goes down as it begins to take up the cold.

"Point of rest" is the graduation at which the water remains for some time, following the fall, with no apparent sign of motion.

"Rise" is, similarly, the graduation to [CLIV] which the water is brought by its expansion from the lowest point of its fall, with a very slow and uniform motion, quite similar to the earlier motion with which it contracted.

"Jump upon freezing" means the graduation to which the water is thrown with very great speed at the moment of freezing.

[200] Without recognizing the fact, they had supercooled the liquid. The phenomenon of supercooling was first described by Fahrenheit in *Phil. Trans.* 33 (1724): 78–84.

[201] Nevertheless, what had certainly happened was that a network of crystals had formed throughout, for the bulb could not have been cooled nearly enough to remove all the heat of fusion. On p. 186 they describe this ice as being like a sherbet.

[202] See p. 206. In his comments on Draft "A," Borelli says that Prince Leopold was the first to notice this effect, and that this was in October 1657. Borelli then saw the cause, but people doubted his explanation, and so the experiment with the heated ring was done, and Leopold wrote to Borelli at Pisa to tell him about it. See *G.* 267, 22v; *DIS*, 335.

It was remarked that after this flight the water does not stop moving for an instant, but goes on rising with a motion that is also very rapid, although incomparably less so than that which preceded it. We have not paid any attention to this sequel, as it has its origin in nothing but the progress of the expansion of the ice already formed, or more correctly of that which has begun to form inside the bulb, as it gradually goes on hardening after the violence of its first impetus. We say "ice that has begun to form" because (as we have found when bulbs have been broken), it is at first very soft, and similar to a sherbet when it is a little too much hardened, as it is really nothing else but the first solidification of the liquids. From this it results that this way of freezing does not make clear the ultimate expansion of liquids that are firmly frozen, for, to save the bulb from being broken, we were unable to leave it to freeze completely, or until the ice acquired its greatest hardness.

We shall also relate how, in order to use all possible care, we wished to compare the thermometer and the pendulum clock in each freezing.[203] Thus we could see, with the thermometer, at what degrees of cold, and with the clock, at what times each of the above-mentioned changes occurred in the liquids. (See figure XIII.) For this purpose a 400-degree thermometer was supported beside the bulb in the same vat; but having found very great disparities both in [CLVI] the degrees of cold shown by the thermometer and in the intervals of time given by the vibrations of the pendulum we realized that all our care had been in vain. This was because of the impossibility of applying either to the bulb or to the thermometer the same environment of heat or cold at all times. The irregularity of the pieces of ice and the various doses of salt made it impossible to distribute them always in the same way. The reason for this is that in freezing a liquid artificially we wish to have snow or ice that, either by crushing or by pounding, are reduced, we might say, to powder; but as they receive the salt the fragments at once stick together and become as hard as rock, so that it is impossible to distribute them around the vessels in any way whatever, or to be sure that they are surrounded on all sides. Yet for completeness we have put in the tables both the readings of the thermometer and the vibrations of the pendulum, leaving it to the discerning judgment of the reader to make use of them, with due regard to this warning.[204]

[203] Borelli recommended that they should do this, when he commented on the first (or at least an early) draft of the *Saggi*. See *G.* 267, 22r; *DIS*, 334. See also p. 68 above.

[204] It must be confessed that the experiments that follow are of no great importance in the history of science. Yet these "quantitative" experiments

The first freezing of spring water

	Degrees in vessel	Differ- ences	Degr. of thermom.	Differ- ences	Vibra- tions	Differ- ences
Natural state	142		139		—	
		1½		6		23
Jump upon immersion	143½		133		23	
		23½		64		232
Fall	120		69		255	
		—		20		75
Point of rest	120		49		330	
		10		16		132
Rise	130		33		462	
		36		—		—
Jump upon freezing	166		33		—	

It is to be understood that the vibrations indicated in this freezing and the four that follow went at 65 to the minute.

[CLVII]
The second freezing of the same water

	Degrees in vessel	Differ- ences	Degr. of thermom.	Differ- ences	Vibra- tions	Differ- ences
Natural state	144		141½		—	
		2½		23½		25
Jump upon immersion	146½		118		25	
		27		80		255
Fall	119½		38		280	
		—		10		135
Point of rest	119½		28		415	
		11½		11		467
Rise	131		17		882	
		39		—		—
Jump upon freezing	170		17		—	

appear in the diary (G. 262) on twenty-nine occasions, and there are also almost a score of experiments without tables. I shall make very few annotations on them.

The third freezing of the same

	Degrees in vessel	Differences	Degr. of thermom.	Differences	Vibrations	Differences
Natural state	143		141½		—	
		2		16½		23
Jump upon immersion	145		125		23	
		25½		74		346
Fall	119½		51		369	
		—		7		196
Point of Rest	119½		44		565	
		10		6		368
Rise	129½		38		933	
		39½		—		—
Jump upon freezing	169		38		—	

From these three examples of freezing of the same water it can be seen that although the natural state of the water was not at exactly the same degree on all the three occasions, because of their different temperatures, changed from one time to the next by extrinsic accidents of heat and cold, so that all the other alterations of the water will not keep to precisely the same degrees; nevertheless, if we reduce the natural state to 142 degrees, reducing all the other levels by the same amount, it will be seen that they differ from those observed in the first freezing by very small and almost unobservable amounts.

[CLVIII]

First freezing of myrtle-flower water distilled in lead

	Degrees in vessel	Differ- ences	Degr. of thermom.	Differ- ences	Vibra- tions	Differ- ences
Natural state	145½		141½		—	
		1½		8½		31
Jump upon immersion	147		133		31	
		38		83½		316
Fall	109		49½		347	
		—		4½		40
Point of rest	109		45		387	
		16		19⅓		538
Rise	125		25⅔		925	
		105		—		—
Jump upon freezing	230		25⅔			

Second freezing of the same liquid

	Degrees in vessel	Differ- ences	Degr. of thermom.	Differ- ences	Vibra- tions	Differ- ences
Natural state	146		142		—	
		3½		11		18
Jump upon immersion	149½		131		18	
		41½		96		442
Fall	108		35		460	
		—		2½		58
Point of rest	108		32½		518	
		18½		13½		809
Rise	126½		19½		1327	
		106		—		—
Jump upon freezing	232		19½			

In the following freezing experiments the clock was changed, taking one that went at exactly 60 vibrations per minute.

First freezing of rosewater distilled in lead

	Degrees in vessel	Differences	Degr. of thermom.	Differences	Vibrations	Differences
Natural state	140½		142		—	
		2½		4		20
Jump upon immersion	143		138		20	
		27		88		331
Fall	116		50		351	
		—		4		38
Point of rest	116		46		389	
		11½		20		356
Rise	127		26		745	
		67		—		—
Jump upon freezing	194		26			

[CLIX]

Second freezing of the same water

	Degrees in vessel	Differences	Degr. of thermom.	Differences	Vibrations	Differences
Natural state	140½		141		—	
		1		16		21
Jump upon immersion	142½		125		21	
		27		86		333
Fall	115½		39		354	
		—		9½		168
Point of rest	115½		29½		522	
		11½		11		735
Rise	127		18½		1257	
		67		—		—
Jump upon freezing	194		18½		—	

First freezing of orange-flower water distilled in lead

	Degrees in vessel	Differences	Degr. of thermom.	Differences	Vibrations	Differences
Natural state	137		142		—	
		2		12		14
Jump upon immersion	139		130		14	
		28		83½		297
Fall	111		46½		311	
		—		2		64
Point of rest	111		44½		375	
		16		24		505
Rise	127		20½		880	
		123		—		—
Jump upon freezing	250		20½		—	

From the tables for the second freezings of all the above liquids it can be seen how much longer they took to freeze the second time than the first. Having observed this, we wished to determine whether this was because of some intrinsic change in the liquids after they had had their first freezing, or an external change in the ice after the passing of the increase of cold obtained from the salt. Therefore the vat was emptied and new ice and salt put in, after which we made a

[CLX]
Second freezing of the same water

	Degrees in vessel	Differences	Degr. of thermom.	Differences	Vibrations	Differences
Natural state	137½		142		—	
		2½		22		29
Jump upon immersion	140		120		29	
		28½		74		337
Fall	111½		46		366	
		—		2		18
Point of rest	111½		44		384	
		15½		12½		523
Rise	127		31½		907	
		121		—		—
Jump upon freezing	248		31½		—	

So the difference in the time of the first and second freezings cannot be attributed to the liquids, but certainly to the ice. Because this has formed much water, and perhaps because the intensity of the cold given it by the salt has weakened, it needs a longer time to operate. And that this is true can be seen from the fact that the whole difference between the first and second freezings of the orange-flower water turned out to be only one minute and 46 seconds, while when the ice was not changed it sometimes reached 7'29" and 13'29",[205] as in the first and second freezings of the rosewater, and the first and third of the water from the spring. That the small difference of 1'46" found in the second freezing of the orange-flower water is purely accidental and not a result of any reluctance to freeze again acquired in the first congelation of the same water is clearly evident from the second freezing of the strawberry water, in which the ice was similarly renewed and which was completed in 3'15" less than the first.

[205] These values cannot be reconciled with the tables.

[CLXI]

First freezing of strawberry water distilled on a water bath

	Degrees in vessel	Differ- ences	Degr. of thermom.	Differ- ences	Vibra- tions	Differ- ences
Natural state	137		143		—	
		2		23		30
Jump upon immersion	139		120		30	
		28		83		405
Fall	111		37		435	
		—		1		15
Point of rest	111		36		450	
		15		17½		538
Rise	126		18½		988	
		89		—		—
Jump upon freezing	215		18½		—	

Second freezing of the same water

	Degrees in vessel	Differ- ences	Degr. of thermom.	Differ- ences	Vibra- tions	Differ- ences
Natural state	139		143½		—	
		2		9		18
Jump upon immersion	141		134½		18	
		27		92½		402
Fall	114		42		420	
		—		1		7
Point of rest	114		41		427	
		15		20		446
Rise	129		21		873	
		86		—		—
Jump upon freezing	215		21		—	

It should be noted that the jump on freezing is higher or lower and also more or less rapid in different liquids; and it appears that in those that freeze harder it is higher and also swifter.

Freezing of distilled cinnamon water

	Degrees in vessel	Differences	Degr. of thermom.	Differences	Vibrations	Differences
Natural state	139½		141		—	
		1½		7½		13
Jump upon immersion	141		133½		13	
		29½		88½		347
Fall	111½		45		360	
		—		6		60
Point of rest	111½		39		420	
		9		12		300
Rise	120½		27		720	

When the water had arrived at 120½ degrees with the same very slow motion [CLXII] with which it had risen after the point of rest, in place of taking a leap, it did nothing but set itself for a time into another, somewhat faster motion. When we observed this, we at once took the bulb out of the ice and found the water coagulated into a jelly that was so soft that it melted as soon as it was exposed to the air.

It should be noted that some of these artificial ices come out softer, like those of cinnamon water and rosewater; others harder, like those of orange-flower water and myrtle-flower water, which, it appears up to now, harden more than any other liquid in the first instant of freezing.

The repetition of this and the following freezings is omitted, as the correspondence between those of each liquid in the examples cited has been seen to be sufficient.

Freezing of water from melted snow

	Degrees in vessel	Differ- ences	Degr. of thermom.	Differ- ences	Vibra- tions	Differ- ences
Natural state	136½		141		—	
		2½		9		27
Jump upon immersion	139		132		27	
		28		80		318
Fall	111		52		345	
		—		4		32
Point of rest	111		48		377	
		5½		8		
Rise	116½		40			

And here, after it had accelerated somewhat, although slowly in comparison to the other liquids at the point of freezing, it began to freeze where it touched the glass and then in the outer parts successively, coagulating gradually as far as the center of the vessel, always with the same slow expansion and motion of the level above. The frozen mass was not at all uniform, like [CLXIII] the others, but interrupted and streaked with random veins intersecting each other in every direction.[206] Repeating the experiment a second time, it turned out exactly as the first, and when it was made later with the same water after this had been boiled, we did not find any great difference.

[206] This behavior was no doubt due to the presence of an ample supply of nuclei on which crystals could form, thus preventing supercooling.

Freezing of fig water

	Degrees in vessel	Differ- ences	Vibra- tions	Differ- ences
Natural State	98		—	
		2		19
Jump upon immersion	100		19	
		29		269
Fall	71		288	
		—		75
Point of rest	71		363	
		12		453
Rise	83		816	
		117		
Jump upon freezing	200		—	

Freezing of the red wine of Chianti

	Degrees in vessel	Differ- ences	Degr. of thermom.	Differ- ences	Vibra- tions	Differ- ences
Natural state	141		141		—	
		2		4		15
Jump upon immersion	143		137		15	
		65½		109½		585
Fall	77½		27½		600	
		—		4		95
Point of rest	77½		23½		695	
		4		7½		340
Rise	81½		15		1035	

From 81½ degrees the motion of its surface accelerated noticeably as it gradually froze, without making any different motion.

Freezing of white muscatel

	Degrees in vessel	Differences	Degr. of thermom.	Differences	Vibrations	Differences
Natural state	140		139		—	
		2½		7		16
Jump upon immersion	142½		132		16	
		65½		108		664
Fall	77		24		660	

[CLXIV]

When it had reached this point it began, without standing still at all, to rise somewhat more quickly than those liquids have so far been said to rise that freeze in an instant and take a second leap. Taken out of the ice it was found to have begun to freeze over in its outermost parts.

Freezing of white vinegar

	Degrees in vessel	Differences	Degr. of thermom.	Differences	Vibrations	Differences
Natural state	141		140		—	
		2		14		11
Jump upon immersion	143		134		11	
		68		110		724
Fall	75		24		735	
		4		5		440
Rise	79		19		1175	
		194		—		
Jump upon freezing	273		19			

With a speed less than that of water, and much greater than that with which muscatel, cinnamon water, and undistilled vinegar leap up.

Freezing of lemon juice

	Degrees in vessel	Differ- ences	Degr. of thermom.	Differ- ences
Natural state	142		143	
		2		9
Jump upon immersion	144		134	
		160		102
Fall	84		32	

When it had reached 84 degrees it began to rise again with a very slow motion, freezing little by little.

[CLXV]

Freezing of spirit of vitriol

	Degrees in vessel	Differ- ences	Degr. of thermom.	Differ- ences	Vibra- tions	Differ- ences
Natural state	140½		140½		—	
		1½		7½		15
Jump upon immersion	142		133		15	
		52		95½		405
Fall	90		37½		420	

It did not stop at all, but after falling to 90 degrees began to rise again very slowly and uniformly, at the same time freezing here and there on various levels, as ordinary water is seen to do when put in a glass vessel to freeze in the open air.[207]

[207] Niccolò Aggiunti had tried without success to freeze spirit of vitriol on January 31, 1634 (G. 128, 104r). The Academicians must have been using rather dilute acid.

Freezing of olive oil

	Degrees in vessel	Differences
Natural state	140	
		18
Jump upon immersion	122	
Fall	—	

The whole of it went down into the bulb, where it congealed without the least expansion. This is perhaps why frozen oil goes to the bottom in liquid oil, while all the other sorts of ice, being made by expansion, float in their liquids.

Spirit of wine is condensed a surprising amount by cold, but does not then expand or freeze.[208]

[CLXVII]
EXPERIMENTS ON NATURAL ICE

Although we have given the title of artificial to the kinds of ice that we have dealt with up to this point, this does not prevent them being made entirely by the hand of nature also. Now as she makes them with a different skill and perhaps with the simple ingredient air, we wished to see whether, in obtaining the same effect by different means, some variation would appear in the progress of the operation.[209] And inasmuch as we had this [CLXVIII] material at hand we managed to extract from it some further knowledge, as will be seen in the following account.

The First Experiment

It has already been said in the preceding experiments that the various sorts of artificial ice, in the kinds of vessel we have mentioned, are very soft when first formed, particularly in comparison with those made in the winter air. Although the latter do not harden as quickly, beginning with a very thin sheet and invisible capillary veins, nevertheless these veins and sheets, apart from the fragility that comes from

---

[208] As Niccolò Aggiunti had also found (G. 128, 104r). Aggiunti made a number of general statements about the freezing of various liquids, roughly those used by the Academy (G. 128, 104r–5r).

[209] In G. 262, 49v (December 19, 1657): "Natural ice is different from that made artificially by the power of ice and salt."

their extreme thinness, are of a harder material, one might say of a more crystalline and dry ice. It is indeed a wonderful freak of nature that we have seen for many years in our observations of natural freezing. For, when we put water obtained from the same spring into various vessels of earthenware, metal, and glass; into deep glasses and wide cups; some half empty, others full; some closed, others open; and in jars and narrow-necked bottles of various kinds, some simply stopped with cotton and some sealed with the flame; all in the same place in the open air, even placed side by side on the same table: sometimes a small amount of water froze before a large amount, sometimes the large quantity before the small; and so for all the rest, with no regard for the form of the vessel or how full it was. As to the material, we believe that we can say dogmatically that earthenware works more quickly than metals and glass.[210] For the rest, we have found nothing else as constant as the perpetual irregularity of all the phenomena; among other examples, there were vessels, next to those that froze at the end of an hour, which stood [CLXIX] the whole long night without even beginning to form a crust.

Moreover, whether the same assortment of vessels has been placed to the north, south, east, or west on the same night, the same strange phenomena have been observed. Thus, sometimes the vessels facing south will freeze before those facing north, even though cold ordinarily comes to us from this direction. Similarly, either the east or west ones have been first, and these have beaten the north and south ones, or been beaten by them.

The process of freezing in this way is very interesting. The water at the top begins to coagulate around the edge, and from this first ribbon of ice that goes around the circumference of the vessel, it begins to send a few delicate threads towards the middle, after which they are sent through the entire depth, confusedly, in every direction. Gradually these threads appear as if crushed together, still remaining thicker on one side and sharper and keener on the other, like the blade of a knife. From the backs of these, other threads begin to run out, very delicate but closely spaced, as in a feather or a palm leaf. From this first network is made what we might call a disordered and confused web, until, the process everywhere growing, the fabric is completed with the total freezing of the water. The surface of this then all appears scratched in every direction, like a crystal engraved with a very fine burin.

At first the surface of the ice always appears flat, but when the freez-

---

[210] It would permit less supercooling. Supercooling was the key that the Academicians lacked and could not have been expected to possess.

ing is finished it becomes convex, but without retaining any regular shape. This effect reminded a few of us of the first experiment described [CLXX] under the heading of artificial freezings, in which the second cover of the silver vessel was found broken open and all covered with a thin layer of ice formed from the water that had come out through the crack at the moment of freezing. Now, in the same way, they maintained, the first crust that forms on the surface of the water seals the inside of the vessel better than any cover, so that the water remaining underneath has no room to expand when it wants to freeze and breaks out wherever it can, and, usually finding less resistance in the ice than in the walls of the vessel, floods over and collects more in one place than in another, according to the inclination of the planes into which the first crust breaks when the water bursts out. Freezing here as time goes on, this finally forms the little rise mentioned above.[211]

There have also been times when it has broken the vessels. This (some asserted) can very probably have occurred because the water at the bottom has delayed its freezing so long that the crust above had become so thick that it had become easier to break the vessel than the cover. But it is impossible to give any rules for these things, for there can be innumerable cases in which either the vessel or the cover alone may break, or first one and then the other, or both together, according to how they are affected by the external circumstances of the air and the cold, of calm or of the winds, of the evenness or nonuniformity of the strength of the vessels, or the internal disposition of the liquids themselves.

Before leaving this subject we must not remain silent about a trifling matter observed this year[212] which, trifling as it is, did not fail to lend some support to the opinion of these people. In a beaker placed in the open during the evening we found in the morning that all the [CLXXI] water was frozen, and on the highest part of its surface it had a point of ice a finger high, as thin and sharp as a splinter of rock crystal. This was apparently nothing but water that had come out onto the first crust when the beaker was freezing, and remained there, caught between this crust and the first layer that was formed by the cold in beginning to freeze this same water. This layer then breaking forcibly, water, entirely ready to freeze, spurted out into the very cold air and froze instantly without having time to fall back.[213]

[211] Boyle made the same observation in his *New Experiments ... Touching Cold* (1665), p. 228 (*Works*, ed. cit., II, 538).

[212] 1667.

[213] From the formal diary that was copied in G. 262, one would infer that all this information was gathered on one very cold night when the

The Second Experiment

We have also tried freezing water in the vacuum made with mercury, and to have a comparison with the freezing done in air we put water in a vessel similar to that in the vacuum. Leaving them thus all night, we found in the morning that both the samples of water had frozen; but with this difference, that the ice formed in the vacuum seemed to us harder and more uniform, and less porous and transparent than the other. Determining which of the two had the greater specific gravity, we found it to be the one in the vacuum. We obtained information about this by putting into spirit of wine pieces of each, turned in the shape of a cylinder and of approximately equal volume. Pouring in red wine, we saw the ice made in air rise from the bottom before that from the vacuum, and when they had both risen it floated more lightly and actively than the other, as the wine engulfed it much less.[214]

[CLXXII]

The Third Experiment

Having put distilled natural water to freeze in various flasks, we found that in all of them it froze more limpid and transparent than ordinary water. Only in the center was there as much as a kernel of whiter and more opaque ice than the remainder, and around it in all directions pieces of ice of similar quality radiated like so many strings. In short, to give an exact simile, in each flask there appeared a chest-

---

temperature went down to about −11°C, the night of January 4–5, 1667 (G. 262, 163v–65v, 166v–69r; *TT* 2, 505–8). An array of vessels was exposed "on the balcony of the Pitti that gives on the square to the west." There were "four boxes [*cassetti*], in each of which there were six small flasks, sealed with the lamp, containing various quantities of common water, with a glass of the same water in the middle, and a 50-degree thermometer for each box." It appears later that some of the flasks contained distilled water. The four boxes faced north, south, east, and west respectively.

The late date of these experiments suggests that Leopold must have had a special interest in the process of freezing, for at this time the *Saggi* was nearly ready for the press. But very similar experiments had been done as early as December 30, 1648 (G. 259, 6r), and February 23, 1654 (G. 259, 7r–9v); that is to say, long before the "foundation" of the Academy. And F. Massai, *Rivista delle biblioteche e degli archivi* 28 (1917): 134–38, has published a letter dated January 15, 1665, *stile fiorentino* (i.e., 1666) in which are described almost identical experiments superintended by Leopold after dinner on several evenings. I should suspect a mistake in the year of the later experiments if it were not that in G. 262, 164r, it is clearly given as "L'anno 1666. ab Incarnate." (i.e., 1667 in our notation).

[214] November 27–28, 1666 (G. 262, 162v–63r); January 4, 1667 (G. 262, 167r).

nut husk frozen in a piece of rock crystal, in the way that flies or worms or butterflies are sometimes seen caught in yellow amber, or blades of grass or straws or other matter in crystal[215] itself.

The Fourth Experiment

One evening, in order to observe the freezing of sea water, we put out two beakers filled with it, at a time when the 50-degree thermometer was at 9 [ca. −7°C]. After an hour we found that the less full of the two had begun to freeze, but in a manner somewhat different from that of ordinary water, for it appeared as if there had been mixed in it a large number of scales of talc, very finely ground. This took away the transparency of the water and gave it a soft consistency like the iced sherbet that we get in the summertime, when it melts if it lacks snow around it. When it was observed again a short time later it was found somewhat firmer, according as the multiplication of the scales had [CLXXIII] diminished the fluid part of the water. In the morning it was harder still, although it did not attain a fraction of the hardness of ordinary ice, since if it was agitated just a little it turned into water. The shape of the scales was rather long and narrow, and between them there was fluid for the most part; therefore the mass was quite detached from the vessel, turning around freely in it. The surface was flat without any protuberance; and in conclusion, the difference consisted in a looser structure and a much finer texture than that of ordinary ice.[216]

The Fifth Experiment

It is common knowledge that ice exerts its cold more effectively when some kind of salt is scattered on it. In this connection we have made the further observation that sal ammoniac strengthens it more than all other salts. We saw this by putting equal amounts of the same water at the same temperature into glass vessels of the same shape, capacity, and thickness, surrounded by equal quantities of powdered ice so that they might remain equally immersed. When the ice around one was strewn with sal ammoniac and the other with the same quantity of niter they did not freeze in equal times. For when a 100-degree thermometer, put in the water that was to be frozen with niter, was at 7½ degrees, another similar one immersed in the water to be frozen with sal ammoniac, put there at 20 degrees like the other, was already below 5 degrees and the water had begun to freeze over.[217]

[215] *Cristallo* can scarcely mean lead-glass here.

[216] January 4–5, 1667 (G. 262, 168r, 168v).

[217] I have not identified this experiment in G. 262. The numerous experiments noted by Targioni Tozzetti in this connection (*TT* 2, 511–15) really belong to the section on cooling by solution (see p. 247 below).

We have already said on other occasions that besides salts, spirit of wine has a wonderful power to help the action of ice. This will become extremely effective if salt is added as well [CLXXIV] as spirit of wine. Sugar also does something, but not much in comparison with common salt, niter, and sal ammoniac, which, more than the others, succeeds marvellously well in the operation of freezing.

The Sixth Experiment

Ice was put in vessels made of different metals in order to see which one would preserve it the best, but no certain results were obtained. Yet if we had to say in general terms what seemed to come out of a large number of observations, we should say that it was preserved best in lead, very well in tin, not very well in copper or iron, less well in gold, and less still in silver. Yet sometimes the ice in the tin and the lead disappeared before that in the silver and the gold, so that, as we pointed out, not much faith is to be put in this experiment, which is put forward rather to induce others to repeat it in surer ways, than to relate anything of which our observations have made us certain.

The Seventh Experiment

Gassendi wrote,[218] and it is quite true, that a strip of ice strewn abundantly with salt on its upper surface attaches itself very strongly to the table on which it is placed. We wanted to do the same with niter, but we did not succeed in seeing it even begin to stick. In those attached with common salt, we have indeed observed that it is much easier to detach them by lifting them perpendicularly to the horizontal plane, or levering them up as we do to loosen a nailed-down plank, than by [CLXXV] pushing them parallel to the same plane. Moreover, the water that is on the bottom is salty. The lower side of the strip remains opaque, and clouded by a white fog made up of innumerable particles of salt, very finely divided. Looking through it at the light it appears rough and as if prettily engraved with a diamond point, with elegant workmanship, so that it is very much like the crystal of those drinking glasses that are commonly called "iced" because they are contrived to resemble ice.[219]

The Eighth Experiment

The mist that gathers on the outside of glasses filled with cold water or ice at times freezes. This happens when the ice or snow contained in them is altered with spirit or with salt. At such times a damp, foggy smoke arises, which for the most part appears to come from the lower part of the glasses, from whence there moves a breath of frozen air.

[218] Gassendi, *Opera omnia*, etc., 6 vols. (Lyons, 1658–75) II, 83.
[219] June 25, 1657 (G. 262, 5v–6r; *TT* 2, 515–16). See also p. 45 above.

Besides being easily noticeable by bringing the hand near it, this appears even more clearly in the agitation that it produces in the flame of a candle that approaches it.

We have repeated this experiment by putting ice, sprinkled with spirit or strewn with salt, in other vessels of different shapes as well as of various materials, in order to observe whether either would make any change in this smoking, and we have observed that as far as the material is concerned it makes no difference whatever, be the cups made of glass or earthenware or wood or metal or gemstones. As to the shape, it seemed to us that whereas beakers and every kind of slender vessels at once begin to [CLXXVI] smoke at the bottom, wide cups, on the other hand, before smoking at the base smoke strongly around the top for some brief space of time.

With a wide golden cup we observed an effect that must be universal in every other vessel, although in some it is made less easily observable because of their shape. This is, that when the fuming stopped, the crust of ice began to rain down very fine particles of ice, like powdered glass, after the manner of dew;[220] and this went on until, the ice in the cup being melted, the liquefaction of the thin layer frozen on the outside ended also.

The smoke that is said to rise from ice seems very different from that produced by something that burns; in fact it is very similar to the fog that rises in the morning.[221]

The Ninth Experiment

The desire came to us to find out by experiment whether a concave mirror exposed to a mass of 500 pounds [170 kg] of ice would make any perceptible reverberations of cold in a very sensitive 400-degree thermometer located at its focus. In truth this at once began to descend, but because the ice was nearby it remained doubtful whether direct cold or that which was reflected cooled it most. The latter was taken away by covering the mirror, and (whatever the cause may have been) it is certain that the spirit immediately began to rise again. For all that, we shall not dare to affirm positively that this could not at that moment have resulted from something else besides the lack of reverberation from the mirror, as we have not repeated all the tests that would be needed to be perfectly sure about the experiment.[222]

[220] Until the nineteenth century it was generally believed that dew falls.
[221] June 9 and 17, 1660 (G. 262, 74r, 79r–v; TT 2, 516–17).
[222] Targioni Tozzetti (TT 2, 518) ascribes this to September 11, 1660, but it is not certain (TT 1, 411) that there was a meeting on that date. The experiment seems to have been made half a century earlier by Giovanni Antonio Magini, who died in 1617, for Sir Christopher Heydon (*An astrological discourse*, etc. [London, 1650], p. 21), who also reports having made

[CLXXVII]

EXPERIMENTS ON A NEWLY OBSERVED EFFECT OF HEAT AND COLD, RELATING TO CHANGES IN THE INTERNAL CAPACITY OF METAL AND GLASS VESSELS

In the experiments on artificial freezing it was said that the first motion that seems to be made by the liquids contained in the vessels used for freezing is a small rise that we therefore named the "jump upon immersion," for it happens the moment the vessels touch the ice. It should now be said that the [CLXXVIII] contrary happens when they are plunged into hot water, since the levels of the liquids go down perceptibly, and as if they were taking their time to rise, like one who wants to make a leap, they are seen very soon to come back to the graduation that they occupied before being immersed in the hot surroundings, and afterwards go on rising as the heat they take up causes them to expand, become lighter, and go up. So on the other hand, although they are raised in that first plunge into cold water or ice, they not only return to the graduation from which they started, but go many degrees lower until either after a long rest or with no pause whatever, they all (leaving out oil and spirit) rise until they become completely frozen.

Seeing this effect, it occurred to one of us[223] to attribute it to a cause that various later experiments appeared to favor admirably. The idea was that the appearance of these sudden movements in water and

---

it, adduces "the testimonies of Maginus, who in the representations of his [burning] glass sent to the Emperor [Rudolph II] doth confirm the same in these words: *Species esse sensu tactus perceptibilis, ut apparet ex lumine candelae, item ex nive & glacie infrigidante per suam imaginem remotè admodum.*" I have not been able to find the original of this, but it must date from before 1611, the gift to the Emperor being mentioned in Magini's *Breve instruttione sopra l'apparenze et mirabili effetti dello specchio concavo sferico* (Bologna, 1611), p. 4.

In an undated letter to Viviani (Florence, Bibl. Riccardiana, ms. 2487, fol. 7r), Magalotti tells how he, with everyone in his house, tried holding a glass lens between the hand and some ice. They seemed to feel something like "an almost imperceptible little breeze" at "the union of the lines of cold." Magalotti thought he felt it himself, "but less, I believe, than all the others, for I know that I am biased." The experiment was certainly not designed to convince a twentieth-century psychologist.

[223] The diary (G. 262, 47r) ascribes this to Borelli, and notes that "someone" [*qualcuno*] disagreed, whereupon the experiment related below (p. 210) was performed to convince him. It seems to have been difficult to accept the fact of the thermal expansion of solids. See also note 232.

other fluids is not due to any intrinsic change of rarity or density produced in their natural states by conflict with the opposite quality in the surroundings, which some justify with the famous word antiperistasis.[224] On the contrary—dealing first with the fall that follows the immersion of the vessel in hot water—they would rather have it that this occurs by the thrusting of the flying corpuscles of fire that evaporate from the water into the external pores of the glass, forcing it apart as if they were so many wedges. In this way the internal volume of the vessel is necessarily expanded, even before the corpuscles are transmitted through the hidden passages of the glass to the liquid inside. Cold too, contracting the same pores, makes the vessel too small for [CLXXIX] the volume of water that is in it, before this volume, still lacking the new cold, becomes smaller. In a word, they believe that the vessel, as the first to be attained by the heat or cold, and thus expanding or contracting first, is the true cause of the apparent rise or fall[225] as it becomes more ample or more restricted for the liquid that is still untouched by the properties of its surroundings. Such a conjecture was made to seem more probable to us by the following experiment.

AN EXPERIMENT FROM WHICH IT IS INFERRED THAT AT THE INSTANT
WHEN THE EXTERNAL HEAT OR COLD IS EXPANDING OR CONTRACTING
THE VESSEL, THE NATURAL TEMPERATURE OF THE LIQUID INSIDE
IT IS NOT CHANGED

Several small hollow enamel balls, hermetically sealed, were enclosed in a glass bulb full of water, as shown in figure I (plate CLXXX). By leaving air inside them, these were all made to have approximately the same specific gravity as water, so that those that were floating descended through it with any rise in temperature, and those at the bottom rose with any small accession of cold.[226]

Having suspended this instrument in the air and left it until the balls settled down, we began to bring basins of water up to it from below, either hot water, or cold water mixed with finely pounded ice. Although when these were applied to it the usual effects—a depres-

---

[224] Antiperistasis in this sense was an imaginary phenomenon dear to the Aristotelian philosophers and consisting of the "compression" of heat or of cold by being surrounded by its opposite. By being compressed the quality was of course intensified. The concept was used by Aristotle in many places, e.g., to explain the formation of dew (*Meteorologica*, I, 10) and of hail (*ibid.*, I, 12). In the latter place it is given as the reason why caves are cold in hot weather and warm in cold weather—a fallacy that succumbed only to careful thermometry.

[225] Thus in the text.

[226] As in the fifth sort of thermometer (see p. 98 above).

CLXXX

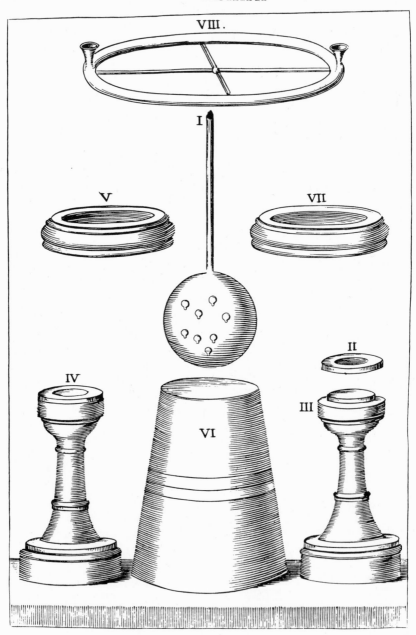

sion on entering the hot bath and a rise with the cold one—were observed, yet during the time taken for these effects to be produced, we never saw the submerged balls rise and float when the water seemed to contract, nor the floating ones go to the bottom when it appeared to expand. [CLXXXI] Such descent and ascent was only seen when the water was beginning to rise again after being depressed by the first ingress of the heat, or to fall after being raised at the entry of the cold. This phenomenon really appears to suggest more strongly that water, and other liquids as well, do not move of themselves in these first motions, but merely in obedience to the changes of the vessels.

Nevertheless it could still be said that these first movements proceed from an intrinsic change in the liquids, for while this is large enough to be seen with the help of the very narrow neck, it is not therefore sufficient to manifest itself by a change in the equilibrium of the balls. It can also be believed that at this instant these really begin to move, although the eye does not take in this first very slow movement from a state of rest.

The answer to this is that the same expansion and contraction of the water that suffices to make it rise or fall through the very short space through which it rises or falls on entering the ice or the hot water is more than enough to disturb the former equilibrium between the water and the balls, making this visible to the eye. And certainly, when the water really rises or falls because of a veritable expansion or contraction, the balls are seen to move somewhat, before it arrives at the graduation that it reached at the moment of its first immersion, while the balls stayed motionless.[227]

The discovery of this effect should not give us the slightest cause to doubt the reliability of our thermometers, since the whole of this contraction or expansion, in vessels containing an ounce and a half [42 ml] will come to a grain [0.05 ml] at the most. It may therefore be seen how much this can amount to, proportionately, in [CLXXXII] a bulb holding a few grains, such as those of the 50-degree thermometers, which are the most convenient, the most reliable, and in consequence the most widely used for measuring the changes of the air.[228]

---

[227] I have not found this in G. 262; nor does Targioni Tozzetti date it; but it is in Draft "A" (G. 265, 43r–v; *DIS*, 317–18), and this, as I have shown on p. 69, must have been written before July 1662.

[228] There seems to be a confusion of thought here. Although they recognized the thermal dilatation of the glass, it does not seem to have occurred to them that the thermometer really made use of the difference between the dilatation of the liquid and that of its container. Even Borelli missed this point (G. 267, 22v–23r; *DIS*, 335).

Now to make the reality of this phenomenon[229] manifest to the senses in various ways, the following experiments were made, which were first founded on theory and confirm it by their results.

THE FIRST EXPERIMENT, SHOWING THE ALTERATION OF A BRONZE RING PLACED IN THE FIRE AND IN ICE, ITS SHAPE REMAINING UNCHANGED

We had a small cylindrical bronze ring cast, as in figure II, and turned so that it was an exact fit on a plug of the same metal (figure III). When the ring was put in the fire for a short time and put back while hot onto its plug it was visibly loose (figure IV), being dilated by the heat into a ring of the same shape but so much larger that the expansion of its concave surface came to 9/100 of its diameter.[230] When it had been on the plug for a little while and warmed it with its heat, between the expansion of the plug and the contraction of the ring as it gradually cooled they not only went back to fitting as at first, but finally bound themselves together so tightly, that even before they were quite cold a considerable force was required to separate them.[231] The opposite happened when the ring was strongly cooled with ice.[232]

[CLXXXIV]

THE SECOND EXPERIMENT, SHOWING THAT A BODY MAY BE DILATED NOT ONLY BY THE ENTRY OF HEAT BUT ALSO BY THE ABSORPTION OF MOISTURE

A conical ring of boxwood was made, as in figure V, its concave surface being turned and polished with the greatest care. An arbor or conical frustum of steel (figure VI) was also made, turned in the lathe, polished to a perfect finish, and accurately divided with many circles parallel to its base. The ring being fitted over it, we observed which

---

[229] *Quest'accidente*; i.e., the thermal expansion of solids.

[230] *Nove parti centesime del suo diametro*. The coefficient of linear expansion of bronze being about 2×10^{-5} per °C, this seems to be an order of magnitude too high.

[231] They probably became covered with oxides.

[232] December 3, 1657 (*G.* 262, 47r; *TT* 2, 520–21). The next day, "In confirmation of the above-mentioned opinion of Sig. Borelli, it was seen that the glasses of spectacles fall out of their horn frames with heating, and tighten in them in the cold" (*G.* 262, 47v–48r; *TT* 2, 521). And on January 10, 1662, "A bulb of the sort generally used in freezing experiments [i.e., of glass], filled with water and set to freeze, was found notably reduced in size when taken out immediately after the freezing. This was determined with the calipers [*le seste*] for measuring the diameters of cannon balls" (*G.* 262, 129v). Nevertheless, it was by no means obvious that the *inner* diameter of a ring would be increased by the expansion of the material, and we find Viviani sending

of the inscribed circles coincided with the bottom of the ring, which was then taken off and put in water. After it had been there for three whole days, so that the water would have time to penetrate the entire thickness of the wood, it was again put on the cone, and we saw clearly that the concave surface had expanded, for the base of the ring went down a considerable distance below the former circle.

This ring was made in two ways (figure VII): in one we took care that the fibers of the wood came out perpendicular, and in the other, parallel to the plane of the base. In the expansion resulting from the absorption of moisture the first one kept its circular shape perfectly; the other became an ellipse and went down less than the first when placed on the arbor.[233]

For making the rings, note that hard and uniform wood must be taken, that is to say without knots and not composed of parts that are notably different in hardness, particularly for the first ring, so that each of the fibers will absorb moisture and reinforce each other, making the enlargement that much greater and more visible. Note also what [CLXXXVI] was said at the beginning of this account, that the rings should be left in the water long enough for it to penetrate their whole thickness, because if they are put on the arbor when only slightly wet on their surfaces, the effect will appear different, for they will go on considerably less than when they are dry. So let them be impregnated with moisture and really full of it, so that their dilatation may show more clearly.

THE THIRD EXPERIMENT, EXHIBITING MORE CLEARLY THE ABILITY OF GLASS TO CONTRACT AND EXPAND BY THE POWER OF HEAT AND COLD

A hollow glass ring was made, as in figure VIII. It was an ell [58 cm] in diameter and had two funnels, so that when liquid was poured into one, the air could escape more conveniently through the other. Over this we adjusted a cross made of two rods of enamel so that their extremities just touched the ring. Then the ring was filled with hot

---

a proof of this to Rinaldini with a letter of November 26, 1657 (G. 256, 36r–38v).

On December 19, 1657, Rinaldini wrote to Leopold (G. 275, 90r–v) recommending as an improvement that the rings should be enclosed in a bladder before they were put in the ice, in order to prevent changes due to the absorption of moisture. This suggests that wooden rings may also have been used; or perhaps the "second experiment" grew out of Rinaldini's idea.

[233] Not in G. 262, except for the figure (fol. 47v). In the diary of G. 261, fol. 61v, the experiment is described under the date December 6, 1657. The importance of these two experiments to the Academicians was that taken together they were thought to support the hypothesis that heat, like moisture, is a substance.

water and as it expanded we saw clearly with the naked eye that one or other of the rods was moving away from it, since they could not all reach it at the same time; until when the support was removed from each one the cross fell on the table inside the ring. The hot water being poured out and replaced with the water that dripped from salted ice, the cross was again placed over the ring and was not only held up once more but more securely than at first.[234]

[CLXXXVIII]

THE FOURTH EXPERIMENT, TO CONFIRM THE SAME EFFECT IN METALS

A thin strip of tin was taken, in the shape of a stirrup (figure IX), and was suspended in such a way that its extremities were very near a flat surface beneath, on which were made two short lines exactly where the two ends would have fallen if they had been prolonged. Then we put a lighted coal on the curved part of the stirrup, and, carefully watching one of the points, saw the line gradually appearing and withdrawing to the inside. This was during the time that the heat, expanding only the convex surface of the stirrup, was making the concave one contract; but when the heat had penetrated (which was in a very short time) through the entire thickness of the tin, expanding it all equally, we saw the point not only return to the line but pass beyond it more or less, according to the various degrees of heat communicated to the bend in the stirrup by the fire.[235]

THE FIFTH EXPERIMENT, TO OBSERVE BY MEANS OF SOUND A SIMILAR EXPANSION IN A GLASS STIRRUP

We tuned a string stretched across a large stirrup of glass, as in figure X, to the octave of a guitar. We applied the heat, as with the tin stirrup, and until it reached the concave surface the sound became lower in pitch because the narrowing of the opening of the stirrup relaxed the string. But when the heat had penetrated the stirrup, the [CLXXXX] cord was so much tightened that the sound rose above its original pitch.[236]

THE SIXTH EXPERIMENT, WHICH REVEALS THE SAME EFFECT MORE CLEARLY TO THE EYE

To the same string was attached a thread with a small lead ball, as in figure XI, and a hand-mirror was placed below this so that it almost

[234] Only the figure is in G. 262. The experiment is described, with a figure, in G. 261 under the date December 6, 1657.

[235] July 29, 1660 (G. 262, 98v; TT 2, 523). This description does not seem to be self-consistent.

[236] August 9, 1660 (G. 262, 102v–3r; TT 2, 524).

CLXXXIX.

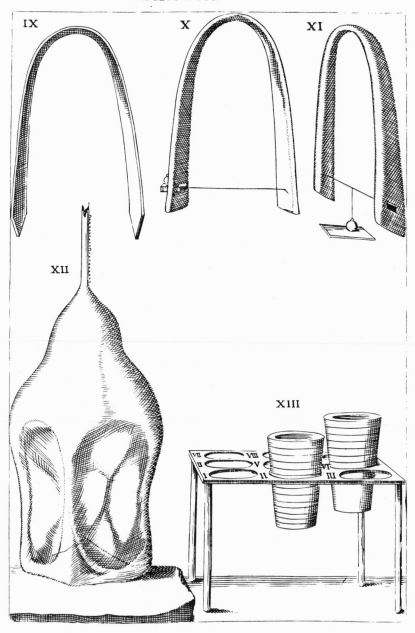

touched it. Then the heat was applied in the usual place. The effect on the stirrup was the same as before, for it first narrowed, the cord loosened, and therefore the ball touched the mirror; and at last, the opening of the stirrup becoming wider, the cord stretched and the ball rose again. Ice used instead of the coal produced the contrary effects, but very much less noticeably, in proportion to its lesser activity compared to fire.[237]

THE SEVENTH EXPERIMENT, WHICH DEMONSTRATES THE SAME EFFECTS IN A COPPER WIRE

A lead ball, attached to an annealed copper wire and hanging at a very small distance over a mirror, came down to touch it whenever the wire was heated by bringing a lighted candle near the wire and went up again whenever the wire was stroked with ice.[238]

In the same way, when two brass wires are tuned in unison and one is struck, the other resonates. But if a small burning coal, or a piece of ice, is brought near one of them they become out of tune. [CLXXXXII] The former, loosening the wire, makes the sound lower in pitch; the latter, drawing it tighter, sharpens the sound.[239]

THE EIGHTH EXPERIMENT, IN WHICH THE APPEARANCE OF A CONTRARY EFFECT CONFIRMS THE OPINION THAT THE FIRST MOVEMENTS OF LIQUIDS ARISE FROM THE CHANGED CAPACITY OF VESSELS WHEN THEY ARE FIRST IMMERSED IN VARIOUS SURROUNDINGS

It may sometimes happen that when vessels are first immersed in cold or hot surroundings an effect contrary to that stated above is perceived in the levels of the liquids; i.e., that they rise at once in hot surroundings and fall in the cold. This will take place whenever the vessel is made in the shape of the one shown in figure XII. In this, then, the liquid is seen to rise the moment it touches the hot water, because the angles between the sides are very robust and thick in comparison to the hollow sides, and the heat, acting first on the outer

[237] August 12, 1660 (G. 262, 104r; TT 2, 524).

[238] G. 252, 31v–32r, except that the mirror is merely a flat glass plate. This was by Viviani (G. 269, 73r–v; DIS, 430, in which it is a mirror.)

[239] This experiment was described by Viviani in a letter to Rinaldini (G. 252, 31r–34v) dated 17 November 1657: "I stretched so that they would sound in unison two copper wires of equal length and thickness (although this is of little importance) and far enough apart. Under one of them I brought a brazier with a small fire, so as to heat it, and striking both strings at the same time, I and a number of others always observed that the pitch of the heated cord was lowered, more or less according to the greater or lesser heat . . . " (fol. 31v).

surface, closes the angles of the corners, as was seen in the glass stir-rups described above. Consequently the thinnest part of the hollows must necessarily be stretched, and thus expanding inwards, diminish the internal volume of the vessel in this first moment, so that the liquid rises in the tube. It then descends to fill the new space when the heat penetrates the whole thickness of the vessel and the vase comes to expand uniformly, coming back to a shape similar to the first and more capacious; and finally the liquid starts to rise again when, re-ceiving the particles of fire into itself, it begins to expand.

It is clear that the opposite will be observed with cold, [CLXXXXIV] the same causes acting in the contrary way; and it should be remarked that the volume of the vessel is seen to be reduced by simple com-pression with the hand in two opposite hollows. The rise of the liquid that immediately follows this compression cannot be attributed in any way to expansion produced by bodily heat, because if the compression is done again with two small pieces of ice the liquid rises just as much in the same way.[240]

The use of the instrument shown in figure XIII can easily be under-stood from the illustration alone, as it is nothing but a steel draw-plate pierced with round holes of various sizes, for measuring the different expansions produced by different degrees of heat, either in the instru-ment itself or in various conical metal rings.[241]

THE NINTH EXPERIMENT, TO SHOW THAT A VESSEL CAN BE EXPANDED NOT ONLY BY HEAT, OR BY THE ABSORPTION OF MOISTURE, BUT ALSO BY WEIGHT

Two glass vessels, one part of a cone and the other of a pyramid, are fitted in mortises in a thick table, as shown in figure XIV. Around the outside of each is marked the intersection with the plane of the table, and they are then taken out. When they are filled with mercury and put back they will not enter as far as the marks, for the weight expands them.[242]

[240] July 28 and 29, 1660 (G. 262, 97r–98r; *TT* 2, 525–26). In the diary this experiment is credited to Borelli. Something of the sort seems to have been suggested to Prince Leopold in a note attached to a letter dated November 14, 1657 from Pisa (*LIUI*, I, 92–94). The original is in G. 275, 84r–85r. In this letter Borelli states his belief that the increase in dimensions of solids when they are heated is due to wedged-shaped "corpuscles of fire" entering their pores. Thus the experiment disposes at one blow of the "frigorific atoms" of cold and the peripatetic idea that heat and cold are separate "qualities."

[241] August 30 and September 5, 1662 (G. 262, 149v, 152v; *TT* 2, 526).

[242] Date unknown, but according to Targioni Tozzetti there was a docu-ment in the Real Segreteria Vecchia describing this experiment, which he believes to be due to Carlo Rinaldini. This is printed in *TT* 2, 527.

CLXXXXV.

Fig·XIV·

꒰꒱

[CLXXXXVII]
EXPERIMENTS ON THE COMPRESSION OF WATER

Even if we do not always reach the truth by experiment, this is often not because the first ideal concept of the experiment is not well-suited to produce such a result, but sometimes because it may be the fault of the material substances and the corruptible mechanisms that have to be made use of in putting it into practice. These, although they cannot themselves contaminate the purity of theoretical speculations, [CLXXXXVII] nevertheless are not always able to support them because of the faults of the materials.

But the experimental way of inquiring into natural occurrences is not on this account to be deemed fallacious, because even if at times we do not succeed with it in getting to the bottom of the truth that we primarily seek, it will be strange if it does not give us some gleams of it or show the falsity of some contrary supposition about it.[243]

Precisely this happened to us in our endeavor to find out whether water suffers compression as air does. In this attempt, although we did not reach a complete knowledge of the truth because of the weakness of the glass instruments, used because they had to be transparent, we were at least taught that water cannot be compressed by a very great force. We have learned that a power not only thirty but a hundred and perhaps a thousand times that needed to reduce a volume of air into a space thirty times less than it first occupied, does not compress a volume of water even by as much as a hair or other lesser observable space, below what its natural extension requires. The ways in which we have gone about getting this information are as follows:

The First Experiment

Let there be two glass bulbs, one larger than the other, at the ends of the two glass tubes *AB*, *AC* (figure I). Fill both vessels with ordinary water up to *D* and *E*, and join them together with the lamp, being careful to leave a passage for air at the joint, and to draw out the spout *AF* as long as possible, leaving it open. Then apply to the two bulbs two beakers full of crushed ice, [CC] in which they will be buried so that, the water contracting, as much air may come into the bores of the tubes as possible.[244] Furthermore, the whole tube *DE* is

---

[243] They may have derived some suggestions from the curious pamphlet of Raffaello Magiotti, *Renitenza certissima dell'acqua alla compressione dichiarata da vari scherzi*, etc. (Rome, 1648). This was reprinted by Targioni Tozzetti (*TT* 2, 182–91).

[244] The water would have taken even less volume at the temperature of maximum density, about 4° C.

CLXXXXIX.

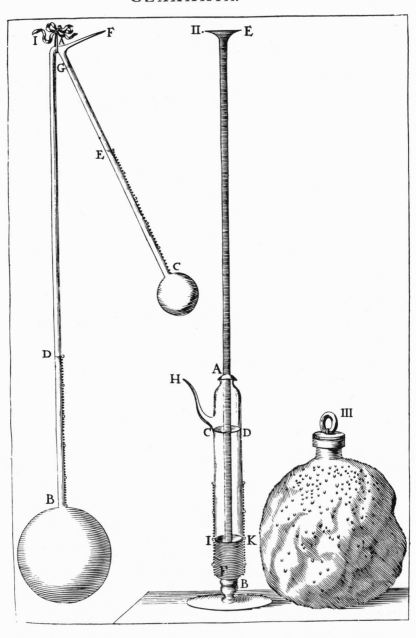

stroked with pieces of ice for a time in order to increase the amount. Thus as the air that comes in through the orifice F is gradually contracted by the action of the cold, new air follows it, so that it remains tightly compressed and held in when the orifice is sealed with the flame.

When this has been sealed, the bulb B is taken out of the ice, and, after being first warmed in tepid water, it is put in hot, and finally in boiling water, the bulb C being kept in the ice all this time so as to maintain the water in it in a state of maximum contraction. Let this water come to the point E, beyond which the cylinder of air GE tries to compress it, and this air is made extremely dense by the force of the water that has gone up to G, because of the expansion produced in it by the heat of the water, which is supposed to be boiling inside the bulb B at the time.[245]

Now if the water suffers any compression it ought to yield some degrees to the cylinder of air that presses on it, going down below the point E. But with us it has turned out otherwise, for when the water in E was really reduced to the state of its greatest contraction, the compressive force of the air GE had made no gain and burst the bottom of the bulb C before moving the level E a hair's breadth. And when we made the two bulbs of copper so as to strengthen the instrument, nevertheless the water in the bulb C, held between the solidity of the metal and the irresistible pressure at E, first broke the tube, which had to be made of glass to show the internal movements of the water, and was fastened [CCII] perfectly to the copper with mastic or with the usual hot cement.[246]

The Second Experiment

Take a vessel such as AB in figure II, holding about six pounds [2.04 kg] of water, and with a mouth that will pass a glass tube reinforced externally with a wrapping of lead, enclosing it exactly to prevent it from breaking. Fill the vessel with water up to CD and immerse the tube EF in this, open at top and bottom, and cement it in the mouth A with the usual cement, taking care to fix it a little above the bottom, as F, B, so that a liquid poured into it can flow freely into the vessel. Then begin to pour quicksilver in through the

[245] The bulb B is in boiling water; but because the internal pressure is well above atmospheric, the water in B would not boil. Some of it would, however, distill over into CE, a possibility that does not seem to have occurred to the Academicians.

[246] September 10, 1657 (G. 262, 32r–v; *TT* 2, 529) with the glass bulbs; October 16, 1657 (G. 262, 40r–v; *TT* 2, 530) with metal bulbs. In the diary the idea is ascribed to Paolo del Buono, who was in Vienna at the time.

tube. This will run into the vessel and raise the water above it until this entirely fills the vessel (since the air has an exit through the spout *CH*), overflowing through the orifice *H*, which is then sealed with the flame, noting at the same time what graduation the mercury has reached with its surface *IK*.

The tube is then to be completely filled by pouring in new mercury. If the water is compressed by such a force, the level *IK* will be seen to rise as the height of the mercury gradually goes on increasing and as the water gives way because of the compression. With a weight of 80 pounds [27 kg] of mercury[247] extended in 4 ells [234 cm] of tube (all that our instrument could stand without bursting) we did not see the level of the mercury *IK* gain by as much as a hair's breadth, the water obstinately resisting the energy of that great force.[248]

[CCIV]

The Third Experiment

We had a large but thin ball (figure III) cast of silver, filled it with water cooled with ice, and closed it with a very tight screw. Then we began to hammer it lightly all over, denting the silver (which because of its hardness could not be thinned and stretched as refined gold or lead or another softer metal would be) so that it was contracted and its internal volume reduced, without the water suffering the least compression, for at every blow this oozed out through all the pores of the metal, looking like mercury pressed through some leather and coming out in a gentle spray.[249]

This is as much as we have been able to learn from these three experiments. Whether by repeating them in stronger vessels, increasing the expansion of the water and thus the pressure of the air in the first experiment, the height of the cylinder of mercury in the second, and the thickness of the silver in the last, water might finally be compressed, we are unable to say. This much is certain, that in comparison with air, water may be said to resist compression infinitely more strongly; confirming what we said at the beginning, that although experiment may not always attain the ultimate truth that we seek, it would be too bad if some small light were not shown by it.

[247] Note that they did not realize that they could have obtained the same effect with a much narrower tube and much less mercury. Nor is it clear how the air in the space *AD* was got rid of.

[248] November 3, 1657 (G. 262, 44v–45r; *TT* 2, 530).

[249] Date unknown. This phenomenon was known to Francis Bacon, according to Targioni Tozzetti (*TT* 2, 531 n). It had been mentioned by Raffaello Magiotti, in his *Renitenza certissima dell'Acqua alla compressione*, etc. (Rome, 1648).

ᵛᵗᵨᵻᵍᵥᵥᵛ

[CCVII]

EXPERIMENTS TO PROVE THAT THERE IS NO POSITIVE LEVITY

It is an ancient and famous question whether those things that are commonly called light are so because of their nature and go upwards on their own account, or whether this rise is really due only to their being pushed away by heavier materials which, having more strength and vigor for descending and placing themselves lower, squeeze them out, so to speak, and force them to go upwards. The latter doctrine, which more particularly seems to have [CCVIII] gained ground in modern times, was not at all unknown to the ancients; in fact it was asserted on reasonable grounds by many philosophers of those centuries, and among these most clearly by Plato in the *Timaeus*.[250] Plato promoted the likelihood of such a concept so much further that he not only maintained that heavier things are able to displace less heavy ones upwards, as air does to fire, but also heavier ones, such as water may be in comparison with air whenever it is lightened by the admixture of heat.[251] He seems to suggest exactly this in the above-mentioned dialogue, the *Timaeus*, when he says that fire, escaping from the hot interior of the earth but not reaching the void, pushes against the air next to it. Not only does this not yield up its place but it even takes away those moist masses that accompany the fire and pushes them up and up into the sphere of fire; and only because, thanks to their union with the fire, the natural heaviness of the moisture is tempered with new lightness.

However this may be, we here adduce only two experiments in confirmation of this opinion. Perhaps the force of these compensates for their small number.

The First Experiment

In figure I (plate CCIX) let *ABC* be a wooden cylinder, the base *BC* of which makes perfect contact with the horizontal plane *DE*. So that the surrounding air may not spoil the exactness of the contact by leaking in between the two surfaces, the base of the cylinder is

---

[250] *Timaeus*, 62c–63c. "Plato's whole point," says E. E. Taylor in his *A Commentary on Plato's Timaeus* (Oxford, 1928), p. 444, "is that every body which is light in some region is heavy in some other, and vice versa."

[251] The fact that water vapor is lighter than air, and hence a moist atmosphere weighs less than a dry one, was clearly stated by Sir Isaac Newton in the 1717 edition of his *Opticks*, but was not given general credence in the eighteenth century.

CCIX.

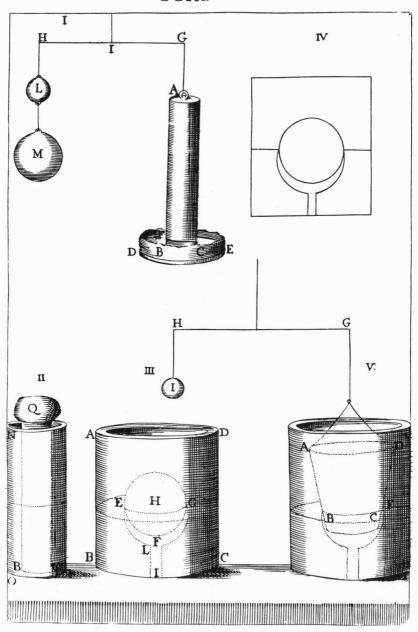

lined with a metal plate, well flattened and polished, and another similar one is laid down on the plane. A bank of wax or clay is made [CCX] around the cylinder *ABC*, and between it and the cylinder is poured quicksilver up to *F*, so that the circle of contact remains quite covered and protected from the ingress of air. Next the extremity *A* is tied to the end *G* of the equal-armed balance *GH*, which has its center at *I*, and to the other end *H* is attached the weight *L*, equal to the absolute weight of the cylinder *ABC*. It is easily understood that in order to detach the cylinder *AC* from the plane below it, the force of the weight *L* will not be enough; so that we go on adding more and more weight to the end H, until the two weights *L* and *M* raise the cylinder *AC*. This resists being raised with a double force, that of its own weight, equal to *L*, and that of contact or repugnance to the vacuum or however this force may be interpreted. The remaining force of the weight *M* will not only equal but exceed the force of attachment of the two surfaces.

When this force (which in our instrument was about three pounds [1.0 kg]) has been measured, put the cylinder *ABC* in a cylindrical vessel *NOP* (figure II) of wood or glazed earthenware, as high or higher, and let it down until the base *BC* makes contact with the base of *OP*, which has also been covered with a thin plate of metal or glass, flattened and polished. Then pour quicksilver into the vessel *NP*, up to any height, even to cover the cylinder *ABC*, and this will never detach it. But if finally the base *BC* is pulled away from *OP* by hand, leaving the cylinder *AC* at liberty, this will be seen to rise with great force and float on the mercury.

We have now to find how great is this upward force that is supposed to be that of levity. We found it [CCXII] thus: We loaded the end *A* of the cylinder with a weight *Q*, sufficient to send it to the bottom and therefore to prevent it floating. In our experiment this weight was about five pounds [1.7 kg] and so we concluded that such was the measure of the force we were seeking. Now we considered that the resistance to the detachment of the two bases was not greater than three pounds, as was noted, and the force of the supposed levity in the cylinder was found to be five, so that in such a case that of the levity was greater than that of the adhesion. Therefore, considering now the cylinder of wood *AB* attached by its base *BC* to the surface *OP*, there are two opposing forces: one of three pounds, which is that of adhesion and holds it back, the other of five, which is that of levity and would lift it. Therefore the lesser force should be overcome by the greater and the cylinder should rise; but this does not happen, since it is not detached. Thus it seems that we must say that what causes the cylinder to float is something other than levity.[252]

The Second Experiment

Let *ABCD* in figure III be a wooden vessel, in the thick bottom of which a hemispherical depression *EFG* has been turned with a lathe. This is of exactly the same size as an ivory ball *H* which fits it at its greatest perimeter *EG*. The vessel is then filled with quicksilver so that the entire ball is submerged in it. It seems clear that as the weight of the quicksilver is borne by the bottom of the vessel, and because it is prevented by the exact fit of the ball in the circumference *EG* from running down between the lower convex surface of the ball and the concave surface of the vessel it will not be able to detach the ball with its impulsive force; [CCXIV] but the natural levity of the ivory, if there really is such a thing, could raise it to float in the very heavy surrounding mercury. But this is not observed to happen, the ball remaining motionless in its socket under any depth of quicksilver.

Nor can it be objected that the abhorrence that nature has for the vacuum (which might arise from the detachment of the hemisphere of the ball from the concave surface of the vessel) opposes the effect of the natural levity of the ball; for even when a hole such as *FI* was made in the bottom of the vessel, through which the air might come in to fill the space that would be empty after the detachment, the ball still did not rise.

And because it could also be said that the ball, touched by air at the bottom, is not lighter, but heavy, the hole was again closed[253] and the cavity in the vessel expanded into a form like *ELG*, so that only the rim and upper circle *EG* remained equal to a great circle of the ball, but the hemisphere *EFG* no longer fitted the concave surface *ELG*, as appears more clearly in figure IV. The space *ELG* is then filled with mercury and the ball dexterously submerged until its great circle fits the edge of the depression, although it is not to be pressed strongly into the upper circle *EG*, but can turn in this with a very small and insensible force. Nevertheless, when the vessel is filled to the top with quicksilver the ball will not move.

Finally, so that there may be no suspicion that the mercury that

252 Targioni Tozzetti "found nothing about this experiment in the diary" (*TT* 2, 533), but in G. 262, 83r there is the notation, "After lunch today [June 28, 1660] the experiment against positive levity was made before the Grand Duke," etc., and this is copied in *TT* 2, 580. He prints a long account found in some detached papers (now in G. 268, 117r–18r; ascribed to Borelli). Borelli described a very similar experiment in his *De motionibus naturalibus*, Cap. IV, Prop. XC, but did not claim it as his own.

253 It is shown open, however, in fig. IV.

rests above the ball, pressing it down with its weight, prevents it floating, a glass vessel *ABCD* (figure V) was taken instead of the ball *H*.
The surface of this is a portion of a cone and near its smaller end fits
the rim *EF*. When it is surrounded with quicksilver [CCXVI] this too
is held motionless. And to make it clear whether the supposed tena-
cious union between glass and quicksilver and the aversion of nature
to permitting an empty space can overcome the force of the levity of
the beaker *ABCD*, the force of the supposed attachment was measured
by removing the mercury from around the glass and, having attached
this to the end *G* of the equal-armed balance *GH*, weight was added
to the end *H* until the glass was detached from the rim *EF*. Let this
weight be *I*, which with us was one pound [0.34 kg]. Then the vessel
was again filled with quicksilver, and the glass put to float on it was
loaded (as in the other experiment) with as much weight as would
send it slowly to the bottom and hold it there. This weight (which
with us was about two and one-half pounds [0.85 kg]) will be an exact
measure of the force that was believed to result from the levity of the
glass *ABCD*. It will therefore be greater than that with which the
vacuum is resisted, which was found to be one pound. Therefore if
levity is what makes the glass float, it must use its effect to detach it,
since its force exceeds the force of attachment that resists it. But it
does not. It therefore appears that the conclusions of the other experiment are confirmed by this second one, that is to say, that something
different from levity raised the ivory ball and glass.[254]

[CCXVII]
EXPERIMENTS CONCERNING THE MAGNET

Although the wonderful operations of the magnet may be a wide
ocean where, however much may have been discovered, much more
probably remains to be found, yet up to now we have not had the

---

[254] Borelli, *De motionibus naturalibus*, Cap. IV, Prop. LXXXII, reports
this experiment as having been made in the Academy, but does not claim to
have suggested it. He has the ball of wood, rather than ivory. However, in
a letter to Viviani dated January 11, 1658, (G. 283, 35r–36r), he describes
the experiment as having been made before the Grand Duke, Prince Leopold,
and numerous "peripatetics" at Pisa, where the Court then was staying.
It is also described in G. 261, 65v.

To us these experiments have an air of unreality about them, but there is
no question that they were taken very seriously, as may be inferred from the
fact that Borelli's long fourth chapter is entitled "Positivam levitatem in
rerum naturam non dari," and is entirely devoted to a refutation of this
doctrine.

courage to engulf ourselves in it, being very well aware that to attempt new discoveries in that region requires an entire and very long study, not interrupted by other speculations. Let no one believe, therefore, that with these two or three observations on the subject we pride ourselves on having lit some [CCXVIII] great light in magnetical philosophy, for we perceive only too well that these reports are very ordinary and perhaps not at all new. For they have not resulted from any aim at a determined application to the study of the magnet, but have either been found by accident or sought out by some one of our Academicians for his own purposes. However, we have not desired to conceal them, such as they are, having no other intention than to communicate everything that has a semblance of truth to us, little though it may be.[255]

THE FIRST EXPERIMENT, TO FIND OUT WHETHER THERE IS ANY SOLID OR LIQUID, EXCEPT IRON OR STEEL, THAT WHEN PLACED BETWEEN IRON AND A MAGNET BRINGS ANY ALTERATION IN, OR ENTIRELY PREVENTS THE PASSAGE OF, ITS FORCE

At one end of the wooden box $ABCD$ (figure I) let there be placed a compass with its needle directed towards the point E. From the other end of the box let the magnet be moved slowly closer until the needle moves a certain distance, for example from E to F. Let the magnet then remain fixed, and into the space that remains empty in the box between the magnet and the compass let there be put glass vessels of quicksilver, or wooden ones full of sand or of metal filings, as long as they are not iron or steel, or solid parallelipipeds made of the same metals or of various stones or marbles; and the needle will always be seen to remain motionless at the point F. Finally, let these vessels be filled with spirit of wine and set on fire; and not even the passage through the flame [CCXX] will dissipate the force that holds the needle at F. Only by a thin sheet of iron or steel, as already noted, is it seen to be released and to return to E.[256] And not only the above-mentioned things fail to interrupt the magnetic activity. Having piled fifty plates of gold one upon the other, we saw a needle placed on the top plate obey the motion of a magnet moved close to the under surface of the bottom one.[257]

[255] The experiments of the Academy in this field were indeed few in comparison with those of William Gilbert reported in his *De Magnete*. Among the papers of the Academy there is a folded sheet (G. 268, 195r–96v) headed "Ex Gilberto," and containing twenty-two short paragraphs that at least show that he had been carefully read. This is printed in *TT* 2, 544–46.

[256] July 10, 1657 (G. 262, 15r–v; *TT* 2, 540); September 2, 1658 (G. 262, 61r–v; *TT* 2, 542).

[257] This was one test that Gilbert probably could not afford.

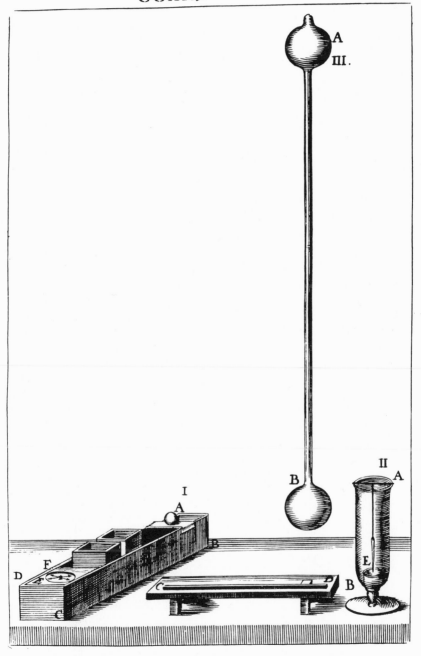

THE SECOND EXPERIMENT, TO SEE EVEN MORE EXACTLY WHETHER
THE FORCE OF THE MAGNET UNDERGOES ANY VARIATION
IN PASSING THROUGH VARIOUS LIQUIDS

Let a needle that has been touched by a magnet be hung by a fine thread on the axis of a glass vase *AB*, as shown in figure II. Let a small cylinder of lead be placed at the bottom of this vase, and let there be two points made of brass, or other metal except iron or steel, on its upper surface, one fixed in the center and the other about the thickness of a piaster[258] from the first. Then let the needle be adjusted so that it comes vertically over the one fixed at the center; and, placing the magnet at such a distance that it does not move the needle, it is brought nearer in such a way that its pole always directly faces this.

To be more certain of this condition, the stone[259] is slid along with one of its faces grazing the straight edge *CD* fastened in the middle of a small plank placed level and in the plane passing through the two points, of which the one not in the center is also turned directly towards the pole of the magnet. Meanwhile, bringing this nearer the needle, it will finally reach it with its force, and when it feels this the needle will slowly begin to move [CCXXII] towards the magnet. The observer will not stop then, but will push the latter forward very slowly until the needle has come out of the perpendicular and reached the second point nearer the magnet. This will at once be stopped and a mark made on the straight edge at the distance that there was between the stone and the needle when the point of the latter was above *E*.

The magnet will then be removed and when the needle has been surrounded by natural water will be brought towards it again in the same way, drawing it forward until the needle returns to the point *E*.[260] This distance being marked, the water is to be poured out and various liquids put into the vase in its place, taking the distances from which the needle is attracted by the same application of the magnet.[261]

It will appear from these that the magnetic force is neither destroyed nor strengthened by any of the liquids through which it penetrates.

[258] Perhaps 2–3 mm.

[259] *La pietra*. The lodestone is usually referred to as *la calamita*, which I have translated "magnet."

[260] June 19, 1657 (*G*. 262, 2r; *TT* 2, 541). This was the first day on which experiments were performed and recorded. They also found that a needle immersed in water would be picked up vertically from the same distance as in air.

[261] July 21, 1657 (*G*. 262, 17r; *TT* 2, 542). Also August 23, 25, and 26, 1660 (*G*. 262, 107r–8r), not August 29 as stated in *TT* 2, 541.

It indeed attracts from various distances, but this happens according as the lighter or heavier medium lightens more or less the needle swimming in it, so that the same force moves it farther or not so far; since it was observed that the various distances from it to the magnet were in reciprocal proportion to the specific gravities of the fluids, that is to say, to the amounts by which the needle was lightened.[262] Thus, among the liquids with which we made experiments, it was attracted from the greatest distance in salt water, from less in ordinary water, less still in spirit, and least of all in the common medium of air.

It should be noted that in repeating this experiment at various times it might happen that these distances would vary from one occasion to another. But it must be considered that this could arise from adventitious phenomena, as for instance the various temperatures of the air, the needle being rougher or smoother, or the accidental proximity of some iron that might alter or deviate the direction of the magnetic force in some manner; [CCXXIV] or other similar phenomena. However, this experiment was made by us on a large table all held together with strong glue and with wedges and dowels instead of nails; and the observer, and anyone else who might remain in the vicinity, was always careful to remove all iron that he had on him, as it was clearly recognized that if anyone came near the table with keys or knives in his pockets this at once altered these effects, which always remained constant when every kind of iron was kept away. As to what may depend on the other phenomena referred to, such as the differing temperatures of the air or others impossible to remedy, we have found that even if the distances change, that is to say that those from which the needle was attracted yesterday disagree with those from which it is drawn today in the same media, nevertheless the differences found at such various times are found to be approximately proportional among themselves.

THE THIRD EXPERIMENT, TO SEE WHETHER THE ACTION OF THE POLES
OF THE MAGNET IS CHANGED BY TURNING THEM TOWARDS
THE OPPOSITE POLES OF THE EARTH

Although in this experiment we have not as yet been successful in giving a satisfactory account of many particulars that still remain unsettled, at any rate we shall give a brief general explanation of the little that it seems to us we can assert with some reasonable grounds of certainty. This is that the north pole turned to the north attracts a needle suspended in air from farther away than if it is turned to the

---

[262] It would be proportional to the square root of the difference between the specific gravities of the needle and the liquid.

south or the west; and towards the west somewhat more than towards the south, and little less than towards the north. As to the south pole, [CCXXV] on the contrary, not only does it seem to us that it attracts towards the south as the north pole does towards the north but also that when it is turned towards the north it continues to attract in the same way as towards south. Towards east and west it also becomes weaker in comparison with north.[263]

[CCXXVII]

EXPERIMENTS ABOUT AMBER AND OTHER SUBSTANCES WITH ELECTRIC VIRTUE

As everyone knows, the electric virtue reveals itself in all those bodies where there is a source of it when these are delicately or vigorously rubbed. Yellow amber is rich in it above all others, after which, it seems, comes the finest sealing wax. After this, it also appears, follow the high-cut diamond, the white sapphire, the emerald, the white topaz, the spinel and the [CCXXVIII] balas-ruby.[264] After these are all the transparent gem-stones, white as well as colored, some of which show more attractive power, some less. It really does not seem that in this they keep to the order of their hardness;[265] for we observe that the soft spinel and the balas-ruby yield nothing in attractive virtue to the hardest diamond and the sapphire. Near the gems come glass, crystal, white and black amber, between which substances there is no great difference in strength and force, all being very weak in their action. For the rest, neither the lapis lazuli nor the turquoise nor jasper nor agate nor other similar sorts of nontransparent gem-stones, nor other stones nor the most noble marbles nor marine jewels such as corals and pearls, nor metals nor crystals of salt attract, as has been asserted by some people. Perhaps this mistake may have arisen from seeing that when such materials touch tiny bits of straw, paper or other things, these adhere to them. We have also observed this, but it may perhaps occur, some say, because certain very small roughnesses

[263] August 30, August 31, and September 1, 1660 (G. 262, 109r–v; TT 2, 544). The field of the earth would be so weak in comparison with that of the lodestone that it would have been difficult to be sure of this effect.

[264] A rose-red variety of spinel.

[265] In G. 262, 70v, there is a classification of gems according to nine degrees of hardness, diamond alone being in class 1. Targioni Tozzetti also found, among the loose papers of the Academy, a list of minerals and other substances with an indication of whether they were found to be "electric" (TT 2, 548–49).

are found on those bodies, and when they are pressed down on the little pieces these remain slightly impaled, and so are carried away. Wishing to avoid this fallacy, we resolved to give credit only to those materials that, when they have been rubbed and presented to very small and light bodies at some distance, attract them; and we have found this to be done only by the materials listed above.

We have also observed that the alterations that amber receives by external accidents such as heating, freezing, and anointing with various liquids, are also exactly repeated in gem-stones and in every other material that has the faculty of [CCXXIX] attraction. However, it is true that in amber, which contains more of the virtue, these are more clearly observed, and for this reason we shall discourse of it only, leaving the rest aside.

Amber, then, of all the substances that are presented to it, refuses to attract only flame, no matter if Plutarch[266] says that it does not attract things steeped in oil, and millet, or as others will have it, sweet basil; we have found this to be untrue. Smoke is also attracted; indeed it is very interesting to see how, when amber already rubbed and warm is brought near the smoke coming from a candle just put out, the smoke at once bends towards the amber. Then part of it is captured and part rises as if reflected by a mirror, while that which remains collects like a little cloud. As the amber cools off[267] this disperses again into smoke and goes away.

On the contrary, not only does flame not let itself be attracted, but if after the amber is rubbed it surrounds it at all it extinguishes its virtue so that it needs to be rubbed again to make it attract. And if the amber is brought up to the flame after it has attracted a small piece of something, it will at once let go of it.

The heat that comes from burning coals is not so inimical to the virtue of the amber, indeed it sometimes suffices to excite it without any rubbing. It is true that with no stimulus but heat it acts very weakly, but if it is then rubbed it becomes more vigorous.

Ice by itself does not harm amber, but if altered by salt and spirit diminishes its virtue to such an extent that sometimes it needs several hours and long and powerful [CCXXX] rubbing to get it back again. Because of this it was believed by some that such a loss of force was not due merely to the increased cold usually brought about in ice by mixing in salt and spirit of wine, but rather to some very slight stain or some film taken up by the amber from the finely powdered salt, or

---

[266] Plutarch, *Quaest. Conviv.*, II, 7, 1. I am indebted to Professor Harold F. Cherniss for this reference.

[267] It was considered that the warming of the amber by friction released the "electric virtue."

indeed from the absorption of spirit, which is one of those liquids that damage its power of attraction.

Not all materials are suitable for revealing the virtue of amber. If the amber is rubbed on bodies that have smooth and polished surfaces, such as glass, crystal, ivory, burnished metal, and gems, the virtue remains unawakened and does not come out. Consequently they need to have some small inequalities and roughnesses in their surfaces, such as cloth and linen have, and a thousand other things that need no mention. Even human flesh is good for bringing out the virtue of amber. But it is true that some people's is better than that of others, and indeed some have been found on whose hands you can rub it as much as you will, and never find a way of making it attract.

It is commonly believed that amber draws bodies to itself; but this is a reciprocal action and in no way more proper to amber than to the bodies, by which it is also attracted; or at least it sticks to them. We have experimented with this and have seen that when the amber is hung on a thread so that it hangs in the air, or balanced like a magnetic needle, and rubbed till warm, it goes towards bodies that are presented to it at a suitable distance and readily obeys their motion.[268]

Liquids also feel the force of amber, which attracts very small drops of them, even of [CCXXXI] quicksilver. It is true that unless these last are very small it has not the force to hold them up, so that they are let fall at once after being attracted. When the amber is presented to the surfaces of still liquids, including quicksilver, it does not pull off even a single drop, but makes such surfaces swell up underneath it and rise towards it after the manner of a falling drop, but in reverse, inasmuch as they tend to unite themselves with it at their most pointed part. This effect is observed better in oil and in balsam than in any other liquid.

When it has been bathed in some liquids amber does not attract after being rubbed, but there are others that do not produce this effect.[269] Those that do are all waters, natural and distilled, without exception; all wines, vinegar, and spirit; all liquids that are distilled within the bodies of animals; all acid liquids and the juices of citrus fruits; balsam and all artificial liquids, such as sweet syrups, essences,

---

[268] It is difficult for us to remember that there was a time, before Newton, when the equality of action and reaction was far from being axiomatic.

[269] Gilbert (*De Magnete*, Book II, chap. 2) had said that amber washed in spirit of wine does not attract, so on July 3, 1657 they tried this and found the statement entirely false (G. 262, 12v; *TT* 2, 551). This seems to contradict the statement about spirit in the next sentence. A loose paper was found by Targioni Tozzetti, listing the liquids that "impede attraction" and those that do not (*TT* 2, 554–55).

and the spirits and oils that are extracted by distillation. On the contrary it is not produced by petroleum, olive oil, sweet almond oil, that pressed from bitter almonds, tallow, lard, and finally butter, either pure or perfumed with the odors of flowers, or incorporated with amber or musk, provided that no essences or oils are mixed with it.

We have observed a very singular effect with diamonds. Of these the high-cut ones, as we said before, are counted among the gems richest in electric power,[270] but table-cut diamonds are so weak and sluggish in attraction that they sometimes appear entirely without this virtue. Indeed it appears to some of us that their flat surface has nothing to do with this effect, seeing [CCXXXII] that when diamonds have depth, even though rounded and smoothed on the wheel, they attract very efficiently, whereas the tables that have no depth, such as those that are commonly called mirrors, used to adorn necklaces, will not attract no matter how large they may be, nor if they are rubbed very hard for a long time; or even if they do attract, they do it with such little energy that one needs, so to speak, to make them touch the little crumb of paper or straw that one wants to make them attract. There is no doubt that at times some are met with that have a little force, but we have found these to be at least very rare. We once had one in our hands which, in many tests made over a number of days, could never be made to attract. A year later, wishing to show some phenomenon or other, we took the same ring in which it was set, and rubbing it very lightly on cloth in the usual way, it was at once brought near some shredded paper, which it attracted marvellously well. All those who had tried so many times in vain to make it attract a year earlier came to see this phenomenon with amazement several times. But on the contrary, high-cut diamonds (as we said at the beginning), that is to say those that are worked in their natural octahedral shape, rarely or never fail.

Finally, so that amber and all other electric substances may not attract, it needs only a very thin barrier put between them and the body that is to be attracted. In fact, when we had made some little windows in a piece of paper, the first made like a shutter with closely spaced hairs, the second veiled with delicate fluff scraped gently from a very fine cloth, and the [CCXXXIII] remaining one closed with a piece of gold leaf, the virtue of the amber did not get through them.[271]

---

[270] Here *potenza*, instead of *virtù*.

[271] These electrical experiments were made chiefly in May, June, and August 1660, especially in the period May 20 to 29, which seems to have been entirely given up to the subject (G. 262, 66r–70v), an unaccustomed procedure for the Academy. This was the first activity recorded in twenty months, and with Magalotti as secretary. Many details are given in *TT* 2, 546–56.

꧁꧂

[CCXXXV]

EXPERIMENTS CONCERNING SOME CHANGES OF COLOR IN VARIOUS LIQUIDS[272]

Nothing is more frequent among the subtleties of the chemists than the extraordinary mutations of colors. In truth we have not professed to meddle in this affair, and if we have done anything about it, this was because we had occasion to use some liquids suitable for examining the quality of natural waters.[273] [CCXXXVI] Concerning this we shall relate what little has come to our notice, again reminding the reader that by this name of *Saggi* is meant that we do not pretend to have examined these subjects with all the experiments that can be invented to deal with them, but simply to give an indication of those things on which we have had the greatest inclination to work.

The First Experiment

Water distilled in lead makes turbid all the waters from rivers, hot springs, fountains, and wells with which we have so far mixed it, taking away their transparency and making them white like whey. Only water distilled in glass, and among natural waters that of the aqueduct of Pisa,[274] remain limpid and transparent.[275] It is true that every water soiled in this way is again made clean by a few drops of strong vinegar, for when it is beaten up with this the obscurity dissipates and clears.[276]

The same waters are changed by the infusion of oil of tartar[277] and of oil of aniseed, which cause a cloud to appear, either higher or lower, which is diffused throughout the water by agitation. This whitening again disappears with a small dose of spirit of sulphur,[278]

---

[272] This is only a small selection from a number "approved by the Academy" and contained in G. 265, 118r–19v; printed in *DIS*, 351–52.

[273] A rare example of an interest in applied science.

[274] This was the aqueduct, begun by the Grand Duke Ferdinand I and finished in 1613 by Cosimo II, that brought water from the springs of Asciano. It still exists in part but is out of use. I have to thank Dr. Carlo Maccagni of the Domus Galilaeana, Pisa, for this information.

[275] Redi, *Esperienze intorno a diverse cose naturali*, etc., pp. 17–18 (bound at end of vol. II in the 1778 Naples ed. of his *Opere*), says that the effect on the Pisa water depends on various conditions.

[276] June 30, 1657 (G. 262, 11r–v; *TT* 2, 556).

[277] A concentrated solution of potassium carbonate. Regarding these names I have consulted M. P. Crosland, *Historical Studies in the Language of Chemistry* (London, 1962).

[278] Dilute sulphuric acid.

which, giving rise at once to bubbling, reduces the water to its first natural transparency.[279]

It is to be noted that all waters are not made turbid indiscriminately, not even by the above-mentioned oils. Indeed oil of tartar and oil of aniseed leave exactly the same waters transparent that are not altered by water distilled in lead. Thus it is that spirit of wine, water distilled in glass, and that of the aqueduct of Pisa do not alter at all or change from their natural [CCXXXVII] limpidity. It is found that in the waters commonly reputed to be lighter, more excellent, and clearer than others, the little cloud produced in them is observed to be smaller and higher up, and only in those that are heavier and full of minerals and impurities does it entirely obscure them and turn them milky. Some people have claimed to test water on this basis with one of the above-mentioned liquids, in order to reveal their most hidden nature and so to disclose their goodness or badness.

If for any reason the water should sometimes be heavily loaded with impurities, so that the ordinary dose of the clearing liquid does not work, it can be increased by a few drops and the water shaken while it is put in. This will be seen to become limpid once more.

The Second Experiment

Oil of tartar produces its effect not only in water but also in wine, since by its natural faculty (as we remarked) of clearing every extraneous impurity from all liquids, it divides by its presence the pure substance from that which is mixed in it. So it happens that what in water is a little white cloud, higher or lower according to its quality and lightness, appears as a thin layer, blood-red in color, in all the white wines that we have tried. When the wine is agitated, this layer loses its natural place of equilibrium and spreads uniformly through it. However, in red wine it makes no change other than to tinge it a deeper color, still darker near the bottom.

On the contrary, spirit of sulphur not only [CCXXXVIII] leaves the natural transparency of wines unchanged but restores to them that which oil of tartar took away.[280]

The Third Experiment

Tincture of red roses extracted with spirit of vitriol[281] becomes a very beautiful green[282] when mixed with oil of tartar. A few drops of spirit of sulphur make it all bubble up into a bright red foam, and

---

[279] June 22, 1657 (G. 262, 4r–v; *TT* 2, 557).
[280] October 19, 1661 (G. 262, 114r; *TT* 2, 558).
[281] Dilute sulphuric acid made by the distillation of ferrous sulphate.
[282] This is an indicator of pH, like litmus.

it finally returns to a rose color without ever losing its scent and can no longer be changed by oil of tartar poured into it.

We have found that the best way of obtaining tincture of roses for this experiment is as follows: Take as many petals from dried red rosebuds as one hand can hold by gentle pressure. Detach them and put them in a glass bottle with one ounce [28 g] of strong spirit of vitriol and shake them with this for a quarter of an hour. By that time the spirit will have acquired the color of the roses, and these will be perfectly macerated.

Half a pound [170 ml] of spring water is then added in three or four installments, shaking the bottle each time until the deep color of the spirit becomes lighter and tints the water. When this has been done, let it stand for an hour, and you will have a brilliant and extremely beautiful tincture of roses. Now in half an ounce of this, ten or twelve drops of oil of tartar, and then the same quantity of spirit of sulphur, serve to produce the effects described.[283]

[CCXXXIX]
The Fourth Experiment

Water charged with saffron, diluted with a little extract of roses, but not enough for it to lose its golden color, is made green by oil of tartar and golden again by spirit of sulphur.

The Fifth Experiment

Water saturated with lily green becomes wine-colored with spirit of sulphur, and regains its own color with oil of tartar.[284]

Lily green is a tincture extracted from the petals of purple lilies[285] which, prepared with a mixture of lime, throws down a very beautiful and vivid green, much sought after by painters of miniatures. It is put to dry in shells like ground gold and silver.

The method of making similar extracts may be seen more fully described in the *L'arte vetraria* of Antonio Neri, printed in Florence in 1612, Book VII, chap. 108, 109, and 110; and also how lakes are extracted from various flowers.

The Sixth Experiment

Lemon juice, spirit of vitriol, and spirit of sulphur change the purple color of *lacca muffa*[286] and that of tincture of violets into bright

---

[283] June 27 and 30, 1657 (G. 262, 8r–v, 11r–v; *TT* 2, 559).

[284] July 2 and 4, 1657 (G. 262, 12r, 13v; *TT* 2, 560).

[285] Possibly *Iris* sp.

[286] According to the *Vocabolario della Crusca* this is a color prepared from *Croton tinctorum*.

red, which oil of tartar again renders purple. Vinegar also makes them reddish, but of a less brilliant color.[287]

꧁꧂

[CCXXXXI]
EXPERIMENTS ON THE MOTION OF SOUND

Sound, a most important phenomenon in the air, keeps such an invariable rate of speed in its movements that the greater or lesser impetus with which the sonorous body produces it cannot alter its velocity. This wonderful property of sound has been referred to by Gassendi,[288] who firmly asserts that all sounds, whether great or small, cover the same distance in the same time. He shows that he has made an experiment with two sounds, one considerably louder [CCXXXXII] than the other, namely a musket shot and the discharge of a piece of artillery. In repeating this experiment with exactly the same results we succeeded in observing some peculiarities about which we did not believe we should be silent, as it might happen that the same idea might not have occurred to everyone, and that if it had, not everyone would have had the facilities for gaining information about it and satisfying himself by experiment.

The First Experiment

This experiment was repeated by us at night with three different kinds of gun: an arquebus, a falconet,[289] and a demi-cannon,[290] situated at a distance of three miles [4.96 km] from the place of observation, from which the flash made by the powder when the piece was fired could easily be seen. Then from this to the arrival of the sound, an equal number of vibrations of the pendulum of the clock was always counted, whether the arquebus or the falconet or the demi-cannon was fired, and this in whatever direction the barrel of the piece was pointed.[291]

We think we should consider here how pleased Gassendi[292] was with the simple analogy adopted from the Stoics to give a life-like representation of how sound is propagated unseen through the air. The Stoics said that just as we see circular ripples made when a

[287] October 29, 1661 (G. 262, 117r; *TT* 2, 560).
[288] Gassendi, *Opera omnia*, etc., 6 vols. (Lyons, 1658–75), I, 418.
[289] A small "four pounder."
[290] A gun of $6\frac{1}{2}$ inches [16.5 cm] bore (O.E.D.).
[291] This experiment was made by Borelli and Viviani on October 10 and 12, 1656 (G. 268, 155r–58v; *DIS*, 449–52), and thus, strictly speaking, does not belong to the Accademia del Cimento.
[292] Gassendi, *Opera omnia*, I, 420–22.

pebble is thrown into still water, these ripples being propagated on and on in ever larger circles until they reach the bank exhausted and die there, or, striking it with force, are reflected back; precisely in this way, they assert, the subtle air around sonorous bodies travels over immense distances in [CCXXXXIII] fine ripples, and meeting our organ of hearing in the form of such waves, and finding it soft and pliant, imprints on it a certain trembling which we call sound. So far the Stoics, without going farther; but to Gassendi such an example appeared to square so wonderfully well with the facts that he really wanted to adapt it to the whole phenomenon and, indeed, to make it sufficient to explain even the particular properties of sound, one of which, as we said, is the unchangeable velocity of its motion. However, he says that this imperturbable constancy of the speed of sound resembles another similar one that is observed in the above-mentioned rippling of water. This, according to him, does not move faster or slower, but approaches the bank with the same speed whether the stone is large or small, or whether it falls into the water only with the energy of its own weight or is flung down with the greatest force. This, it must be said with all respect to that great man, we have found to be untrue, for we have observed in repeated experiments that the larger the rock and the greater the force with which it is thrown into the water, the greater the speed with which the circles reach the bank.[293]

The Second Experiment

Another marvellous thing happens in respect of the movement of sound, as Gassendi[294] himself relates. This is that it is neither retarded by a contrary wind nor made to go faster by a favorable breeze, but always with imperturbable pace covers the same distance in an equal time. We also wished to compare this with experiment, and found it to be quite true, in the following way.[295] [CCXXXXIV]

[293] Discovered on January 18, 1662 (G. 262, 132r; *TT* 2, 562).

[294] *Opera omnia*, I, 418.

[295] It is quite untrue, as was shown half a century later by the Rev. W. Derham, who made careful experiments and observations for three years on Blackheath. See *Phil. Trans.* 26 (1708): 2–35. The effect of wind had begun to be suspected in Italy by that time, Mario Tornaquinci having suggested such experiments in 1706. (See Florence, Bibl. Marucelliana, ms. A.68, fol. 311v–12r and 314v.) But at the Royal Society there seem to have been serious doubts about this as soon as the *Saggi* arrived on March 12, 1668. Of only two experiments in the entire book was it suggested that they might be repeated, and this was one of them. On March 26, "It being mentioned again [the first time had been a week earlier], that the Florentines had affirmed sounds to move equally swift against and with the Winde, it was suggested by the President, that the Experiment might be conveniently enough made,

At a time when the west wind was blowing two guns were discharged, one situated to the east of the point of observation, the other to the west, and at equal distances from it, so that one was favored and the other opposed by the wind. Nevertheless each transmitted its sound to the observers in the same time, measured by the same number of vibrations of the same clock, although the one to the east sounded considerably weaker than the one to the west.[296]

The Third Experiment

On the occasion of the above-mentioned experiments it occurred to one of our Academicians that, apart from the motion of all sounds being equally swift, it might also be uniform. Meditating later, he acquired various notions, not less curious than useful, on the basis of this conjectured truth. But first, to be certain of the existence of such uniformity, the following experiments were made.

At the distance of one of our miles, exactly measured, equal to 3,000 ells of the sort commonly called land ells,[297] we had several shots made, six with an arquebus and six with a saluting gun. In each of these, from seeing the flash to the arrival of the sound we counted about ten whole vibrations of the pendulum of the clock, each of which was half a second. When the same shots were repeated at half a mile, i.e., at half the distance, the clock also gave precisely half the time, counting about five of the same vibrations at each shot.[298] From this it seemed to us that we could be assured of the supposed uniformity.

The consequences, then, that are claimed to result from this uniformity are, among others, that [CCXXXXV] by using the flash and the sound of various firings we shall be able to get the exact distance of places, and especially at sea, of ships, rocks, and islands, where one cannot take up various positions at will, such as would be necessary if we wished to use the ordinary instruments.[299] Also, with a simple

---

between Deal and Dover, and that he would desire the Governor of Deal Castle to take care of it." (Royal Society, *Journal Book* III, p. 195.) I have been unable to trace this further.

[296] Neither this experiment nor the next appears in G. 262, and Targioni Tozzetti does not provide a date, but G. 261, 65v, tells us that it was performed near Leghorn on January 19, 1658. Thus it may not strictly belong to the Academy (see p. 61 above).

[297] The *braccio a terra* was 55.12 cm, or 17/18 the length of the ell we have heard about up to now. See also p. xiii.

[298] This works out at 331 meters per second, almost exactly the accepted value at 0° C. The close agreement is, of course, fortuitous.

[299] The speed of sound varies not only with the relative wind speed but also with temperature.

blow on wood, stone, metal, or other resonant body, we shall be able to deduce the distance of the man who strikes the blow, counting the vibrations from the fall of the tool with which the blow was made, until the stroke is heard. With a favorable wind this will be audible at a distance of some miles. It will also be easy and interesting to know how far away the clouds are and at what distance from the earth thunder is made, measuring the time from when the lightning is seen until we hear the thunder. Then if we desire the distance of places that are not visible from one-another, either because of the roundness of the earth or because of the interposition of mountains or other similar obstacles, we can still easily obtain it by means of two shots, arranging that another shot from the distant point will at once answer ours. Taking half the time from our signal until the arrival of the reply, we shall have precisely half the path of the sound, that is to say the whole distance of the other place, which we were seeking.

By using sound in this way we shall be able to correct the plans of particular places and draw maps of various countries, first taking bearings on towns, castles, and villages in order to locate them neatly in their places; and to do other similar interesting things, perhaps very useful and not to be entirely despised.

In order to determine each unknown distance the time taken by sound to cover a known distance of one mile will serve us as a scale. We have found this time to be five seconds.

[CCXXXXVII]
EXPERIMENTS CONCERNING PROJECTILES

Galileo believed that if a culverin were set level on the top of a tower and shots fired with it point blank, i.e., horizontally, with large or small charges in the gun so that the ball might fall a thousand ells away, or four or six or ten thousand, etc., all these shots will make their journey in equal times, each equal to the time that the ball would take to go from the mouth of the gun to the earth [CCXXXXVIII] if it were simply allowed to fall straight down with no other impulse.[300] This would happen if it were not for the fortuitous resistance of the air, which may somewhat retard the very rapid motion of the shot. We wished to put this opinion to the test of experiment, and it seems to us that it holds up very well. Therefore we

[300] *G.OP.*, VII, 181. *Dialogue Concerning the Two Chief World Systems —Ptolemaic and Copernican*, trans. Stillman Drake (Berkeley and Los Angeles, 1953), p. 155.

shall relate what little we can say that we have certainly seen in this subject.

The First Experiment

On the tower of the old fortress at Leghorn, fifty ells [29 m] high, with a falconet that takes an iron ball weighing 7⅓ pounds [2.48 kg] and 4 pounds [1.35 kg] of fine powder, several shots were made point blank towards the sea with banded shot, and these were seen to hit the water at a distance of about two-thirds of a mile [1.1 km], in a time of four and a half vibrations, the going and coming of each of which signified half a second. The fall of other similar balls from the above-mentioned height of fifty ells was observed and found to take place in four of the same vibrations.[301]

The Second Experiment

With a small culverin taking fourteen-pound [4.75 kg] balls, also of iron, and ten pounds [3.40 kg] of fine powder, the banded shot arrived on the water in five of the above vibrations, and the bare shot in five and a half; and it appeared that they carried farther than the banded ones.

[CCXXXXIX]
The Third Experiment

Concerning projectiles, Galileo writes these exact words:

From a height of one hundred or more cubits fire a gun loaded with a lead bullet, vertically downwards upon a stone pavement; with the same gun shoot against a similar stone from a distance of one or two cubits, and observe which of the two balls is the more flattened. Now if the ball which has come from the greater elevation is found to be the less flattened of the two, this will show that the air has hindered and diminished the speed initially imparted to the bullet by the powder, and that the air will not permit a bullet to acquire so great a speed, no matter from what height it falls. For if the speed impressed upon the ball by the fire does not exceed that acquired by it in falling freely, then its downward blow ought to be greater rather than less.

This experiment I have not performed, but I am of the opinion that a

---

[301] Not in G. 262, but in G. 260, 183r, this experiment is referred to in exactly these terms and dated April 2, 1662. It was also performed at Leghorn in January 1658, by Carlo Rinaldini, according to his letter to Viviani dated February 1, 1658 (G. 283, 39r), no difference of time being observed. This experiment is also described in G. 261, 66v–67r. It ought to be noted that the Genoese nobleman G. B. Baliani had found this result before 1646, when he published at Genoa his *De motu naturali gravium solidorum et liquidorum*. Regarding Baliani's mechanics see Serge Moscovici, *L'expérience du mouvement*, etc. (Paris, 1967).

musket-ball or cannon-shot, falling from a height as great as you please, will not deliver so strong a blow as it would if fired into a wall only a few cubits distant, i.e., at such a short range that the splitting or rending of the air will not be sufficient to rob the shot of that excess of supernatural[302] violence given it by the powder.[303]

We have made this trial with a rifled arquebus, not indeed firing it against a stone in order to observe the flattening of the ball but against an iron breastplate. Now we observed that the shots fired from a lesser height impressed a much deeper dent in this than those [CCL] from a greater, because (as some said, following Galileo's opinion in this) in the longer journey that the ball makes, dividing the air, the supernatural impetus and force given it by the fire is continually abated.[304]

The Fourth Experiment

In confirmation of what Galileo also asserts in several places,[305] that the power impressed on projectiles is not destroyed by a new direction of motion, some of us proposed the following experiment:

A small cannon firing a one-pound [0.34 kg] iron ball was mounted on a six-horse wagon so that it stood perpendicular to the horizontal. Various shots were fired with this, all with the same charge, three scruples [3.5 g] of musket powder. Some of these were fired with the wagon standing still, others while it ran at full gallop over a very uniform plain. In the first the balls fell back around the mouth of the gun. In the second, after the wagon had travelled 74 ells [43 m] between the shot and the return of the ball, this came down only about four ells [2.4 m] behind the gun, and the times of both turned out to be approximately equal.[306]

The Fifth Experiment

The same experiment was made with a large crossbow of the kind that are bent mechanically. The three-ounce [85 g] lead balls fell behind the wagon only six ells [3.5 m] in a course of 78 ells [45 m] (always meaning from discharge to return). [CCLI] Those of ordinary

[302] *Soprannaturale,* i.e., greater than its natural speed of fall.

[303] Galileo, *Discorsi e dimostrazioni matematiche intorno à due nuove scienze,* etc. (Leyden, 1638.) See *G.OP.,* VIII, 279. I have used the translation of H. Crew and A. de Salvio (New York, 1914), p. 256.

[304] December 29, 1661, and January 5, 1662 (G. 262, 127v–28r, 129r; *TT* 2, 566).

[305] In the *Discorsi* and in the *Dialogo.*

[306] In the *Dialogo,* second day (*G.OP.,* VII, 200–1), Galileo suggested a similar experiment, but on a moving ship. He also described an experiment (*ibid.,* p. 194) in which a crossbow was fired from a wagon at a proper elevation to produce a maximum range. The projectile would fall farther away if fired in the direction towards which the wagon was moving.

clay, in 100 ells [58 m], fell behind 17½ [10.2 m]. From this some of us were still firmer in Galileo's opinion that the air takes not a little from the impetus of heavy bodies that cleave it, more notably of those that are lighter.[307]

[CCLIII]
MISCELLANEOUS EXPERIMENTS

Although in our Academy we have always done our best to maintain a continuous course of experimentation upon some subjects, this has not prevented the occasional admission of some particular observations outside of these, suggested by the Academicians from time to time, each according to the requirements of his own studies. Now as this has produced a mass of disconnected experiments having for the most part little or no connection with one-another, a few more items have been selected from these. To give a sample of them, as of the others, we have kept them in this last position in order to complete the book.

[CCLIV]
AN EXPERIMENT FOR FINDING THE ABSOLUTE WEIGHT OF AIR IN RELATION TO THAT OF WATER

A lead ball was taken, completely closed and full of air.[308] Because this did not sink when put in water, it was weighted externally with enough additional lead to make it go to the bottom. The whole assembly was weighed in air with a very accurate balance, and found to be 31,216 grains [about 1.53 kg].

When the same assembly was put in water, hanging from the same balance, its weight was reduced to 4,272 grains,[309] so that the differ-

---

[307] The fourth and fifth experiments are not represented in G. 262; nor does Targioni Tozzetti provide further information.

[308] An experiment on "the weight of air with the metal ball of Dr. Uliva" was tried unsuccessfully on June 21, 1660, and again on June 30 (G. 262, 81v and 84r; *TT* 2, 572). Not enough detail is given for us to be sure that this was the same as the experiment described here, which was made on October 25, 1660 (G. 262, 110v–12v; *TT* 2, 569–71). Borelli claimed to have invented it (*De motionibus naturalibus*, Cap. V, Prop. CXX).

[309] The weights given in the 1667 edition are not consistent, and the attempt to make them so in the 1841 edition was not successful. I have thought it best to reconcile them with those in G. 262, 110v–12r, which are given in pounds, ounces, scruples, and grains (see p. xiii). The weights in the 1667 edition were plainly "cooked," as is clear from the fact that the crushed ball in air was made to weigh less, rather than more, than it weighed before being flattened.

ence, 26,944 grains, was the absolute weight of a volume of water equal to that of the above-mentioned assembly.

The ball was then crushed by compression as far as its thickness would allow,[310] and when it was weighed in air again, with all the lead, it turned out to weight 31,223 grains [which is 7 grains more than the weight of the same ball in air before it was crushed].[311] This was concluded to be the absolute weight of the volume of air, not compressed, that there was in the space occupied in the ball by the volume removed in the crushing.

In this state the whole assembly was put back in water and weighed, giving 12,518 grains, which, subtracted from 31,223 (the weight of the crushed ball [and its attachments][311] in air) yields the remainder, 18,705 grains, the weight of a volume of [CCLV] water equal to that of the assembly after the crushing. Then this weight, 18,705 grains, subtracted from the other weight, 26,944 grains, leaves the remainder, 8,239 grains, which is the weight of a volume of water equal to the volume of the air that weighed 7 grains. Therefore it is concluded that the weight of the kind of air that was weighed by us was in the proportion of 7 to 8,239, i.e., 1 to 1,177, to the weight of an equal amount of water.

When we repeated this experiment at various times, the proportion did not always turn out the same. It is true that the variations were not very great, being of the order of one, two, or three hundreds of grains more or less, which is as much as one can expect in making a comparison between something which never changes in weight, so to speak, and another that is never the same.[312]

[CCLVI]

EXPERIMENTS ON SOME EFFECTS OF HEAT AND COLD

The First Experiment

When two small steel rods of equal weight, one red-hot and the other cold, were put on the assay balance, the cold one seemed heavier

---

[310] In G. 262 this phrase reads "To compress the air inside, the ball was crushed until a notable resistance was observed." The success of the experiment of course depends on the air-tightness of the lead ball after it was crushed.

[311] The clauses in square brackets are not in the 1667 edition, but are in G. 262, fol. 111r. The first one was put in the 1841 edition. Without it the paragraph makes nonsense.

[312] It does not seem to have occurred to them that a grain more or less in the small difference (7 grains) would throw the result widely out. Several other variants were suggested by Borelli (G. 268, 63r–64v, 94r–v, etc.; DIS, 488–94), but they all depend on a small change in a relatively large weight. See also p. 269 below.

than the other; but when a burning coal or a red-hot iron was brought very near it, it at once returned to equilibrium with the hot one. The same thing will occur if the rods are of gold, silver, or any metal whatever. Even holding a burning coal above one of the empty pans raises it, and holding it below makes it fall.[313] But in spite of this, none of us were in a hurry to believe that simple heating, as such, could in any way change the ordinary weight of the metal. In fact some considered that as much as any other cause, the pressure of the air might have its part in this phenomenon.[314]

[CCLVIII]
The Second Experiment

The vessel *AB* (figure III, plate CCXIX), with a neck an ell and a half [88 cm] long and two closed bulbs of equal capacity, was half filled with spirit of wine, and the bulb *A* [*sic*] was then put in a beaker of oil placed on the fire. The spirit began to show its usual rarefaction and to rise in the tube, but when the oil was boiled very strongly, the spirit all passed over little by little into the upper bulb, leaving the lower one, and the lower half of the tube, entirely empty. But for this effect to be produced it is necessary not only to have a vigorous fire but to blow continually on the coals that surround the beaker; and one should be careful to do this through a hole in a plank that serves as a parapet for the man using the bellows, and behind which also stands the observer looking through a sheet of glass.

This is because, when the spirit has all gone into the upper bulb, it explodes it; and sometimes not only the upper one but the lower one as well bursts with such a great downward force that on one occasion, when a brass vessel was being used instead of the glass beaker, it ruptured the bottom of this, blew the bottom out of an iron brazier, though this was made of thick plate, and cracked one of the stones of the floor. Oil and glass were then selected because their transparency showed the progress of this wonderful phenomenon, although wax, pitch, lard, and perhaps all oily substances may produce the same effect.[315]

[313] This was probably misreported.

[314] July 12, 1657 (G. 262, 15v; *TT* 2, 574). The experiment was due to Candido del Buono (*TT* 1, 419). This was acknowledged by Borelli (*De motionibus naturalibus*, Cap. IV, Prop. LXI), who also claims to have explained it to the Academy. However, the explanation given by Borelli in a letter to Paolo del Buono at Vienna, October 10, 1657, is quite incorrect. The letter is in *LIUI*, I, 94–100. It is amusing that Fabroni suggests, in a footnote, another incorrect explanation.

[315] This highly dangerous and no doubt entertaining experiment was performed on July 23, 27, and 28, 1658 (G. 262, 50v–52r; *TT* 2, 574–75).

The Third Experiment

In order to do something in the matter of antiperistasis[316] we filled a leaden vessel with finely powdered ice [CCLIX] and, putting into this a 50-degree thermometer, let it come to a steady state, which was about 13½ degrees.[317] Then we plunged the vessel into a bowl of boiling water, paying attention to whether, at the moment when the ice became surrounded by its contrary, the thermometer gave any sign of increasing cold by going down. But, however often the experiment was repeated, this was never seen to change by a single hair's breadth. Nor was it ever seen to rise when, on the contrary, the vessel was filled with hot water and plunged into ice water; indeed it was then very soon seen to begin to fall, as the quality of the surroundings arrived more quickly through the liquid water than it did through the ice in the first experiment. And it is not that all precautions were not taken so that the air around the thermometer would not receive any alteration from the various surroundings when the lead vessel was immersed in them; for this vessel was fitted into a board which, extending around it in every direction,[318] cut off all communication between the bowl below, in which the vessel was immersed, and the air above; but with all this, we never managed to see anything more than what has been related.

The Fourth Experiment

To throw some light on the question, whether the cooling of a body results from the entry of some kind of special atoms of cold,[319] just as it is believed that it is heated by atoms of fire, we had two equal glass flasks made, with their necks drawn out extremely fine. These were sealed with the flame, and we placed one in ice and the other in hot water, where we let them stand for some [CCLX] time. Then, breaking the neck of each under water, we observed that a superabundance of matter had penetrated the hot one, blowing vigor-

[316] Boyle performed these experiments in exactly the same way and obtained exactly similar results (*New Experiments . . . Touching Cold*, pp. 785–88. *Works*, 1772 ed., II, 682), but as the Academy did them on August 7, 13, and 16, 1658 (G. 262, 55r, 55v, 56v–57r; *TT* 2, 577–78), the two sets were probably entirely independent. Boyle's demolition of antiperistasis is worth reading for its own sake. See also p. 207 above.

[317] The Duke's glassblowers seem to have made these thermometers come quite consistently to 13½° in ice.

[318] Apparently it extended out for 2 *braccia* (1.17 m) in every direction (G. 265, 133r; *DIS*, 363).

[319] Another peripatetic dogma which, rather surprisingly, was supported by Gassendi (*Opera omnia*, I, 401). Borelli, writing to Leopold on August 24, 1658, emphasized the difficulty of finding "something clear, and sufficient to convince either the Gassendists, or the peripatetics" (G. 275, 109v).

ously out of the flask, as was seen by the bubbling in the water as soon as the flask was opened. It seemed to some of us that the same thing should have occurred when the cold one was opened, should the cooling of the air in it have proceeded in the same way as the heating of the other, i.e., by the intrusion or packing in of cold atoms blown by the ice through the invisible passages of the glass. But it turned out quite the other way, for instead of excess matter coming out, it seemed rather to demonstrate an evacuation or loss of something (if it was not really the contraction of that which was there), much water being sucked in to fill the space.[320]

The Fifth Experiment

Vitriol, its spirit removed, remains like a tartar or incrustation the color of a bright fire.[321] This, with long and continuous heating, distils an oil, almost as black as ink and strongly corrosive. When mixed with water in a certain proportion this oil at once produces heat which, increasing notably without raising bubbles or steam, gets to the point where the beaker containing such a mixture can hardly be carried in the hand. The same effect is produced by mixing it with all other liquids except oil and spirit of wine, the first of which does not change at all from its natural state, and the second, even if it does, may be said to do so imperceptibly.[322]

On the contrary, there is the well-known experiment in which niter, dissolved in water, cools it. Sal ammoniac freezes it to the extent that [CCLXI] if, into the water in which it is mixed in the proper quantity is put a vessel of very thin glass containing water previously cooled with ice, the cold that this salt produces in dissolving is sufficient to make the water freeze.[323]

Now if one-third sal ammoniac and two-thirds oil of vitriol are brought together a very strange effect is produced, for while the salt goes on dissolving in it, it smokes and bubbles up furiously, and all the more if it is stirred with a stick, since then the whole mixture rises more easily in froth, to the extent that it has sometimes occupied a

[320] This experiment appears in G. 262 undated, but at some time between September 7, 1658 and May 20, 1660. During this interval no records were kept, though it is by no means certain that no meetings were held. The experiment has been criticized as being nothing but a demonstration of the expansion and contraction of air, and from their tactful parenthesis it would seem that the academicians were well aware of this. A draft in G. 263, 59r–60r, ascribes the experiment to Prince Leopold (see *DIS*, 392).

[321] The word "vitriol" was applied to a number of substances, many, of course, impure. The chemistry of this is therefore elusive.

[322] June 27, 1657 (G. 262, 7v–8r; *TT* 2, 579).

[323] June 17, 1660 (G. 262, 79v–80r; *TT* 2, 581). Some thermometer readings are given.

volume twenty times greater than the two separate volumes of oil and salt together. But with all this fury of smoking and bubbling, not only is no tendency to heat to be found in this mixture but a marvellous cold is engendered in it, by which the glass of the beaker containing it is frosted over, and the spirit of a thermometer put in it falls rapidly, until the salt has dissipated and vanished and it ceases to bubble and the oil returns to its natural state.[324]

We have found such a production of cold every time we have repeated this experiment. It is true that this, and also the bubbling and smoking, is greater or less according as the salt is stronger or the liquid better refined. We have also observed that a few drops of spirit of wine or spirit of vitriol, put in the oil during the greatest fury of its bubbling, stop this and cause the mixture to warm up at once. Putting in oil of tartar increases the heat and revives the smoke and the bubbling; but with the infusion of spirit of sulphur it immediately begins to cool.

It is worthy of reflection that just as oil of vitriol produces heat when mixed with every liquid, [CCXLII] oil and spirit of wine excepted, so also sal ammoniac mixed into every liquid cools all of them more or less, again excepting oil and spirit of wine, with which alone it does not work; and putting oil of vitriol and the above-mentioned salt together there results the wonderful cold boiling that has been described.[325]

[CCLXIII]
EXPERIMENTS MADE TO FIND OUT WHETHER GLASS AND CRYSTAL ARE PENETRABLE BY ODORS AND MOISTURE

The First Experiment, Concerning Odors

Oil of wax, quintessence of sulphur, and extract of horse's urine, which are considered the most acute and powerful odors in existence, do not transpire perceptibly from a sealed glass ampoule, however much they are shaken or however long the vial is heated. That breath of delicate spirit, too, that evaporates when we cut the rind of a bitter lemon, or is seen to spray in tiny drops when this rind is pressed, does not penetrate to give a scent to water closed hermetically in an extremely thin little crystal jar.[326] Similarly, when a partridge was sealed

[324] June 26, 1660 (G. 262, 82v–83r; TT 2, 580). In the margin of fol. 82v there is a notation, "nuova osservaze. del Sigr. Lorenzo Magalotti."

[325] Experiments on cooling by the solution of sal ammoniac in various liquids were made on about a dozen occasions between June 1660 and January 1662.

[326] This observation was made on August 16, 1657 (G. 262, 26r), and October 19, 1661 (G. 262, 114v). The other does not seem datable.

in a thin glass vessel, and hidden in the corner of a room, a gun-dog that was walked around for a time in the vicinity gave no sign of smelling it out.

[CCLXIV]
THE SECOND EXPERIMENT, CONCERNING MOISTURE

A glass bulb filled with perfectly dry powdered salt and sealed with the flame did not gain in weight after ten days at the bottom of a cistern and the same number in an ice-house. When it was broken open the salt came out so very dry that it poured out like powder.[327]

It once happened, indeed, that in the ampoule of salt a very small part of this was found slightly moistened; but this does not indicate penetration, because if it had really occurred it does not seem that it should be greater on one side than on another. But finding this slight moistening in only one place seems to be enough reason to believe that this small amount of humidity is only what the force of the cold could squeeze out of the air that remained in the vessel by way of the usual misting.

[CCLXV]
EXPERIMENTS ON LIGHT AND ITS PHENOMENA

The First Experiment
In the first dialogue of the treatise about the two new sciences, Galileo suggested a very easy way to attempt to find out whether light moves at a measurable speed or with an infinite velocity.[328] This consists in having two of the company train themselves to uncover two lights in turn, in such a way that one is exposed the moment the other is seen; so that when one man exposes his light, that of his companion comes into view at the same time. When such a procedure had been arranged at a short distance, Galileo desired that the same observers should test themselves at a greater distance, to see whether the answers to their exposures and occultations followed the same tenor as they did from nearby, i.e., without observable delay. At a distance of a mile (which means two, as one light has to go and the other to return) we have not been able to find any. Whether it would be possible to succeed in perceiving some [CCLXVI] sensible delay at a greater distance is an experiment that we have not yet been able to make.[329]

---

[327] June 25 and 27, 1657 (G. 262, 6v, 8v–9r; *TT* 2, 584–85).

[328] *Dialogo*, first day (*G.OP.*, VIII, 88; Crew & de Salvio, p. 43).

[329] They, or at any rate Viviani and Magalotti, tried. On July 24, 1663, they organized a series of rockets and explosions of gunpowder at Pistoia and at Florence, a distance of about 33 km. The flashes could not be recog-

The Second Experiment

Light refracted by a lens of crystal, or reflected from a burning mirror, is unable to set fire to spirit of wine, even if this is made opaque with some sort of tincture. Among other inflammable materials, gunpowder goes up in flames at the focus of the lens or the mirror, but a perfumed pastille, white balsam, storax, and incense liquefy but do not catch fire. The whitest paper and Dutch linen, when they are spread out and exposed to the reflection from a large burning mirror, finally burn. Consequently it is not true that light does not inflame white and shining things, as is commonly believed. It is true that these receive the heat with greater difficulty than things of other colors, and perhaps one may not succeed in burning them with a small mirror or a lens.[330]

The Third Experiment

Apart from the gun-flint, there are other bodies in which it appears that there is a greater store of light, since by striking them together or breaking them in the dark they give out flashes. Such are candied sugar, loaf sugar, and powdered rocksalt, which when pounded in a mortar send out light so copiously that one is able to discern the sides of the mortar and the shape of the pestle. We have not yet succeeded in seeing this same appearance by pounding common salt in blocks, alum, or niter, nor by pounding corals, yellow or black [CCLXVII] amber, granite, and marcasite; but rock crystal, agate, and oriental kinds of jasper, struck together or crushed, give an extremely bright light.[331]

[CCLXVIII]

EXPERIMENTS ON THE DIGESTION OF SOME ANIMALS

The force with which the digestion of hens and ducks operates is astonishing. When these birds were fed with balls of solid glass, gutted by us after a few hours and their gizzards opened to the sunlight these appeared as if lined with a shining coat, which, viewed with the mi-

---

nized (G. 262, 155r–60r; *TT* 2, 587–91). But six years earlier, on April 14, 1657, Borelli had written to Prince Leopold suggesting folding up a light beam by means of an array of five mirrors, so that only one observer would be required (G. 275, 65r–67r; *LIUI*, II, 59–62). As he pointed out, this would permit using the sun as a source of light.

[330] August 11 and 14, 1658; July 12, 1660; September 1, 1660; January 3, 1662 (G. 262, 55r–v, 56v, 91r, 109v, 128v; *TT* 2, 592).

[331] December 23, 1661 and January 4, 1662 (G. 262, 127r–v, 128v; *TT* 2, 593).

croscope, was recognized as nothing else but an exceedingly fine and impalpable glass powder.[332]

In some, fed in the same way with balls of glass, but empty ones with fine holes in them, we happened to see some of these balls already pounded and powdered, others only beginning to break up, and full of a certain white substance like curdled milk, which had got through that very small hole. We have observed that those birds grind better than others, that have in their gizzards a greater quantity of small stones.

So it is less surprising that they crush and break up cork, and other harder woods like cypress and beech, and grind down and finally break into the tiniest splinters, olive stones, the hardest pine seeds [CCLXIX], and pistacchio nuts that they were made to swallow with the rinds on. After twenty-four hours we have found pistol balls considerably flattened, and of a number of small hollow boxes made of tin, we found some scratched and out of shape and some broken in pieces.[333]

---

[332] Francesco Redi, *Esperienze intorno a diverse cose naturali*, p. 49, refers to this passage and disagrees, saying that while hollow glass balls would soon be broken up, solid ones would remain whole for days or weeks. (This work is bound with vol. II of the 1778 Naples edition of his *Opere*, but separately paged.)

[333] Nothing in the diary, but in G. 263, 114v–20r, there are papers describing a large number of such experiments, mostly in January and February 1659 (see also *TT* 2, 594–99).

SUMMARY TABLE OF THE MATTERS DEALT WITH IN THIS WORK

Canon Lorenzo Panciatichi will please examine the present work and report whether there is anything in it that is repugnant to the Catholic Faith or public morality. September 18, 1667.

VINCENZIO BARDI
Vicar-General of Florence

These *Examples of Experiments in Natural Philosophy*, examined by me, contain nothing that offends Christian piety or public morality. Therefore I consider that to publish it would be very useful to lovers of truth, and bring great praise to him who has described them with such propriety and elegance.

LORENZO PANCIATICHI, *Canon, of Florence*

To be printed according to the regulations; VINCENZIO BARDI, *Vicar-General of Florence*

October 5, 1667
The Very Reverend Father Sebastian of Pietra Santa, Minorite Observant, Consultant of the Holy Office in Florence, will examine and report whether in the present book entitled *Examples of Experiments*

in Natural Philosophy there is anything repugnant to the Catholic Faith or to public morality.

BROTHER JAMES TOSINI OF CASTIGLION FIORENTINO,
Vicar-General of the Holy Office in Florence

October 7, 1667

I have examined these *Examples of Experiments in Natural Philosophy,* and have found in them nothing repugnant to the Holy Faith or to public morality, and on that account I consider them very worthy of being printed, for the common benefit of first-class minds, to whom they should be very acceptable, as by these means many philosophical truths are lifted from the darkness of opinion to the light of evident proof.

BROTHER SEBASTIAN OF PIETRA SANTA
Consultant to the Holy Office in Florence

[In Latin]   *October 11, 1667*
In view of the above declaration let it be printed at Florence. On the above date, etc.,

BROTHER JAMES TOSINI OF CASTIGLION FIORENTINO,
Vicar-General of the Holy Office in Florence
JOHN FEDERIGHI

Notice to Booksellers Who Will Bind the Work

The title-page and the dedication are single leaves, but take note that although in the text they are all in two's, yet those bearing the figures are without signature letters. Therefore in binding observe the Roman numeral of each page, and the catchwords at the bottom of those pages that precede the figures, for these correspond with the first words of the pages that follow them. Understand, too, that the dots in the following collation substitute for the letters that are missing in the sections, as explained above.

First register
✠ A B . D E . . . I . . M N . Q R S T V X Y Z

Second register
. . . . Ee Ff Gg Hh Ii . . Ll * *2 *3

5.

THE UNPUBLISHED EXPERIMENTS
AND OBSERVATIONS

1. INTRODUCTION

The boundless curiosity of Prince Leopold and his coadjutors, as well as their empirical attitude towards science, resulted in a very large number of experiments. Of these only a fraction appear in the *Saggi*. Of this remainder, found in the surviving eighteenth-century copy of the diary, G. 262, and in other papers that have come down to us, some are of importance, some interesting as illustrations of the preoccupations of their makers, and a large number quite trivial.

The diary contains many variants of the more important experiments that appear in the *Saggi*, and others are scattered through the papers, especially in G. 263. The most interesting of these last will be found in *DIS*, and I have mentioned some of them in the notes to the translation of the *Saggi*. In this chapter I propose to make a rapid survey of the many activities of the Academy that did not find their way into its only publication. These will be dealt with under the heads of astronomy, pneumatics, heat and cold, other physics, and miscellaneous. One might expect a section on instruments, but upon investigation it seems hard to find any contemporary account of instruments not published in the *Saggi* and certainly devised in the Academy.[1] It is likely that the hydrometers that exist in large numbers mainly antedate the Accademia del Cimento, and at any rate they do not seem to be described in the diary. Neither shall I discuss the routine meteorological observations, for they were part of the first network of weather stations in the world, set up by the Grand Duke more than two years before the foundation of the Accademia del Cimento. The surviving observations are to be found in G. 296 to

---

[1] I refer to *instruments* as distinct from *apparatus*, in the sense that an apparatus is devised for a particular experiment, an instrument (e.g., a thermometer) to perform some function of measurement that may be needed in many.

G. 307 inclusive, and those in G. 306 were published by V. Antinori.[2]

Only a selection—a *saggio*, indeed—from the unpublished[3] experimental work of the Academy will be attempted here. A complete treatment would need another book.

2. ASTRONOMY

It has been well said by the editors of *DIS* that the astronomical part of the work of the Academy appears to be a digression, "desired by Prince Leopold for his own satisfaction on account of his correspondence with foreign astronomers[4] and because of the very notable astronomical phenomena that happened in those years, rather than the execution of a pre-conceived program."[5]

Leopold's interest in astronomy dates from before 1640, for in that year there is a correspondence with the Olivetan monk Vincenzio Renieri on astronomical subjects, especially the satellites of Jupiter which Galileo had discovered and named "the Medicean Stars."[6] Besides these satellites, which were almost considered Tuscan territory in those days, the main astronomical interests of Leopold and his Academy were comets, eclipses of the sun and moon, the planet Saturn, and finally the technical matter of evaluating the performance of telescopes, inspired by the rivalry between instrumentmakers.

Giovanni Alfonso Borelli published his *Theoricae mediceorum planetarum* in 1666 at Florence; but it is evident that the work must have been begun much earlier, because when Michelangelo Ricci sent Leopold some observations of Jupiter's satellites made on August 6, 1664, by G. D. Cassini in Rome,[7] Leopold replied,[8] on August 26, that they had been compared with Borelli's calculations and found to agree very well. The following summer, on July 11, 1665, Leopold and several of his courtiers made many observations of the satellites and the times of their eclipses.[9] When Borelli's book was finally printed, Leopold sent it far and wide and was apparently very pleased with it. Writing to Hevelius at Danzig on May 19, 1666, he says that it con-

[2] *Archivo meteorologico centrale italiano, prima pubblicazione* (Florence, 1858).

[3] Except, of course, by Giovanni Targioni Tozzetti to a great extent in his *Notizie*.

[4] Especially Boulliau, Hevelius, and Huygens.

[5] *DIS*, 54.

[6] A letter from Renieri to Leopold from Genoa, June 1, 1640, is in *LIUI*, I, 73–77.

[7] *TT* 2, 747–49. Cassini later became the first director of the Paris Observatory.

[8] *Ibid.*, p. 750. Targioni Tozzetti speculates that this letter may have been dictated by Borelli for the Prince's signature.

[9] *TT* 1, 391.

tains "very curious speculations" and praises its author's "clear and perspicacious understanding."[10]

The interest of the Academy in comets was part of its anti-Aristotelian program, the idea being to demonstrate that their parallax was very small, so that they must be outside the "orb of the moon"; the question of parallax being therefore paramount. During the 1660's there were impressive comets in 1664, 1665, and 1668, and we find Borelli writing to Leopold from Pisa on December 22, 1664, begging him to try to get observations from Paris so that they might be compared with those made in Tuscany.[11] Leopold wrote to Ismael Boulliau on January 16, 1665, sending him the observations from Pisa and asking for his, "so as to be able to judge the distance of this comet."[12] Plainly the "Copernicans" were in the awkward position of having to prove a negative.

On January 30, 1665, Boulliau wrote, enclosing some observations by the French astronomer Adrien Auzout.[13] The theories of the latter[14] were severely criticized by Borelli, who later wrote a small book about the comet under the pseudonym P. A. Mutoli.[15] There followed a very bad-tempered exchange between Auzout and Borelli, which has been published by Targioni Tozzetti,[16] and at least serves to show that Borelli was a formidable antagonist. There seems to have been a good deal of international co-operation in the matter of comets, apparently organized by the Polish nobleman Stanislaus Lubienietz, who published correspondence with more than forty astronomers in his *Theatrum cometicum*, etc.[17]

As far as eclipses are concerned, it is difficult to find much of particular interest, with the exception of a "horizontal" lunar eclipse that occurred on June 16, 1666, at sunset. Borelli had calculated[18] that because of atmospheric refraction, it would be possible to see both the sun and the partly eclipsed moon at the same time, and Leopold

[10] *BM*, Add. ms. 24,214, fol. 79r. For a full appraisal of this work see A. Koyré, *La révolution astronomique. Copernic, Kepler, Borelli* (Paris, 1961), pp. 461–520.

[11] *LIUI* 1, 118–20, from G. 277, 61r–63v. There are many other letters about the comet in G. 277.

[12] The draft is in G. 282, 84r.

[13] G. 277, 94r–96r.

[14] *L'éphémeride du nouveau comète* (Paris, 1665). The dedication is signed "Auzout."

[15] *Del movimento della cometa apparsa nel mese di dicembre 1664* (Pisa, 1665).

[16] *TT* 2, 766–83.

[17] Three parts in 2 vols. (Amsterdam, 1668, 1666, 1668). One of the correspondents was Otto von Guericke, who published his part of the exchange in his *Nova experimenta magdeburgica* (Amsterdam, 1672), pp. 184–97.

[18] Borelli to Leopold, June 1, 1666, G. 277, 280r–81r.

asked various people to observe the phenomenon. A team on the mountainous island of Gorgona in the Tyrrhenian Sea was successful, and estimated the sum of the two refractions at less than a degree.[19]

Another rather similar phenomenon took place in April 1662. On April 10, Borelli wrote to Leopold from Pisa that he had calculated that on April 21 and 22 Venus would be both an evening and a morning star if the horizon was free.[20] He and his students planned to go up a nearby mountain called La Verrucola, and he hoped that the Prince would make arrangements for other observations. Leopold was apparently in Pisa, or at least had arrived there by the 15th, for we find him sending Lorenzo Magalotti one of those characteristic handwritten notes that show how well he had things in hand:

Signor Lorenzo: As you will see from the enclosed note and calculation of Borelli's, we have a chance to see a fine curiosity in the sky that rarely happens in relation to Venus; but it would be necessary to observe in places where the horizon is free where the sun rises and sets. Away from the sea we have to go up into the mountains. Borelli here wants to go to La Verrucola. I have ordered that another observation should be made at Gorgona, and lent a telescope to the commander of the castle there; but I shall not be able to make this observation. If either you, or Antonio Maria del Buono or others there could do so, it would not be a bad thing. Meanwhile I wish you perfect health and every happiness.[21]

Borelli and his students had no luck with the weather.[22] I have not seen the results of any other observations.

Certainly the most interesting astronomical activity of the Accademia del Cimento, and indeed the best organized of all its operations, was the study of the planet Saturn made in the late summer of 1660, a time, incidentally, of great activity in experimental physics.

Saturn had been a very difficult object for the first imperfect telescopes, and Galileo had been surprised by its appearance, or rather its appearances.[23] Sometimes it seemed to have two large satellites in fixed positions, one on each side, and at other times these entirely vanished. Other people, with somewhat better telescopes, thought that they saw "handles" of various sizes and shapes.[24]

[19] *TT* 2, 753–56. I have not found the original.

[20] *LIUI*, II, 330–32. The original is in *G*. 276, 161r–62r.

[21] Florence, Bibl. Laurenziana, *ms. Ashburnham* 1818, fol. 12r.

[22] *LIUI*, II, 332–34, from *G*. 276, 167r–v.

[23] Galileo, *Istoria e dimostrazioni intorno alle macchie solari*, etc. (Rome, 1613). See *G.OP.*, V, 110–11: 237–38.

[24] See *Oeuvres complètes de Christiaan Huygens, publiées par la Société Hollandaise des Sciences*, 22 vols. (The Hague, 1888–1950), XV, 227, 274–75, and especially the plate at the end of vol. XV. I shall not go into detail here.

The problem appealed to the great intellect of Christiaan Huygens. Unable to buy a good enough telescope in Holland or in Paris, he taught himself the art of lens-making, and by 1656, besides discovering a small satellite of Saturn, he had been able to see that instead of having large companions that came and went, or handles, the planet was surrounded by a ring, somewhat inclined to the plane of its orbit. He had worked out precisely how such a ring could produce the changing appearances that had baffled earlier observers. This was such a revolutionary idea that he referred to it only by way of a cypher at the end of a short tract,[25] dated March 5, 1656, describing his discovery of a moon of Saturn. He did not publish his idea about the rings until 1659, after trying it out in 1657 and 1658 on several friends in Paris, notably Ismael Boulliau, Jean Chapelain, and Pierre Petit.[26] When he did have his *Systema Saturnium* printed[27] it was elaborately dedicated to Prince Leopold, who had been praised to him by Boulliau. By way of Nicolas Heinsius and Carlo Dati, a copy was sent to Leopold, who read it and, reasonably enough, suspended judgment and decided that the matter should be investigated by his Academy. Observations were made in the late summer of 1660, and sent to Huygens by Leopold, again via Dati and Heinsius, on 4 October, together with memoranda by Borelli and—rather surprisingly—Magalotti.[28] Using the best telescopes available in Tuscany, including one belonging to the Grand Duke that was said to have been made by Evangelista Torricelli, the Academicians made observations that supported the picture drawn by Huygens. But it became important to explain how Galileo and others had seemed to see what they had reported; and in the late spring of 1660 this was underlined by the appearance of a tract attacking Huygens' observations and theory and maintaining that Saturn really does consist of separate globes. This book,[29] pur-

[25] *Christiani Hugenii de Saturni lunâ observatio nova* (The Hague, 1656). *Oeuvres*, XV, 165–77. This tract is so rare that the Société Hollandaise had to use an edition published in 1725 at Leyden.

[26] See for example Huygens' letter to Boulliau, December 26, 1657 (*Oeuvres*, II, 108–10).

[27] *Cristiani Hugenii . . . Systema Saturnium, sive De causis mirandorum Saturni phaenomenôn, et comite ejus planeta novo* (The Hague, 1659). *Oeuvres*, XV, 209–353, with French translation. The dedication to Leopold is dated July 5, 1659.

[28] See Huygens, *Oeuvres*, III, 158–67. In a letter to Falconieri dated October 26, 1660, Magalotti referred to this as "a rigmarole of mine, sent only in obedience to the Prince's command" (*Lettere familiari*, I, 66).

[29] *Eustachii de Divinis Septempedani Brevis annotatio in systema Saturnium Christiani Eugenii ad Serenissimum Principem Leopoldum Magni Ducis Hetruriae fratrem* (Rome, 1660). *Oeuvres*, XV, 404–37, with French translation.

porting to be the work of the instrumentmaker Eustachio Divini, but really written by the Jesuit Honoré Fabri,[30] reached Huygens in August. Huygens at once wrote a reply,[31] treating "Divini's" work as it deserved.

Borelli's memorandum,[32] written on behalf of the Academy, is a most interesting document. Pointing out that a complete series of observations showing all the various phases of Saturn could not be made in less than eight or nine years, he describes an experiment to determine what a system such as that postulated by Huygens would look like. This was "according to the custom of Your Highness' Academy, which is to find out the truth by means of experimental tests."[33] What was done was to observe a small model of Saturn according to the proportions stated by Huygens (Plate 6). The sphere and the ring were painted white, and the model was set up at the end of a gallery 128 braccia (75 m) long—no doubt in the Pitti Palace—and illuminated by four torches, located so that they were hidden from the eyes of the observer. The results were what Huygens predicted; and they found that when the plane of the ring made only a small angle with the line of vision, and if the model was observed with an imperfect telescope, the parts GF and CD vanished, and the ends B and E looked exactly circular, just as Galileo had seen them in 1610. With a better telescope the ring was clearly seen.

The observations were repeated in daylight from 37 braccia (23 m) with the naked eye, "at which distance Saturn appeared between two stars B, E, separated from their planet, and round."[34] A very small telescope restored the proper shape.

This was a good experiment, but the sequel was even better. Fearing that they might be deceiving themselves, they called on "many people, among them illiterate persons, who had not seen from nearby the arrangement of the machine that they were to observe"[35] and had them make sketches, one at a time, of what they saw from 37 braccia. Almost all drew the disk of Saturn between two round bodies separated from it by a short distance. I think that this may be considered

[30] Fabri (or Fabry) was a rabid Aristotelian who wrote a great deal, often under pseudonyms or under the names of others. See also p. 262 below.

[31] *Christiani Hugenii Zulichemii Brevis assertio systematis Saturnii sui, ad Serenissimum Principem Leopoldum ab Hetruria* (The Hague, 1660). *Oeuvres*, XV, 439–67. To make this more readily available in Italy, an edition was at once printed in Florence at the instigation of Prince Leopold.

[32] Huygens, *Oeuvres*, III, 152–58.

[33] *Ibid.*, p. 152.

[34] *Ibid.*, p. 154.

[35] *Ibid.*

PLATE 6.   Borelli's model of Saturn and its ring (*courtesy of the library of Leyden University*).

an excellent early investigation in experimental psychology. Its excellence was appreciated by Michelangelo Ricci in Rome, to whom it had been sent by Prince Leopold. Writing on August 22, 1660, Ricci says, "I have been greatly delighted by the experiment which showed the ring around the globe made in the shape of Saturn, sometimes in the form of two separate globes, sometimes in its natural shape. It is the most novel and ingenious idea that I ever heard of."[36] He goes on to say that he had told Fabri about it, and that the latter had been very defensive. Eight days later Ricci sent Magalotti a "thick envelope" containing a long apologia by Fabri, which had been written in a few hours.[37]

By this time Leopold was convinced of the truth of Huygens' description of Saturn,[38] and it is pleasant to report that by 1665 even Fabri publicly acknowledged that he had changed his opinion,[39] although he still managed to oppose the heliocentric theory. Meanwhile, the actual structure of the ring was unknown, though Borelli[40] and Magalotti[41] speculated at length. It is interesting that the writer and critic Jean Chapelain was the first to guess that it might be formed of a large number of small solid bodies.[42] Huygens was quite annoyed with himself for not having thought of this;[43] but that this is the real constitution of the ring was proved only in the nineteenth century. The account of the whole affair, magnificently edited in the second, third, and fifteenth volumes of the *Oeuvres complètes*, can be recommended as fascinating reading for anyone with a grasp of the three languages involved.

A little-known activity organized by the Academy was the comparison of various telescopes made by the rival instrumentmakers Campani and Divini. This was done at Rome in the autumn of 1664 by the use of what would now be called visual acuity charts, and Paolo Falconieri acted as "liaison officer" for the Academy. This activity gave rise to a good deal of correspondence which was in the Ginori-Conti archives, but has been sold abroad. It is being published by Silvio Bedini and Maria Luisa Righini Bonelli, so that I merely mention its existence here.

[36] *LIUI*, II, 93–94.
[37] Ricci to Magalotti, from Rome, August 30, 1660. *LIUI*, II, 94–95. The correspondence on the subject continued throughout the autumn of 1660 (*LIUI*, II, *passim*).
[38] *Ibid.*, p. 151.
[39] See Huygens, *Oeuvres*, XV, 401.
[40] *Ibid.*, III, 159–62.
[41] *Ibid.*, III, 162–67.
[42] Chapelain to Huygens, March 4, 1660 (*Oeuvres*, III, 34–37).
[43] Huygens to Chapelain, April 28, 1660 (*Oeuvres*, III, 76).

3. PNEUMATICS

No branch of physics developed more rapidly during the period in which we are interested than that dealing with the properties of air. The barometer had been made possible by experiments in the 1640's, and the study and extension of these experiments spread quickly to France, and then to England, in the years before 1657.[44] The interest in these experiments was first of all in the demonstration that a vacuum could actually be produced, and in the defense of the proposition that it really was a vacuum. We have seen that the first experiments described in the *Saggi* were designed for this purpose. These were made mainly in the last half of 1657. At this time, or at any rate before the end of 1659, Robert Boyle in England was making his beautiful pneumatic experiments with the air pump that had been invented by Otto von Guericke and improved by Boyle's assistant Robert Hooke.[45] There is no doubt whatever that the *New Experiments* was eagerly read in Florence,[46] for some of Boyle's pneumatic experiments are referred to in the *Saggi*.[47] We can get an idea of when it arrived from a letter written to Boyle by Henry Oldenburg, the secretary of the new Royal Society, and dated merely "October 1661,"[48] which begins: "I cannot yet obey you in delivering y^r letter for Sir John Finch, nor ye book to be sent Prince Leopold, because ye one is not yet returned to London, nor ye other come into my hands." In a postscript we find out what this book was: "When I was going to seall this, a porter brought me from Warwick lane two Copies of y^r Latin pneumaticall book, whereof I shall immediately deliver one to Mr. Herringman, to have it Princely bound up, the other I know not, whom it is for seing you left one, with me, very handsome bound up for Sr Viviani."[49] Thus, assuming that either Viviani's or Leopold's copy arrived safely, it would probably have been early in 1662. The second edition (1662), which contained "Boyle's law," cannot have reached Tuscany until 1664, for on February 4 of that year Borelli

---

[44] For details, see Middleton, *The History of the Barometer* (Baltimore, 1964), chapter 3.

[45] Boyle, *New Experiments Physico-Mechanicall touching the Spring of the Air, and its Effects*, etc. (Oxford, 1660). In *The Works of the Honourable Robert Boyle . . . A New Edition. Edited by* Thos. Birch, 6 vols. (London 1772), I, 1–117. The 1660 edition is dated on p. 399, "Becons-field the 20th of December, 1659" [Fulton, no. 13].

[46] They evidently had the Latin edition. *Nova experimenta physico-mechanica*, etc. (Oxford, 1661) [Fulton, no. 19].

[47] See pp. 157, 159, and 162 above.

[48] *The Correspondence of Henry Oldenburg*, ed. and trans. A. Rupert Hall and Marie Boas Hall (Madison and Milwaukee, 1965–[in progress]), I, 440–42.

[49] *Ibid.*, p. 440.

wrote to the librarian Antonio Magliabecchi asking him to procure a copy "as soon as they arrive at Venice."[50]

Nearly all the experiments on the vacuum were done in June, July, or August 1660, but the ones that they frankly admit they copied from Boyle, the ebullition of tepid water and the solution of pearls and coral in the vacuum, were done on August 7 and 8, 1662,[51] and it must be remembered that in 1662 there were no recorded meetings between January 18 and July 27. Neither Draft "A" nor Draft "B" mentions Boyle, and these experiments are not in Draft "A" at all, which had been finished by this time, as I have shown.[52] There are two more references to Boyle's pneumatics in the *Saggi*, one to the transmission of sound, the other to his experiments with animals under the receiver of the air pump.[53]

If then they had the *New Experiments* by the summer of 1662, why did they not have an air pump made, especially as it is clear from the *Saggi* that they realized the advantages of such a device? I find this extremely difficult to understand.

It can scarcely be maintained that they had no workmen who could have made one, for "Monsu Filippo," i.e., Philip Treffler of Augsburg, was the Grand Duke's turner and clockmaker. His presence is directly referred to in 1656,[54] 1657,[55] 1660,[56] and 1665,[57] although by November 21, 1665, we find Viviani writing to him at Augsburg.[58] Targioni Tozzetti also mentions another turner, Teodoro Sengher.[59] There were also the rival telescope makers, Giuseppe Campani and Eustachio Divini, who competed fiercely for the favor of the Grand Duke. After seeing Boyle's illustrated description, any of these four could surely have made an air pump.

The problem is made even more difficult by the existence of a document in the handwriting of an amanuensis, undated, but containing the following passage:

These gentlemen have decided that it would be well to carry out many proposed ideas for various experiments to be made in the vacuum and to verify all those that Mr. Boyle has made in his instrument. For this two

---

[50] *FBNC*, Cod. Magl. VIII, vol. 518, fol. 35.

[51] G. 262, 136r–v; *TT* 2, 441.

[52] See p. 69 above.

[53] *Saggi*, 1667 ed., p. LXXXXVII and p. CXVI.

[54] G. 268, 158r.

[55] Florence, Bibl. Riccardiana, Cod. 2487. fol. 14r (Magalotti to Viviani, March 20, 1657).

[56] *TT* 1, 426.

[57] *BM*, Add. ms. 24,215, fol. 61r–62v, (Viviani to [?], 2 Feb. 1665).

[58] G. 252, 103r–6r.

[59] *TT* 1, 381.

vessels are required: a large one for making the vacuum with water, with a copper tube, silver-soldered, and a small glass one for making it with quicksilver. It remains for His Highness to give orders for both and to delegate someone to superintend their manufacture.[60]

It would seem that they deliberately decided not to use an air pump, but to continue using the Torricellian vacuum. Perhaps—and this is no more than a guess—they were not sure that Boyle's vacuum would show the same effects as theirs; an explanation that is supported by the passage on page CXVI of the *Saggi*, in which they correctly account for the difference in the behavior of birds in their vacuum and in Boyle's.

I cannot find any record of experiments made in the large apparatus using water, or any indication that it was ever built.

Let us look at a few of the unpublished pneumatic experiments that are to be found among the papers, and first of all at one which is in Draft "A," and so at one time was thought worthy of publication. It is known that this experiment was suggested by Marsili,[61] and his motive is at once apparent in its title, as given in Draft "A," from which the following is taken.[62]

An experiment to determine whether the vacuum apparently left in the usual vessels of quicksilver is filled by the evaporation of very thin humors such as might rise from the mercury itself, as some believe.

Around the lip of the glass tube *AB* (fig. 5.1) is tied the mouth of a little bladder such as *AC*, carefully taken from the inside of a fish. This is introduced into the vessel *DF* through the mouth *EF*, and the space left around the opening closed perfectly with hot cement. Then the small tubes that branch off the tube and the vessel are carefully connected by the glassblower where they meet at *G, H*. Next the bladder is filled with quicksilver through the mouth *B*, shaking the vessel delicately so that any small quantity of air that may remain in the folds where the bladder is tied may escape. When the bladder and the short length of tube *FI* are full, the mercury begins to flow through the two tubes *IH, IG* into the vessel *DF*, filling it, together with the rest of the tube *BI*. Finally, closing the orifice *B* with a finger, and plunging it into the mercury *KL*, the vessel is inverted, and, together with the bladder, will be emptied of mercury, the latter directly through the tube *AB* and the former through the tubes *G, H*, down to *M* at the usual level of an ell and a quarter. Therefore, if from the cylinder of mercury that is sustained anything mercurial should evaporate, such as insensible breaths of very tenuous exhalations, the bladder *CA* should be seen to inflate (without any tiny amount of air remaining there, which

[60] G. 269, 83v. Reproduced in *DIS*, 434.

[61] G. 262, 104v.

[62] G. 266, 79r–80r; *DIS*, 305–6. There is a rough draft in Magalotti's writing in G. 263, 205r–v, with a sketch on fol. 204r.

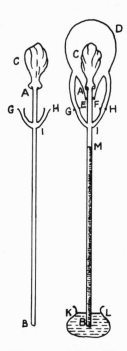

FIGURE 5.1   Diagram of Marsili's apparatus.

would then dilate in the empty space), it could be adduced as a cause of such inflation, since the bladder was at first full of mercury. Nevertheless it remained deflated, just as if it had been in the middle of the exhausted air.

This was of course a very badly designed experiment, as was realized by someone or other; for there exists a note in Viviani's writing in which we are told that:

In his [whose?] opinion it would be necessary that they [the exhalations] should not be able to continue to enter the vessel that surrounds it [the bladder], and that to avoid this either the "handles" could be connected by bladders at G, H, and after the vacuum has been made the passage could be closed by tying them tightly; or [the experiment] made in this other way, perhaps easier, etc.[63]

I do not know what this other way could have been, but this criticism is exactly what would first occur to anyone today, and it is surprising that Borelli, in commenting on Draft "A," did not see this point,[64] but confined himself to the thought that such tenuous exhalations would undoubtedly be able to pass through the bladder.

[63] G. 267, 44v; DIS, 348.
[64] G. 267, 20r; DIS, 332.

But apart from the design of the experiment, it is one that it would never have occurred to anyone to try, if the appropriate concepts, in this case vapor pressure and molecular motion, had been at hand. We know that the space indeed contained mercury vapor, but in an amount that could not inflate the bladder, be it ever so delicate and flexible.

It seems to have been very important to some of the Academicians to refute the claim of the peripatetics that the apparently empty space was full of these exhalations. Early in 1659 Leopold must have asked them to think about it, for there exists a long letter from Magalotti on the subject, dated April 24, 1659,[65] which begins:

In carrying out Your Highness' orders, it seems to me that we should remember that I did not put forward the opinion of the evaporation of quicksilver as being peculiar to any person, for it is the most universal subterfuge of all those who deny the vacuum, to have recourse to these exhalations of the mercury, violently extracted from its bulk in some way, to fill the space left open by its fall.

He goes on to remind the Prince about an experiment made in Florence "two years ago" in which a vacuum was made many times with the same mercury in such a way that if any exhalations had been produced they would have escaped after each experiment; but no diminution of the volume of mercury was observed. He admits, though, that it was done only on one morning, "having been done more as a trial, and to establish the method, than to perform the experiment,"[66] a state of affairs that was all too common in the Florentine experiments, quite understandably in view of the number of things that presented themselves to be tried. I have not identified this experiment in the records, and of course if Magalotti's estimate of two years was correct, it would have been before systematic minutes began on June 19, 1657.

In the same letter, Magalotti suggests another observation which, he believes, ought to dispose of the "exhalations." When a Torricellian tube and its cistern are moved, the mercury librates; but no matter how many times it is allowed to do so, the height at which the mercury is sustained by the pressure of the air does not change, although presumably fresh mercury comes to the surface. In reality the problem was, as we know, an imaginary one.

Galileo had been quite sure about the bubbles that rise through *boiling* water. This was as far back as 1612. They were not "vapors generated from any part of the water,"[67] for if you boiled water in a

[65] G. 275, 147r–51r.

[66] *Ibid.,* 149v.

[67] In *Considerazioni intorno al discorso . . . di Lodovico delle Colombe* (*G.OP.,* IV, 635).

long-necked flask, you saw thousands escape, and yet when the flask was cooled down, the volume of the water was not changed. "And if you performed such an operation a thousand times, you would see millions of such *little spheres of fire* pass through the water, without the water ever decreasing by a hair's breadth,"[68]—surely a fine example of Galileo's independence of real experiment. But we see that he thought that the bubbles were full of fire.

The bubbles that form in water, even in cold water, that is placed in a vacuum were a great puzzle to the experimenters, as they were to everyone else in the seventeenth century; for they could scarcely be full of fire. It seemed impossible that a little water could hold so much "air." This phenomenon had been observed with amazement the very first time a water barometer (as we should now call it) had been set up by Gasparo Berti in about the year 1640;[69] and we know that at least two accounts of that experiment were available to the Academicians. At any rate, after making many experiments on July 1 and 3, 1660, with water floating on the mercury in the Torricellian tube, and arriving at no conclusions, they arranged to have a tube 20 *braccia* (11.7 m) long set up, with the first 16 *braccia* of tin and the rest of glass.[70] I have not discovered the results of this experiment.

If the concept of vapor pressure was far in the future, the existence of water vapor was known to, or rather believed in by, the Academicians. Indeed they were obliged to postulate it in order to overthrow one more Aristotelian shibboleth, the interconversion of air and water. Their experiment, done on September 14, 1657, was very ingenious:

The water that collects in a glass vessel full of air when it is put in ice is not congealed into water, but vapors, distinct from the air. This was proved, because after the water had collected, it was made to evaporate by heat, so that the air was left extremely dry; then putting some gunpowder into the vessel, this was sealed hermetically and returned to the ice; but the air did not become moist, not even enough to make it attack the gunpowder in the concavity of the vessel. The contrary would have occurred if the water that collects were air, condensed by the power of cold.[71]

---

[68] *Ibid.* (my italics).

[69] See my *History of the Barometer*, pp. 10–18.

[70] G. 262, 84r–85r; 86r–87r; *TT* 2, 437–39. A water barometer had been set up at Artimino and then at Florence in September 1657 (G. 262, 35r–v; *TT* 2, 411–12), but the intention had been to see the effect of the difference in elevation of the two places.

[71] G. 262, 33v–34r; *TT* 2, 478. In the eighteenth century the standard test for highly rectified spirit of wine was to put a few grains of powder into a spoon, cover them with the spirit, and set it alight. If the powder finally burned, the spirit was judged to be free from water.

The Academicians tried very hard to find an acceptable substitute for the crude method of measuring the specific gravity of air described on pages CCLIV–CCLV of the *Saggi*, in order to circumvent the problem of measuring a small difference between two large weights. Borelli devised several pieces of apparatus for this purpose, all on the general principle of the ordinary hydrometer, the best of which is probably the one illustrated in fig. 5.2.[72] First of all, a copper wire was to be calibrated in units of the volume of one grain [45 milligrams] of water by attaching it to the top of the vial *AB* of the left-hand figure, ballasted with lead shot so that it submerged to the point *F*. Then by adding ring-shaped one-grain weights at *D* the wire would be submerged a little at a time, and the "grains" could be marked off by painting every other space white. (It does not seem to have struck him that the paint would increase the volume of the wire.)

Next a larger vessel (*ABCD* in the right-hand figure) was to be prepared, the neck of which was longer than the barometric height, 1¼ ells [73 cm]. This would be filled with mercury and inverted into more mercury so that a Torricellian vacuum would be formed in the vol-

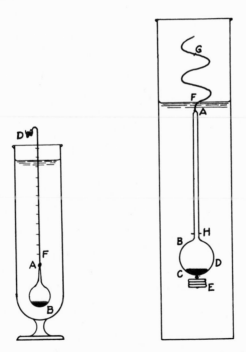

FIGURE 5.2 Illustrating one of Borelli's methods of weighing air.

72 G. 268, 100r–101v.

ume *HBCD*, and while the mouth *A* was under the mercury it would be closed with a little piece of bladder, firmly tied on.

This vessel was then to be inverted as shown in the right-hand figure, and the calibrated wire fastened on and coiled into a helix to save space. The vessel was to be ballasted with some lead plates *E* so that it would come to equilibrium in water at the level *F*. A thermometer would indicate the temperature of the water. Next the apparatus was to be taken out of the water and dried, and a small hole made in the bladder to let the air in, this then being sealed with a tiny piece of wax before the apparatus was again put in the water, where it would submerge to the level *G*, the water being, if necessary, restored to the temperature that it had before. The interval *FG* would thus indicate the number of grains of air that had come into the space *HBCD*, and two weighings of the vessel *ABCD*, one empty and the other with water up to *H*, would give the weight of a corresponding volume of water.

I have found no indication that this experiment, well designed except for the painting of the wire, was ever performed by the Academy.

To end this section I shall refer to an experiment made on November 25, 27, and 28, 1661, in which the Academicians tried to find out to what extent air is compressed by a given head of water.[73] The air was confined in the innermost of two concentric glass tubes. Unfortunately they measured only the change in volume instead of the volume remaining, so that they could not derive a law. It is to be regretted that a long letter from Paolo del Buono, a copy of which still exists,[74] never reached Florence, for he had shown the proper way to do this experiment, though he had no one to make the necessary apparatus. I think that it can be argued that the almost unlimited availability of beautiful glassware sometimes led the Academicians to make experiments before they had thought them out properly.

4. EXPERIMENTS ON HEAT AND COLD

In the second half of the twentieth century, when the astonishing achievements of technology have made *nihil admirari* a way of life instead of the hedonistic maxim intended by the ancients, it comes as a slight shock to read in the *Saggi* of the delight of the experimenters in the phenomena of freezing and the various demonstrations of the thermal expansion of solids. Experiments on heat and cold account for 90 pages in the 269 pages of text and plates in the 1667 edition, and

---

[73] G. 262, 121r–v, 122r, 123v; *TT* 2, 630–31.

[74] *BN*, fonds fr. 13039, fol. 126r–40v. I have published this in *Arch. Hist. Exact Sci.* 6 (1969): 1–28.

yet there are records of experiments in this field over a score of different subjects that are not represented in the *Saggi*. I shall make a
arbitrary selection of half a dozen.

In the first week of their meetings they tested an assertion which
they had heard, namely, that water shaken in a vessel became warmer.[75]
They found that it became cooler, but with our hindsight we can
recognize that such an experiment would be difficult to control. Five
weeks later, on July 27, they observed a jar of water left for three days
in a closed room with a thermometer, in order to see whether it took
up the temperature of the air, which may seem to be a trivial inquiry
until we remember that it was a scholastic dogma that water is naturally cold. In the hot summer weather there must have been a good
deal of evaporation, for they found the water several degrees colder
than the air. This is recorded in the diary without comment.[76]

A more elaborate experiment, indeed a very ingenious one, was
made on September 11, 1657, with the idea of measuring the thermal
dilatation of air between, presumably, room temperature and that of
boiling water.[77] A metal vessel was made in the form of a huge hydrometer, weighted so that it would float with most of its long thin
neck submerged. At the end of the neck there was a screwed stopper
shaped like a cup to hold lead shot. With the apparatus floating in
water, enough shot was added to submerge it up to a mark at the top
of the tube. The stopper was opened and the instrument left for some
time in boiling water, the stopper was then closed and the whole thing
brought back to its former temperature, when it was found to float
higher because the air within was now more rarefied. The difference
in weight could then be found by adding more shot until it floated at
its original level. This is all the information that we are given, no
numbers being vouchsafed us.[78]

On the other hand we have a series of numbers from an experiment
made on August 12, 1662,[79] about the rate of cooling of a large 400-
degree thermometer placed in a freezing mixture and timed with a
pendulum. The Academicians timed it at each 10-degree mark and
calculated differences, but then decided that there was "no rationality

[75] This was on June 22, 1657 (G. 262, 4v–5r; *TT* 2, 624).

[76] G. 262, 19v. A copy of part of the diary in a secretarial hand (G. 260, 5r)
did not specify how long it had been left, but Magalotti has added, over a
caret, "the day before." Long enough, in all probability.

[77] G. 262, 32v–33r; *TT* 2, 705.

[78] In the account given in G. 261, 47r, we are told that 23½ grains of shot
had to be added.

[79] G. 262, 139v–40r; *TT* 2, 628. Both these sources contain an obvious
mistake in copying which can be corrected from the original "laboratory
notes" in G. 260, 196v.

in this progress," blaming the poor contact between the ice and the bulb.

Now as a matter of fact they had an excellent series of numbers, even though they did not know it. If a thermometer at a temperature θ_o is suddenly brought into an environment of temperature θ_e, its temperature θ at any later time t will be given by the equation

$$\theta - \theta_e = (\theta_o - \theta_e)\, e^{-t/\lambda}$$

where e is the base of the natural logarithms and λ is a quantity known as the lag coefficient, having the dimensions of time and depending on the properties of the thermometer and the medium, but constant in any single experiment of this kind.[80]

If we plot $\theta - \theta_e$ against t on semi-logarithmic paper we shall get a straight line. Unfortunately the experimenters did not give us θ_e in terms of their thermometer; but we need not be deterred by this, for we have only to assume various values of θ_e until we find the one that gives a straight line. Figure 5.3 shows a plot of these observations on the assumption that $\theta_o - \theta_e$ is 110 degrees, and it will be seen that the scatter is very small. The lag coefficient works out as about 150 seconds.

I have gone through this argument to make the point that this experiment, like so many others made in the Academy, was completely futile simply because it was completely empirical.

Almost at the end of its period of activity the Academy became interested in what we should now call calorimetry. On January 5, 1667, the experimenters wished to determine "the degree of heat that keeps

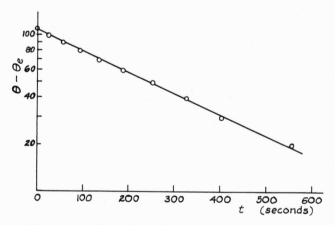

FIGURE 5.3   An experiment in cooling, plotted from the data recorded by the Academy.

[80] See Middleton and Spilhaus, *Meteorological Instruments* (3rd ed.; Toronto, 1953), p. 63.

tin melted;"[81] so they fused 2 lbs [0.68 kg] of it, put it in 3 lbs [1.02 kg] of water, and measured the rise in temperature with one of their "100-degree" thermometers. This was also quite unfruitful, because they could not know that they ought to have distinguished between the heat of fusion and the heat that merely raised the temperature of the tin. On January 28 they made a similar experiment with 2 lbs of iron, heated until it would ignite sulphur.[82] This time they remembered an experiment that they had made on January 4 at the suggestion of Rinaldini, and again on January 26, in which a jar of strong nitric acid was put into a larger jar of water containing a thermometer.[83] They added half a pound of tin filings, and when the reaction was completed they noted the rise in the temperature of the outer jar. Taking the inner jar out, they found that sulphur touched to it was not ignited. Now after doing the experiment with the iron they noted that the solution of the tin filings had apparently produced more heat, as measured with the thermometer, than there had been in the hot iron, although the weight of this had been made equal to the combined weights of the tin and the nitric acid, "so that it seemed possible to conclude that this heat, although more powerful, was not of the same nature as that taken up by the iron held under the lighted coals."[84] The development of the various ideas associated with these experiments was to take nearly two centuries. But even in matters of thermal physics the Academicians could sometimes ask very acute questions, and even guess at plausible answers. As an interesting example I quote a jotting by Viviani, "I am looking for the reason that any instrument [sc. thermometer] that continues to stand in the sun, when it has risen a certain amount, stays there for a while."[85] And he added in pencil, "because it is getting just as much heat as is escaping." I do not intend to imply that this is a foreshadowing of the law of the conservation of energy; but it shows an intuition of some principle of conservation. Viviani also suggested the use of the volume of a mass of air at constant pressure as a measure of the amount of heat in a medium,[86] but this was a vague suggestion that had to wait until the end of the century for Guillaume Amontons to develop it.[87]

[81] G. 262, 167v; TT 2, 670–71.
[82] G. 262, 175v–76r.
[83] G. 262, 165v–66v, 175r; TT 2, 621–22.
[84] G. 262, 175v; TT 2, 620.
[85] Quoted from G. 269, fol. 84, in DIS, 436. The work of Viviani has not had the study it deserves, and I am sure that this is mainly due to the atrocious handwriting in his private notes. There are 104 volumes of his letters and papers in the "Galileo" ms.
[86] G. 269, 99r.
[87] Amontons, Mém. Acad. R. Sci. Paris (1699), pp. 112–26.

5. OTHER PHYSICAL EXPERIMENTS

The diary, G. 262, contains many unpublished experiments on subjects that we should nowadays put under the heading of "general physics." To bring these into a reasonable compass I shall confine my attention to those in which the Academicians took the greatest interest, using as a criterion the number of days on which they were attempted, and leaving out of account those tried on less than three occasions.

The very first experiment registered in the diary concerned the specific gravity of a sample of water into which ashes had been stirred, compared to that of pure water from the same source. They thought they detected a slight increase, but were not sure. Then they dissolved as much salt as they could, and found an increase of 6½ drams per ounce,[88] which is about 27 per cent.[89] But they left it until the next day, when it had become clearer, and it then weighed only 3 drams, 13 grains "more than its natural weight of an ounce."[90] They must have had some salt in suspension on the first day; or the temperature may have fallen.

Altogether they made experiments of this sort on eleven days in June and July 1657, trying various things, such as dissolving a second and even a third salt in a solution of common salt, measuring the specific gravities of solutions of various amounts, and so forth. Of more interest was an experiment made on June 18, 1660, "to find out as exactly as possible whether salts occupy space in water, or adapt themselves to the empty spaces left in the corners where the water-particles touch."[91] This was a particularly well-designed experiment, and deserves description. A vessel such as is shown in fig. 5.4 was made, the mouth B having a wax plug with a flat lower end. This plug could be sealed on with mastic. The vessel was filled with water and the plug put in, raising the level to C in the side tube. After this level had been noted, the plug was removed, a quantity of salt put into the water, and the plug pushed in again and sealed, raising the level to D. To prevent evaporation they then sealed the side tube hermetically at E. Then,

We let it stand for twenty-four hours, so that in dissolving, the salt might have room to adapt itself to the little empty spaces assumed by Gassendi,[92]

---

[88] See p. xiii.

[89] G. 262, 2r; *TT* 2, 638.

[90] G. 262, 2v; *TT* 2, 638.

[91] G. 262, 80r–v; *TT* 2, 644–45. A similar but less careful experiment was made on June 28, 1657 (G. 262, 10r).

[92] *Opera omnia*, etc., 6 vols. (Lyons, 1658–75), I, 212.

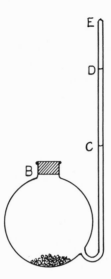

FIGURE 5.4 An experiment on the solution of salt in water.

letting the water return to its first level C; but even after six or seven days it did not go down a hair's breadth below the height to which it had risen immediately the salt was put in.[93]

They recognized that such a vessel might act as a thermometer and kept a close watch on the ambient temperature, which stayed within three or four degrees [centigrade] of its original value during the entire experiment.

The observations of the Academy on the flow of liquids, though numerous, are of no particular importance. Rather naturally, they measured the rate at which a tall vessel was emptied through a hole near the bottom, and we are told that the two tables of readings obtained on July 23, 1657, were by Prince Leopold and Antonio Uliva.[94] First differences were calculated, but no attempt seems to have been made to check the results against Torricelli's well-known rule that the amount of water coming out in unit time is proportional to the square root of the depth.[95] Indeed, numerical comparison of observation and theory was uncommon in 1657, except in astronomy, and the Academy were no innovators in this respect.

On seven occasions in 1657 and 1661 they recorded numerous observations of specific gravity. Some of these were done with a little

[93] G. 262, 80v; *TT* 2, 645.
[94] G. 262, 19r–v; *TT* 2, 661–62.
[95] Torricelli, "De motu aquarum," in *Opere di Evangelista Torricelli*, G. Loria & G. Vassura, eds., 4 vols. in 5 (Faenza, 1919), II, 193.

cage containing hollow glass balls of various specific gravities, called a *gabbiolina*[96] and intended for weighing liquids that are lighter than water; others with a "one ounce ball," weighted to float upright and used by adding small ring-shaped weights until it just submerged; still others with a hydrometer very much like those commonly used today, which float higher in heavier liquids. We are told[97] that these last were calibrated in grains per pound, based on some particular liquid as standard. It is clear that all these instruments antedated the Accademia del Cimento; their invention is ascribed to Grand Duke Ferdinand II.[98] It is difficult to see any system in these observations.

Concerning the experiments with pendulums, these were mainly devoted to the establishment of the fact that for a given pendulum large vibrations take longer than small ones.[99] The experimenters also tried ivory and metal balls of various sizes rolling in semicircular channels; but their results were not very consistent.[100] One interesting experiment was performed on October 4, 1657, in which a small basin filled with water was hung from four cords that were united into one, and set swinging without spilling a drop and with no apparent motion of the water relative to the basin. But when the four cords were attached to separate hooks so that they were parallel, and the basin remained horizontal in swinging to and fro, the water spilled out even though the motion was only small.[101] No discussion of this excellent experiment is recorded.

Besides the experiments on capillarity made in the vacuum and published in the *Saggi*, the Academicians also made ordinary experiments on the rise of liquids in narrow tubes.[102] In the most of these the idea was to investigate a possible relation between the specific gravity of a liquid and its capillary rise. Apart from spirit and vinegar, their samples were mostly wine from various sources. Their entire lack of success will not surprise us; what may astonish us is that although they knew that liquids rose higher in the narrower branch of a U-tube with legs of different diameters,[103] they do not seem to have measured the bores of these tubes.

It is remarkable how little interest the Academy seemed to take in optics. Apart from their abortive attempt to measure the speed of

[96] G. 269, 230v.
[97] G. 269, 227r.
[98] TT 1, 149 and 153.
[99] See also p. 103 above, and appendix E.
[100] G. 262, 127v, 128v, 129r–v; TT 2, 669–70. These experiments were done in December 1661 and January 1662.
[101] G. 262, 36r; TT 2, 669.
[102] G. 262, 24v, 49v–50r; 71r, 71v–72r, 73r, 120r, 123v; TT 2, 652, 657–61.
[103] G. 262, 49v–50r; TT 2, 652.

light, and the experiment on reflection from a glass surface *in vacuo,* both published in the *Saggi,*[104] I can find only two real optical experiments in the diaries. On August 25, 1657, they looked at the flame of a candle through 10 ells [5.84 m] of water contained in a wooden tube with windows.[105] The object was to determine how far light goes down into water. They intended to increase the length of the tube until nothing could be seen, but the project seems not to have been carried out. On June 10, 1660, they tried backing a glass plate with various substances and observed the effect on the "second reflection" (i.e., from the back of the glass) of a candle flame.[106] Besides this there are a diagram and a sketch by Viviani showing a box in the form of a flat cylinder with glass ends for the investigation of the refraction of light by the surface of a liquid;[107] but it seems that this experiment was performed before the Grand Duke and not in the Academy of Prince Leopold.

I shall end this section by mentioning some experiments suggested by Rinaldini on the bouncing of balls[108] and done in the last winter of the Academy's activities. Galileo had clearly stated that a falling body acquires enough "impetus" to carry it back to the height from which it has fallen.[109] This is the principle on which the motion of the pendulum depends. It had of course been noted that balls did not bounce quite up to the place from which they had been released, and the peripatetics claimed that this was due to the repugnance of a heavy body to motion upwards. Rinaldini also desired to find out whether bodies of different specific gravity, and thus affected differently by the resistance of the air, would bounce to different relative heights.

So they set up a vertical scale, rising from a block of marble, and began with a ball of buffalo horn, chosen because of its dryness and hardness, finding that it bounced to between 75 and 80 per cent of the height from which it was dropped. The variations at once led them to suspect that specific gravity had little to do with the matter, it being ascribable rather to variations in the hardness of various parts of the ball. They also confirmed that absolute weight had nothing to do with the question by dropping a larger ball of the same material. This was on January 14, 1667.

The next day they made a cavity in the ball and stuffed lead into

[104] Pp. LXXXII and CCLXV.
[105] G. 262, 28v; *TT* 2, 683.
[106] G. 262, 74v; *TT* 2, 420.
[107] G. 269, 95v and 96r; *DIS*, 444 and 445.
[108] G. 262, 170r–73r; *TT* 2, 666–68.
[109] *Dialogo* (1632), first day. G.OP., VII, 46–47.

it. The results differed little from those of the day before. On January 18 they tried a solid glass ball and observed that it bounced to more than 96 per cent of the height from which it was released, as did a larger, hollow glass ball, until on the fourth drop, this broke.

Finally they poured melted lead into the ball of buffalo horn that had been hollowed out. When this was dropped, it bounced to only 56 per cent of the height of fall, and this surprised them, but they found an explanation and recorded it in the diary:

But it is true that the lead inside was not firmly fixed, so that it could probably be suspected that in the act of bouncing, the heavy lead inside, being loose, did not entirely follow the motion of the bone ball, and resisted it, and in this resistance part of the impetus already created in the fall was extinguished. For this reason the bounce was hindered to some extent.[110]

This was an extremely acute conjecture, and it may be interesting to note that in our own time this principle is used to prevent bouncing in the contacts of large electromagnetic switches.

6. Miscellaneous Experiments

There are some experiments and observations recorded in the diaries and elsewhere that do not seem to fit into any of the above categories. Many of these are mere notes of natural phenomena and need not detain us.

There are some interesting experiments on the effects of the oxides of nitrogen evolved in the reaction between copper and nitric acid. On August 28, 1657, they found that when bubbles of this gas were passed through oil, the oil solidified and became like wax, which, however, melted in the sun.[111] Passed through red wine, they gradually bleached it.

At the very end of the activity of the Academy there was a sudden flurry of interest in the amalgamation of metals with quicksilver, a subject that had only had their attention once, more than eight years earlier,[112] when they found that a brass ring had not changed its dimensions by absorbing a good deal of mercury.

On January 12, 1667, they left cylinders of eight different metals, previously weighed, in quicksilver for nineteen hours, in order to find out which absorbed it most easily.[113] Lead and tin simply dissolved; silver and gold absorbed a good deal, while iron, steel, and—surprisingly—copper and brass seemed unaffected. The gold lost a little

[110] G. 262, 173r; *TT* 2, 668.
[111] G. 262, 28v–29r; *TT* 2, 671.
[112] On August 3, 1658 (G. 262, 54r–v; *TT* 2, 672).
[113] G. 262, 169v–70r; *TT* 2, 672–73.

weight, but the silver gained, and so they drove the mercury off by heat and found both cylinders lighter. On February 9 they did a rather similar experiment, using mercury that had already been contaminated with lead.[114]

Meanwhile, on January 20 and 21, 1667, they had made experiments in the Torricellian vacuum to see whether the presence of the air was necessary to the amalgamation of gold and of tin. They found that it was not.[115] Evidently feeling that this was important, they repeated the experiment with a cylinder of tin on March 5, 1667, taking various precautions.[116] This was the last recorded experiment made by the Accademia del Cimento.

Very few zoological observations are mentioned in the diaries, and, apart from the ones published on the last pages of the *Saggi*, only one real experiment, which was on the effect of snakebite.[117] On June 28, 1660, they caused two cockerels to be bitten by a viper, dissected them immediately after death, and found the blood in the ventricles and the vena cava to be clotted. They suffocated another cockerel as a control. After lunch they continued the experiment before the Grand Duke, using various other animals. I suspect that Francesco Redi may have been the moving spirit on this occasion.[118]

In considering the hundred or so experiments that do not appear in the *Saggi*, we must remember that they can have had little effect on the contemporary learned world, apart from those members of the Academy, such as Borelli, whose own later researches may have profited from them. What effect the work of the Academy may have had on scientists in other countries will be investigated in the following chapter.

[114] G. 262, 176v–77r; *TT* 2, 673.
[115] G. 262, 173v–75r; *TT* 2, 673–75.
[116] G. 262, 177r–78v; *TT* 2, 675–76.
[117] G. 262, 83r, 84r; *TT* 2, 680.
[118] See Francesco Redi, *Osservazioni alle vipere* (Florence, 1664).

6.

THE TRANSALPINE RELATIONS OF THE ACADEMY

1. INTRODUCTION

We have seen in previous chapters that a strict rule of anonymity was enforced in the one publication of the Accademia del Cimento. It may therefore come as a surprise that the Academy as such had no external relations whatsoever; I have discovered only one letter that was addressed to it as a body,[1] and this was after the publication of the *Saggi*, when it was no longer in existence.[2] Magalotti was only a "recording secretary."

Nevertheless, the reader must not jump to the conclusion that this chapter is superfluous, for some of the members of the Academy, above all Prince Leopold, had a large transalpine correspondence, much of it relating to the activities of the group. In the "Galileo" manuscripts at Florence there are preserved about 280 letters of the Prince's transalpine correspondence between June 1657 and the end of 1667, and about 120 similar letters to and from Viviani, some of which relate in one way or another to the Accademia del Cimento. In addition to these, a large number of the letters between Giovanni Alfonso Borelli and Michelangelo Ricci and Prince Leopold have to do with the exchange of information with correspondents to the north of the Alps. The interest of this correspondence for the historian of science lies in the light it casts on the beginnings of international scientific cooperation, and the struggle of this tender plant to establish itself in a soil made infertile—especially in Italy—by a secular distrust of the *forestiero*, the foreigner.

Much of the correspondence was conducted through the resident ministers of the Grand Duke, especially in Paris, as well as through other Italians who were from time to time sent on diplomatic missions; but the most interesting exchanges are those with a few foreign

---

[1] G. 284, 11r–13r.
[2] See p. 326 below.

scientific workers who were known to Leopold either in person or by reputation.

A surprising number of the letters have to do with the purchase, gift, or exchange of books. There are three aspects of this. In the first place, Leopold was an avid collector and sent long lists of *desiderata* to the Tuscan residents and to special envoys. In the second place, he was also extremely generous in distributing throughout Europe the books published in Tuscany under the auspices of the court, a number of which were written by members of the Accademia del Cimento. Thirdly, there was the natural desire of the authors to obtain an even wider distribution for their works, and to do this they were quite prepared to give a good deal of trouble to foreign visitors, who might also find themselves expected to perform many other chores on the way back home. An admittedly extreme example of this, unfortunately too long to present here in full, is a letter written on July 28, 1661, by Vincenzio Viviani to his friend Robert Southwell, later to become a president of the Royal Society, but in 1661, at the age of twenty-six, on his way back to England during his "grand tour."[3] In this letter Viviani asks his friend—"instructs" would be a better word—to promote the sale of his book *De maximis et minimis* in Venice, Vienna, Amsterdam, Aix, and Leyden, quoting the prices he should ask and other commercial details. It is interesting that forty copies had already been sold to "Jacopo d'Alestre" of London, presumably James Allestry, one of the Royal Society's printers. Most of the letter might have been written by a sales manager to his favorite travelling salesman.

For convenience, I shall treat separately the relations between the Academy and England, and those with other countries, chiefly France and Holland. With England the correspondence was scarcely official at all, as also with Holland, but with other countries the Residents of the Grand Duke often came into the picture.

2. RELATIONS WITH ENGLAND

The relations between England and Italy on the intellectual plane are closely associated, at least in the seventeenth century, with the "grand tour" that was almost obligatory for the nobleman or man of means who wanted to round out his education. Milton was in Florence shortly before the death of Galileo, Evelyn in the autumn of 1644 and the spring of 1645, Southwell in 1660. We have already seen that the last-mentioned became a firm friend of Viviani; he was also

[3] The holograph draft is in G. 252, 73r–74r. I am obliged to Signor Giuseppe di Pietro of Florence for permission to consult his typed reading of this very difficult manuscript.

very highly esteemed by Prince Leopold and by the Grand Duke, and there exists a copy, or perhaps a draft, in Viviani's hand of a letter from Ferdinand to "il Ré di Gran Bretagna" [Charles II] dated March 24, 1661 and recommending Southwell to the King's attention as a virtuous man and full of noble qualities,[4] as indeed he was. More than thirty letters, still extant, were exchanged between Southwell and Viviani in the years 1660 to 1670, and they are pleasant to read, if not always as to subject matter, as when on November 7, 1660, Viviani is informed that the Tiber is in flood,[5] and nine days later has to reply that the Arno has also left its banks.[6]

But there is one of these letters, written by Viviani on October 6, 1660, which is of importance to our story, for it shows that overtures were being made to the English with the idea of establishing a scientific correspondence. Southwell evidently sent this letter to Boyle, and it found its way into the archives of the Royal Society.[7] Viviani was at Pistoia, no doubt on his engineering duties:

As I find myself here, uncertain whether I can return to say good-bye in person before you leave for Rome, I do not want to miss reminding you of your intention of making our Tuscany enjoy the inexhaustible treasures of your England, introducing a mutual correspondence on matters of physics between it and the philosophical Academy of His Highness Prince Leopold; a correspondence founded on experiments made or to be made for finding out natural effects and for the discovery of the truth. By such means could be established a useful and perpetual friendship between our Assembly and some of the first scholars of that kingdom. This is a thing greatly desired and coveted by us Academicians, and not least by His Highness the Prince, who, as you well know, has let you see by his very gracious requests how much he would enjoy this admirable interchange, and especially with that Mr. Robert Boyle whose many titles to excellence you have so often held up to us. His Highness anxiously awaits a suitable juncture to reveal his generous benevolence towards such rare and sublime intelligence. He assures you that as soon as this correspondence is begun with your help, you will see it also perpetuated, so that the satisfaction of these English gentlemen will perhaps not be less than the benefits that we promise ourselves from the novelty of their experiments and their sound doctrine.

It is therefore entrusted to your incomparable prudence and skill, in the hope that we may happily see this embryo, so to speak, soon born and come to perfection, or rather, being picked up, nursed, and educated by your favorable genius, quickly grow, taking on an ever greater force and vigor.

---

[4] C. 258, 40r–v.
[5] G. 254, 168r–69v.
[6] G. 157, 99r.
[7] Ms. BL. 5, fol. 166r–67r.

To this end some merchant in Lyons will be appointed by us, who, receiving letters from Florence and London, may deliver them when required.

So, wishing you (in case I am absent) a very happy journey, I remind you to favor me in all the other matters about which I asked you; and I beg you to make my excuses to Mr. Boyle for having ventured to send him my work *De massimi et minimi,* and also for not having accompanied it with a letter. Please perform a similar office with the other gentlemen, Barlow, Wallis, Ward, and Oldenburg. I desire that all these should be made aware of my particular reverence for the great fame of such illustrious people, and of my great desire to be honored with the title of their servant, as I am a humble admirer of their lofty doctrine in the disciplines of philosophy, geometry, and astronomy.

I trust that by this time you have sent to Leghorn, for London, not only those books of mine, but also that portrait of my so greatly revered master Galileo, which you have been pleased to receive. I shall be very glad to hear that by your liberality and intercession, Mr. Barlow has been willing to adorn his famous library at Oxford with it, honoring the portrait not less than such an agreeable nation will honor and esteem the works of Truth.

In addition I beg you to signify specially to Mr. Boyle how pleased His Highness the Prince would be to hear some particulars of his progress in that comparison that you told His Highness he had in hand, about the ideas of atoms and chemical experiments, which he is meditating on all the time and on which he has already written any number of pages. Ah! blessed Mr. Robert [Southwell], if with your help I succeed in finding favor with so many heroes! but more if some day it may fall to my lot to be able to sign myself to them with the claim that I am

Your most faithful friend and most
devoted and grateful servant

Pistoia, Oct. 6, 1660   VINCENZIO VIVIANI

This letter tells us a good deal, and at the same time leaves us with some problems because of the loss of some of the correspondence. Behind its baroque verbiage is evidence that Southwell had convinced the Tuscan scholars of the great merit of some of their English contemporaries, especially Robert Boyle. As to Boyle, it is interesting but not important that he had spent the winter of 1641–42 in Florence. He was probably one of the brightest fourteen-year-olds that the city had seen.[8] Viviani knew Barlow, Wallis, and Ward by their works, and of Oldenburg (the first secretary of the Royal Society) probably from Southwell's conversation. It will be seen that Viviani was not aware that the Royal Society was in the process of formation, although Viviani and the Prince knew by the next spring that it had been formed, as we shall see.

Four days later Southwell sent Viviani's letter to Boyle, enclosed in

[8] Thomas Birch's *Life* in Boyle, *Works,* 6 vols. (London, 1772), I, xxiii–xxiv.

one of his own that fairly bubbles with enthusiasm for Florence and its court.[9] Here are some extracts:

At my first arrival here . . . I thought it requisite, for the arriving unto some kind of admission in the court, wanting the benefit of friends, to settle my reputation in a lower class. And obtaining entrance into a meeting of the virtuosi, I also put in an oar, and shot my bolt among them; and then got not only the occasion of choosing good acquaintance, but heard, that at the court some favourable words passed of me; so that thence I took hint to go and make my reverence to the great duke, and some time after to prince Leopold his brother; who certainly is a perfect Maecenas, not only the patron to all that pretend to letters, but also admirably knowing, and illuminated himself . . . . his highness' darling study is that of natural experiments, for the prosecution of which he has a select company, that with him every morning make a private academy. He was pleased to ask me of those in England, that have the vogue, as to that kind of application. And having named two or three, I finally represented unto his highness, in those slender terms I was able, the imperfect character of Mr. Boyle;[10] for whom I can assure you his highness expressed so great a passion to be acquainted, that letting him know he was as much a master of civility as knowledge, and that I believed I might be the mediator of a correspondence, his highness told me, that I would engage myself in a very grateful embassy. So that to tell you the naked truth, the inclosed, that I here send you, is written in Florence (though dated from abroad)[11] with design to be sent unto you within this of mine; but with pretext directed to me, that so the prince may a little keep state, and yet secure your correspondence.

There follows some information about Viviani, and Southwell continues:

And being most confident, that if you shall please to embrace this occasion of correspondence with the prince, it will highly redound, not only to your own satisfaction, but even a general advancement of learning; and I shall number it among the felicities of my life, that I was but instrumental unto so great a good . . . .

The prince has correspondence with monsieur Bullialdus and others of Paris, unto whom he punctually repays all their addresses. So that if you please to write to the prince, under cover to signior Viviani, for the first time, taking notice to this of the present of his book, and to the prince, of the notice I gave you, of his courteous inclination to your acquaintance, you will find all things afterward run in a prompt and smooth channel.

It is worth noting that Southwell's visit happened to coincide with one of the most active periods in the history of the Accademia del

9 Boyle, *Works*, ed. cit., VI, 297–98.
10 I.e., an imperfect idea of the character of Mr. Boyle.
11 I.e., from Pistoia.

Cimento, just after its reopening, with Magalotti as Secretary, on May 20, 1660.

For some reason or other, Boyle does not seem to have accepted the Prince's invitation.[12] We shall see on page 294 below the excuse that he himself gave more than seven years later. I incline to the opinion that he thought that he was being unduly favored, and that such correspondence should be the privilege of the new Royal Society. In March of the following year Southwell, still in Rome, wrote again on the 30th:

WORTHY SIR,

Although I have not the happiness to hear from you, yet the fresh resentment[13] that I have of your past civilities, keeps me still to an observance of my duty and gratitude. From Florence, the last summer, I wrote unto you; and at the same time the mathematician of the Great Duke, a person of excellent qualities, sent you one of his books; and I was in hopes, that long before this time there would have been a firm correspondence between you, for it is desired here; and on the other side I know you have no aversion to it, but rather a genius to confer with all persons, that are singular in their kinds.

There is certainly some unhappiness arrived in the business, and I am sorry for it.[14]

Meanwhile the books and the statue had arrived, as Southwell informed Viviani on February 19, 1661.[15] Only on May 17 did Thomas Barlow get around to writing a Latin letter of thanks on behalf of the Bodleian Library.[16]

Before this, things had been taken out of Robert Southwell's hands by a man whose connection with Italy was much closer than that of any tourist, however "grand." This was Dr. John Finch, who on June 10, 1661 became Sir John. Born in 1626, he graduated M.A. at Cambridge in 1649 and soon after went to the medical school at Padua, taking the degree of M.D. and becoming English consul at that place. He attracted the attention of the Grand Duke Ferdinand, who made him a professor of anatomy at Pisa in 1655. A firm and sincere friendship with Prince Leopold began at about this time, and their correspondence is very pleasing. Finch wrote entirely in Italian, signed himself "Giovanni Finchio," and in one of his letters referred to Italy

[12] R. E. W. Maddison published a "tentative index" of Boyle's extant correspondence in *Roy. Soc., Notes and Records* 13 (1958): 128–201. It contains no letters to or from Prince Leopold.

[13] A common seventeenth-century meaning was "a feeling of gratitude."

[14] Boyle, *Works*, VI, 298–99.

[15] G. 161, 298r–99v.

[16] G. 254, 183r–v.

as "the terrestrial Paradise."[17] His lifelong friendship with Dr. (later Sir) Thomas Baines, who was with him on all his travels, was remarkable; in writing of his experiences he almost always used the pronoun "we." Most probably because of his intense chauvinism, of which we shall see a further example on page 300, Borelli took a violent dislike to "these English anatomists," and in many letters to Marcello Malpighi[18] during the years after 1659 he accuses them of all sorts of tricks, in fact of being charlatans.

At the Restoration of Charles II Finch and Baines returned to England, but went back to Italy in October 1662, and on May 16, 1665 Finch wrote to Leopold from Venice with the ecstatic news that he had been appointed the resident minister of Charles II in Florence,[19] an appointment that he filled until May 28, 1671. Evidently Charles II wanted to be properly represented at the Tuscan court, for Finch had £1,000 quarterly for entertaining, "besides intelligence money."[20] Borelli notes the resulting "pomp and very proud display" without enthusiasm in a letter to Malpighi dated July 10, 1665,[21] four days after Finch had presented his credentials.[22] For the last ten years of his life, which ended in 1682, Finch was ambassador at Constantinople.

Coming back to early 1661, we find Finch writing to tell Leopold of the formation of the Royal Society, using terms that he no doubt felt the Prince would understand and enjoy. After saying that he and Baines have only one wish, to return to Florence, he goes on:

By royal command we have in London an Academy, begun on the model of that of Your Highness, for making experiments. Knowing the inclination of His Highness [i.e., Ferdinand] I have already proposed to some of the most eminent of these scholars that they should correspond with Your Highness. As soon as I know that this is your pleasure I shall arrange it, for these people are more than ambitious of such an honor.[23]

The letter ends with political news.

Finch of course knew perfectly well that Charles II had not "commanded" the setting up of the Royal Society, and—although he was not made a Fellow until 1663—that it was not formed on the model of Leopold's Academy. His letter evidently produced a reply that I have not been able to find, but it was more likely to Finch than to the

---

[17] G. 281, 66r.

[18] Bologna University, ms. 2085, vol. IX, *passim*.

[19] G. 281, 71r–72r.

[20] *Calendar of State Papers*, vol. CXVI (London, 1863), entry dated March 1665.

[21] Bologna University, ms. 2085, vol. IX, fol. 160r–v.

[22] FAS, ms. 136, fol. 249v (*Diario di Settimanni*).

[23] G. 276, 2v (January 18, 1661).

Society. There is, however, a small piece of paper, probably in Finch's hand, which may be translated as follows:

I am very glad that those scholars have begun to study in the book of nature in order to get to know the truth about things. As to the correspondence that you propose, it will greatly please me and be valuable in every way, and I offer to reciprocate their desire at once and always with every possible diligence.

PRINCE LEOPOLD[24]

Whether this was the letter referred to or not, the Society's *Journal Book* tells us that on May 8, 1661:

This day a letter was read to the Company from the Duke Leopoldus.[25]

My Lord Brouncker [the President], Mr. Boyle, Sr. Kenelme Digby, Sr. Robert Moray, Dr. Ent, The President in being, & Mr. Croone, appointed a Comittee constantly to correspond with the Duke Leopoldus.

This Comittee to meet on Mundays att My Lord Brouncker's Lodgings and to make a speedy answer to the Duke's letter.

Three of this Comittee at least to be Quorum.[26]

A week later on May 15, "Sr Robert Moray having had occasion to acquaint His Majesty with the Duke of Florences Letter to the Society, had the Kings consent to return an answer."[27] The "Comittee" took the easy and obvious way out, for on May 22, Finch (who was not at this time a Fellow) was requested to draw up the letter to be sent to the Prince.[28] From a private letter from Finch to Leopold dated June 9[29] we learn that this formal epistle was to be in Latin and that it had not yet been written. Leopold replied in a letter of which we have a draft[30] but do not know the date, saying that he will be glad to receive the Latin letter, but whether that letter was ever sent I have not been able to find out. On June 5 it was referred to the Committee, but after that neither Committee nor letter were heard of again, and for some time whatever contact there may have been between the Royal Society and Leopold's Academy was indirect, depending on Oldenburg's correspondence with French scholars. This will be referred to later. Meanwhile Boyle, although he does not seem to have

[24] Royal Society, ms. MM.3, item 84.

[25] Birch, *History of the Royal Society*, 4 vols. (London, 1756–57), I, 22, has corrected this to "prince Leopold, brother of the Grand duke of Tuscany," and made corresponding amendments.

[26] *Journal Book*, I, 16.

[27] *Ibid.*, I, 17.

[28] *Ibid.*, I, 19.

[29] G. 276, 130r–v.

[30] G. 282, 217r.

written, did send to Leopold and to Viviani, through Oldenburg, copies of the Latin edition of his *New Experiments Physico Mechanicall*,[31] and Southwell, on his return to England in 1662, showed Boyle a Florentine liquid-in-glass thermometer.[32]

As I have shown in chapter 4, the publication of an account of the "Florentine experiments" was momentarily expected throughout the learned world. We find Oldenburg writing to Boyle on November 24, 1664 that there is "no account yet" of them; "howsoever they were promised me, to be sent with ye packet received."[33] On December 5, 1665 he was still hopeful, but he now hears, he tells Boyle, that they will not be printed for four or five months yet.[34] He tried to find out by writing to Finch two days later,[35] and on April 10, 1666, not having received an answer, he repeated his request: "We exspect [i.e., await] with some impatience ye Florentin Experiments, and hope, yt as soon as they are come to light, some Exemplars of ym will be hastened hither."[36]

The failure to maintain a direct correspondence between the Royal Society and the Italian experimenters is a mystery that seems deeper the more I read and the more I consider it. We have seen that Prince Leopold was persuaded to favor the project, and on the English side one finds nothing but praise for His Highness. Latin was a common language. Oldenburg was a most industrious letter-writer. The only explanation that I can imagine is that the Fellows believed that their books would speak for them, but only Boyle seems to have done much to ensure that his works reached Italy. However, when Thomas Sprat's *The History of the Royal Society of London for the Improving of Natural Knowledge*—which, of course, is not a history but an *apologia* —was published in 1667, the Council of the Society sent one to Leopold through Finch, accompanied by a letter in Latin composed by Oldenburg.[37] This letter, the original of which bears the date No-

---

[31] See p. 263 above.

[32] I have documented this in a *A History of the Thermometer* (Baltimore, 1966), pp. 38–39.

[33] *The Correspondence of Henry Oldenburg*, ed. & transl. by A. Rupert Hall and Marie Boas Hall (Madison & Milwaukee, 1965–[in progress]), II, 319.

[34] *Ibid.*, II, 629.

[35] *Ibid.*, II, 631–33.

[36] *Ibid.*, III, 87.

[37] Royal Society, *Council Book I*, p. 140 (November 5, 1667). They also sent copies of *The History* to Hevelius in Dantzig, to "Mr Winthorp"— presumably John Winthrop the younger, governor of Connecticut, who was a Fellow—and to Auzout and Petit in Paris, apparently jointly. A peculiar distribution-list.

vember 16, 1667, is in Florence,[38] and the beginning is worth quoting in the Halls' translation:[39]

As the designs for the cultivation of a sound and useful philosophy that have been framed both by our august King and by your Highness are so well matched, it seems entirely fitting that there should be a close and friendly relationship between our two societies. And indeed, when one examines the matter deeply, it seems that the successful execution of so arduous and burdensome a plan requires rather the joint labors of the industrious and wise men of the whole world in mutual co-operation, than those of this region or that alone.

They did not know that the Accademia del Cimento had ceased to function.

In his covering letter to Finch,[40] Oldenburg says that his two last letters [of December 7, 1665 and April 10, 1666] must have miscarried. This one, with the book, was eight months on the way, being acknowledged by Finch on July 24, 1668.[41] But the lack of correspondence can only be explained by the difficulty of communication, if at the same time we can explain why the Society's committee on correspondence was stillborn and why the suggestion about the merchant in Lyons was not taken up and made use of. As we shall see later, communication between France and Italy was fairly regular, and Oldenburg found little difficulty in corresponding with France. The following passage from Sprat's book, which was to some extent an official publication, does nothing to dispel the mystery:

In Italy the Royal Society has an excellent priviledge of receiving and imparting experiments, by the help of one of their own Fellows, who has the opportunity of being Resident there for them, as well as for the King. From thence they have been earnestly invited to a mutual intelligence, by many of their most noble wits, but chiefly by the Prince Leopold, brother to the Great Duke of Thuscany; who is the patron of all the inquisitive philosophers of Florence: from whom there is coming out under his name an account of their proceedings called *Ducat Experiments*.[42] This application to the Royal Society I have mention'd, because it comes from that country, which is seldome wont to have any great regard, to the arts of these nations that lye on this side of their mountains.[43]

---

[38] *G.* 278, 82r–v. It is followed by a table of contents of Sprat's book, in Latin.

[39] *Correspondence of Henry Oldenburg*, III, 619–23.

[40] *Ibid.*, III, 618.

[41] *Ibid.*, IV, 541–42.

[42] We have already seen (p. 74) that people had been guessing at the title. This guess looks like a misprint.

[43] Sprat, *The History*, p. 126.

That the Society intended to have relations with Florence is indeed implied by the election of both Baines and Finch to fellowship on May 20, 1663, when they were in Italy.[44]

At the very time when Sprat's book was being sent off towards Florence, the copies of the *Saggi* were being distributed by the Prince. One of the first to write his thanks was Finch, to whom had been sent four copies: one for Charles II, one for Boyle, one for Finch himself, and one for Dr. Baines.[45] Not one for the Royal Society, which seems strange, but was not; for Lorenzo Magalotti and Paolo Falconieri were on an official voyage around Europe, and were to present it in person. The story of Magalotti's visit to England in 1668 has been dealt with in a very entertaining way by R. D. Waller,[46] and I shall treat briefly the part of it that interests us here. As Miss Crinò has pointed out,[47] the visit was as much social and political as scientific.

The story begins before they reached England, with a letter from Magalotti to the Prince, written from Antwerp on January 6, 1668, in which he says: "I have become greatly disillusioned about my reception by the King, for while I was given to understand that the so effective protection that he gives to the famous Royal Society was the effect, if not of kindness, at least of his esteem for these studies, I have learned that he is accustomed to call his Academicians by no other title than *mes furets* [my ferrets]."[48] At the end of January he was in Brussels, and he did not know what arrangements had been made about the copies of the *Saggi* for England, so he wrote on January 31 to his friend Sir Bernard Gascoigne in London, a Florentine, born Bernadino Guasconi, who had fought with the royalists in the Civil War, spent the period of the Commonwealth in Italy, returned to England after the Restoration of Charles II, had been knighted and, rather incredibly, elected to the Royal Society in 1667. He wanted Gascoigne to delay the execution of any orders from the prince about the books until his arrival, thinking that Leopold might have wondered where he was.[49] He need not have worried, for when he arrived in London about February 10,[50] the copies of the *Saggi* had not yet arrived. On the 11th Oldenburg wrote to Boyle:

[44] They were in a long list of Fellows named by the Council under the second Charter. See Birch, *History*, I, 239–40.

[45] Finch to Prince Leopold, December 11, 1667; G. 278, 112r–v. Autograph.

[46] Waller, *Italian Studies* 1 (1937): 49–66.

[47] Crinò, *Fatti e figure del seicento anglo-toscano* (Florence, 1957), p. 127.

[48] G. 278, 119r–20v. Printed in *LIUI*, I, 295–98.

[49] The letter is in BN, fonds ital. 2035, f. 269r–70, and has been printed by C. Delcorno, *Studi secenteschi* 7 (1966): 141–42.

[50] Old style; hence nearly three weeks later.

Here are arrived two of ye Florentine Virtuosi, whereof one is Signor Magelotti; the others name I have not yet heard. I met ym yesterday accidentally in Westminster-Hall, accompanied with Sir Bern. Gascoyn, who brought us together, and who intends to conduct ym to ye meeting of ye Society on Thursday next.[51] They exspect every day a packet from Paris, containing an Exemplar of the Florentin Experiments, to be by ym presented from Cardinal Leopold[52] to ye Society. I intend this very night yet to give them a visit, and as solemn a welcom as I can. When they return home, I may by their favor purchase a correspondence with all the Ingenious men in Italy, if I please, and am able to entertain it.[53]

But they did not go to the meeting on the 13th. The following day Magalotti wrote to the Prince-Cardinal to tell him that unfavorable winds had delayed the shipment of the books from the Continent. Then:

Yesterday was Thursday and the Academy met as usual, and I think they were expecting me, for there were an extraordinary number of experiments. But, understanding that it is not permitted to go in simply as a curious tourist, I did not want to take my place as a scholar, firstly because I am not one, and secondly because even if I were I would not take it as being the role most advantageous for getting into the Court.[54]

Charles II had a "bad press," then as now. There is also evidence that the bright eyes of the English ladies were more attractive to Magalotti than dull experiments at Arundel House.

On February 18, Oldenburg wrote again to Boyle:

Some of the Florentin Virtuosi are come, but not yet their Book of Experiments. They were pleased the other day to give me a visit, wherein they told me, upon my inquiry, that ye chief subject of their Experiments was the Air and the pressure thereof, wch things if they had been publisht 3. or 4. years agoe, might perhaps appear new, but would hardly doe so now.[55]

Boyle would see what they meant, Oldenburg went on; also the Italians had hoped that Boyle himself would be in town.

In spite of Magalotti's qualms, he and Falconieri did attend the next meeting on February 20, arriving late, after the President had left. Their presence was not recorded in the minutes of the Society, but Magalotti wrote to Leopold the next day, a long letter giving a lively impression of the way the meeting was conducted and a garbled

[51] February 13 (O.S.).
[52] Leopold had become a cardinal on December 12, 1667. See chapter 7.
[53] *Correspondence of Henry Oldenburg*, IV, 170.
[54] G. 278, 145r–v. The letter is printed by Waller in *Italian Studies* 1 (1937): 52–53.
[55] *Correspondence of Oldenburg*, IV, 186.

account of some of the experiments.[56] Oldenburg reported their attendance in a letter to Boyle, noting their tardiness.[57] They had intended to go to Oxford on February 25,[58] but something prevented this, and they were at the next meeting on the 27th, their presence being recorded in the minutes on that date.[59] If Magalotti gave an account of this occasion to the Prince, it seems to have been lost, but the meeting must have been a very interesting one, to judge by the record. However, Magalotti appears to have been so little impressed that in writing to Viviani two days later[60] he did not mention the meeting at all, but made a great thing of his luck in being in London when Prince Rupert[61] came home from Portugal. Magalotti announced that he was going to Oxford the next day and hoped to get Viviani two more books from the list that that compulsive collector had given him. But he did not mention his chief reason for going to Oxford, which was to visit Robert Boyle. He was not disappointed, as we learn from his letter of March 13 to Prince Leopold,[62] which begins:

There was so short a time last week between my return from Oxford and the departure of the post, that I was unable to tell you that I found Boyle so agreeable [garbato] that it seemed to me too little to have gone only fifty miles to see him. It is impossible to say how courteous, considerate and obliging he is, and how pleasant and amiable his conversation. I enjoyed about ten hours of it, spread over two occasions, and had I not feared he might dismiss me, I should not have left his house at all in the two days and a half I spent in that city. He showed me several experiments, some connected with air pressure, others concerning the changes of colour produced by the mixing of different liquids. He spoke of Your Highness with those sentiments of respect and veneration which were appropriate to your virtues and to my own wishes, and I replied in a way I thought proper to the esteem in which you hold this worthy man. I attempted to persuade him to write to you, imagining that he had never done so before, and

---

[56] G. 278, 151r–52v; printed in *LIUI*, I, 298–301, and translated into English by Waller, *Italian Studies* 1 (1937): 53–55.

[57] Boyle, *Works*, VI, 269–70 (Oldenburg to Boyle, Feb. 25, 1667–68). *Correspondence of Oldenburg*, IV, 205–8.

[58] *Ibid.*

[59] Birch, *History*, II, 252.

[60] The letter, in the Biblioteca Riccardiana, Florence, ms. 2487, fol. 92r, is dated 29 February/9 March 1667/8. But 29 February O.S. is 10 March N.S. It is more probable that he would get the O.S. date right, as it was in use in London.

[61] The son of Elizabeth of Bohemia, daughter of James I of England, and hence Charles II's cousin. The Prince apparently knew Viviani and Dati.

[62] G. 278, 154r–55r. Printed in *LIUI*, I, 301–3. I have used the translation by Waller (1937), pp. 57–58.

could not gather very clearly from his reply if I was mistaken. He told me that Sir John Finch had invited him to assume this honour, but that he had refrained, not being able to bring himself to the point of writing with his own hand because of his infirmities, and not having anyone at hand to whom he could dictate in a language other than English. I sought to ease this difficulty as far as I could without incurring the suspicion of having been commissioned to invite him to write. As far as I myself am concerned, I have promised to send him news of anything which is likely to attract his noble curiosity; and since he gave me the Latin translation of one of his books that had not yet arrived in Italy when I left, I promised to let him have a copy of Your Highness's which I humbly beg of you to add to those which are coming to England for sale or presentation.[63]

This letter was written the day after he and Falconieri had formally presented a copy of the *Saggi* to the Royal Society on Leopold's behalf, and it is interesting that he described his visit to Boyle first; but he goes on:

Yesterday, so as to be in time for the Royal Society meeting, I had a copy of the book delivered to Mr. Oldenburg, bound in the most magnificent style, adding in Your Highness's name the highest possible expressions of esteem for so celebrated and virtuous a society. When Lord Brouncker the President had arrived and all were seated, the Secretary translated my compliments into English and read in the same language the titles of the subjects treated in the book. The president, taking off his hat, replied that these matters were among the most essential and the most difficult in the order of natural phenomena, and that having been examined in the presence and under the protection of a prince so great, so splendid, and so wise, they could not be otherwise than extremely well determined and illuminated; that the book would be consigned to the persons best equipped with a knowledge of Italian, so that reporting to the Society they might enable all to enjoy the fruits of such distinguished favour; and that in order to express the deeply humble gratitude which the Society would always feel for so high an honour, he would order the Secretary to send a special letter. This reply was translated into Latin for me by Mr. Oldenburg, who immediately wrote a memorandum of the occasion on the flyleaf.[64] We shall be here till after Easter; with which I subscribe myself, etc.

The "special letter," in Latin,[65] was purely formal and need not detain us. The minutes of the Society[66] agree almost exactly with Maga-

[63] Magalotti could not know that one was on the way.

[64] *Frontespizio*, i.e., title page. It was really at the bottom of the title page, and is still in the Royal Society's copy, as follows: "Munus Serenissi Principis Leopoldi ab Hetruria transmissum Regiae Societati, et traditum à Clarissis Viris Dnis Laurentio Magalotti, et Paulo Falconieri in publico consessu, d. 12 Martij, 1667/8."

[65] G. 278, 158r. It was approved on March 19.

[66] *Journal Book III*, p. 188. Birch, *History*, II, 256.

lotti's account, with the additional information that the Fellows chosen to report on the book were Dr. [Christopher] Merret and Dr. [Peter] Balle.

The next number of the *Philosophical Transactions* (No. 33) came out four days later and briefly noting the presentation, gave thirteen chapter headings, in English, and concluded: "As all these heads are very considerable, and of main importance, so doubtless will the handling of them be by competent judges found worthy of these famous *Academians del Cimento*."[67] That was all the readers were ever told, except for the translation of a few pages that appeared two years later.[68]

Merret and Balle had some difficulty in getting the book away from Robert Hooke and the Gresham professor of astronomy, Walter Pope, who made an unenthusiastic report on March 19.[69] It is amusing that four months later Hooke was "desired to return the book to the doctor [Merret]."[70]

Leopold had also sent a copy for Charles II, and Magalotti had it "bound with all possible richness" and delivered it in person on March 25 to the King, who "turned over the pages while going to bed."[71] This leaves us with a minor mystery, for, as we have seen, Finch had been sent a copy for the King.[72]

Magalotti and Falconieri left Dover on the night of April 6 (O.S.). It is clear that his duties in England were more political than scientific, even including a little espionage,[73] and he had a private cipher.

On May 7, 1668 Oldenburg evidently reported to Magalotti on the Royal Society's opinion of the *Saggi*, in a letter that the Halls were unable to find, which is a pity, for it must have been rather obviously tactful, to judge by Magalotti's reply of May 25.[74] It is quite clear that Magalotti knew that the *Saggi* had been delayed too long.

Meanwhile, on May 5 Cardinal Leopold had acknowledged Brouncker's letter in a formal reply written in Italian[75] and offering no serious promise of future relations. Indeed the relations between the Society and Italy from that time forward, while by no means

---

[67] *Phil. Trans.* 3 (1668): 640.

[68] *Ibid.* 5 (1670): 2020–23.

[69] Birch, *History*, II, 257. See also p. 335 below.

[70] *Ibid.*, II, 308.

[71] Magalotti to Leopold, March 27, 1668 (*G.* 278, 167r–68r). The letter also enclosed the formal letter from Brouncker, the president of the Royal Society.

[72] See p. 291 above.

[73] His political correspondence at the time is in *FAS*, filza Medici del Principato 4240.

[74] *Correspondence of Oldenburg*, IV, 410–12.

[75] There is a copy in Royal Society ms. LBO2, p. 201.

negligible, had nothing to do with the Accademia del Cimento, which no longer existed, or even with Tuscany, if we except a few visits by Alessandro Segni in the winter of 1668–69.[76] The President asked him, in the name of the Society, about the Cardinal's health. Segni's master, Prince Cosimo, visited the Society's rooms but had not bothered to go to a meeting.[77]

There were echoes of the "Florentine experiments" at meetings of the Royal Society on November 23, 1671 and again in February 1684. On the first of these occasions Robert Boyle reported at some length on his failure to confirm the observations of the Academicians that water reached a minimum volume and then expanded before congelation began.[78] The account, read in Boyle's absence, is embalmed in the most opaque Boylean prose. As far as I can understand this, his observations can only be made self-consistent on the assumption that he used water that was already near the ice point.

A dozen years later, on February 6, 1684, "Dr. Croune" [William Croone, M.D.] reported that he had observed the increase in volume of water before the onset of freezing,[79] but Robert Hooke "attributed the rising of the water in the neck of the bolthead [i.e., flask] to the shrinking of the glass."[80] Three weeks later Croone reported further well-designed experiments, the best of which was to put a flask of water and another of spirit in the freezing mixture together. As the water expanded before it froze, the spirit continued to shrink.[81] But it was plain that the Fellows, including Edmond Halley, were unconvinced.

3. Relations with Continental Countries across the Alps

The relations of the Accademia del Cimento, or rather of its members, with the countries on the Continent of Europe began sooner than those with England and were also more specialized. Prince Leopold's early and continuing interest in astronomy resulted in a rich correspondence with Ismael Boulliau, extending at least from 1649 to 1672, and in the 1660's with Huygens, Heinsius, and Hevelius. There is also correspondence with the mathematician Sluse at Liège,

[76] Segni to Leopold, January 20, 1668/9, G. 278, 123r–25r; printed in *LIUI*, I, 288–90.
[77] A. R. Hall, in his introduction to the reprint of Birch's *The History of the Royal Society of London*, 4 vols. (New York and London, 1968), I, xxii.
[78] Birch, *History*, II, 492–95.
[79] *Ibid.*, IV, 253.
[80] *Ibid.*, 254. I noted on page 183 above that Hooke never could believe in the existence of a point of minimum volume for water.
[81] *Ibid.*, 263–64.

with Chapelain and Menage, both professional literary men and amateur scientists, and with Pierre Petit, Melchisadec Thevenot, and Habert de Montmor, who are resistant to classification.

To deal with all this correspondence in a limited space would result in a catalogue of surpassing dullness and would not be germane to our subject, for most of it deals with matters that scarcely touch the work of the Accademia del Cimento. This can be said even more categorically of the extensive correspondence of Vincenzio Viviani. It is astonishing how much of all this deals with books, and in the case of Leopold, works of art as well. The appetite for books seems to have had no limits, and it appears that in Tuscany the only way to acquire foreign books was to have someone buy them abroad. It was different in England; Rhodes has recently shown that in the first half of the seventeenth century several London booksellers specialized in the importation of Italian books.[82] There is little doubt that the difference is to be ascribed to the ecclesiastical censorship, and in this connection there is an interesting letter of April 2, 1666 to Leopold from Alessandro Segni, the original secretary of the Accademia del Cimento, who was in Paris on official business, showing that the Prince had ordered a large selection of Jansenist literature, which Segni was sending by a route other than the usual one.[83]

Leopold also exemplified the maxim that to give is more blessed than to receive, distributing books by Tuscan authors with a liberal hand, often accompanying each gift with a gracious letter, which of course produced a grateful reply. Indeed, there were voluble thanks for books simply given to people by the Tuscan agents on the orders of the Prince.

In view of all this, I shall take the purchase and exchange of books for granted and concentrate on that part of the correspondence that deals with, or attempts to promote, the exchange of information about experiments and observations between the Accademia del Cimento and scholars or groups abroad.

As far as groups are concerned, only France comes into the picture, for the Collegium Naturae Curiosorum in Germany was entirely a medical society and, at any rate, does not seem to have attracted any attention in Tuscany until after the Accademia del Cimento had ceased to exist. It will therefore be necessary to look at the one group in France whose purposes were similar, at least in part, to those of the Florentine experimenters. This was the Académie Montmor, which met fairly continuously in Paris from 1657 until 1664, when the group

---

[82] Dennis E. Rhodes, *Studi secenteschi* 7 (1966): 131–38.
[83] G. 277, 265r–67r. See also page 319 below.

began to meet at the house of Melchisadec Thevenot and became known as the Académie Thevenot. The apparently complicated history of the Parisian academies of the time becomes a little simpler when we realize that they were named after the man who provided the hospitality. In 1666 the Académie Thevenot gave place to the Académie Royale des Sciences.[84]

Henri Louis Habert, "seigneur de Montmor et de la Brosse," was more than rich enough to play host to a distinguished company in his beautiful house which still stands in the Marais district of Paris. Born in 1599, he had been a friend of Mersenne and of Descartes; he was a member of the Académie Française and a collector of rare books and pictures. Gassendi lived in his house from 1653 until in 1655 he died there, and a group was undoubtedly attracted by the presence of that learned cleric—Pascal, Fermat, Desargues, Auzout, Petit, Carcavy, Roberval, Boulliau, Chapelain, Frenicle de Bessy, Claude Perrault, Pecquet, and later that great genius from the Netherlands, Christiaan Huygens, from whose immense correspondence we learn much of what is known about the Académie Montmor. Besides this veritable roster of French science there were two others, Samuel Sorbière, a rather extraordinary adventurer who became secretary and tried to dominate the proceedings; and Melchisadec Thevenot, an indefatigable traveller, diplomat, and amateur scientist who in effect performed the role of liaison officer with the Italian scholars. He spoke and wrote Italian, having been on a diplomatic mission to Rome from 1652 to 1654.[85] He was a remarkable book collector, and when he was put in charge of the very extensive royal library, he found that it lacked more than 2,000 volumes that were in his own.[86]

Unfortunately Montmor was unable or unwilling to keep out fashionable people of no scientific ability, and the meetings quickly deteriorated in spite of Sorbière's efforts. A letter to Christiaan Huygens from his brother Constantyn, written on November 18, 1660, suggests what was wrong and shows that Prince Leopold's Academy was becoming a standard of comparison:

We have had a good laugh at that fine assembly at Monseiur de Montmor's, and what happened in that meeting of fools when you were there hardly makes us respect the intelligence of these academicians, who patiently

---

[84] See Pierre Gauja, "Les origines de l'Académie des Sciences de Paris," *in* Institut de France, Académie des Sciences, *Troisième Centenaire 1666–1966* (Paris, 1967), pp. 1–51. I have drawn freely on this excellent article.

[85] Harcourt Brown, *Scientific Organizations in Seventeenth-Century France* (Baltimore, 1934), p. 135.

[86] Louis Cousin, "Eloge de Thevenot," *J. des Sçavans*, November 17, 1692, pp. 435–37.

listen to pedants jawing for hours on end about nothings. To tell you what I think, it seems to me that those gentlemen in Florence are worth much more than these Parisians and treat things with fore-thought and modesty.[87]

But I have found nothing to show that the formation of the Accademia del Cimento had any influence at all on the formation of the Montmor Academy,[88] or on its program. If there was any influence, it may have come through Boulliau.

We must not conclude from the above letter that nothing of interest took place in the Montmor Academy. With Christiaan Huygens there, this would have been impossible; and indeed his theory of Saturn was first read there in May 1658.[89] We shall get a glimpse of their experimental work in a moment.

In the main the contact between the Montmor Academy, in the person of Thevenot, and the Florentine experimenters, took place indirectly, through Rome. In the autumn of 1658 Borelli was in Rome, helping Abramo Ecchellense with the translation of an Arabic manuscript of Apollonius of Perga. On September 23 he wrote to Prince Leopold[90] that Ecchellense had received a letter, dated August 29, from "a learned gentleman of Paris," who turns out to be Thevenot. Borelli translated a paragraph of this letter, which shows that Thevenot had heard of the translation of Apollonius and was offering to have it printed in Paris on whatever terms they might wish. While Borelli was writing, a Signor Ricardo Bianchi (i.e., Richard White) had come with a letter dated August 1 from James Allestry—one of the "Printers to the Royal Society"—offering to print it in London. Quite naturally neither of these offers was seriously considered, but their interest here is that they show how comparatively rapidly scientific gossip got about. I have not been able to determine how this piece of news reached London, but it is highly probable that Thevenot in Paris heard about it by way of Michelangelo Ricci, who is known to have been in correspondence with Thevenot at this period and indeed for some years afterwards. It is very unfortunate that the letters exchanged between these two seem to have been lost; or, at least, I have been quite unable to find them. It is possible that an undated and unidentified secretarial copy of part of a letter[91] written to someone in Italy by someone in France may belong to this correspondence. There

[87] Oeuvres de . . . Huygens, III, 178.
[88] Harcourt Brown, Scientific Organizations in Seventeenth-Century France, p. 72, suggests "perhaps only an indirect influence."
[89] Ibid., pp. 83–84.
[90] G. 275, 121r–22v.
[91] G. 293, 30r–v.

are a number of letters between Ricci and Leopold and between Borelli and Leopold that give us a good idea of what was going on, but I have been unable to find the exchanges between Ricci and Borelli in spite of an extensive search in Italian libraries.

Correspondence between Leopold and foreign scientists was sometimes carried on in what seems to us nowadays to be a very roundabout way. Although he wrote directly to old friends like Boulliau, the establishment of such a relationship was often done by way of Carlo Dati, to whom he could give orders, or Ricci, whom he trusted implicitly. We shall see some instances of these procedures.

In November 1658, Borelli, back in Pisa, received a letter from Ricci which is now lost, but on November 11 Borelli quoted part of it to Prince Leopold: "Signor Thevenot wrote to me recently about the new French Academy [i.e., Montmor's], the thoughts of which run on the same lines as those of the one that meets under the auspices of Their Highnesses of Tuscany." Ricci then outlines some experiments that had been made in Paris on capillarity and on unannealed glass drops, and continues:

The latter were given consideration in Italy a short time ago. Now if the Frenchmen had found the true cause of all this, then I will say that they would have seized the priority and the glory from Italian scholars. Signor Thevenot, who is as courteous as he is diligent, and desirous of promoting the knowledge of the natural sciences, shows that he is anxious to communicate with you gentlemen. You may therefore see whether you wish to write directly to him, mentioning my name in some way, and I for my part will inform this gentleman of your merit and talent so that the correspondence may turn out more satisfactorily.[92]

But Borelli, with a lively suspicion that is excusable in an inhabitant of a country that has been invaded again and again for a thousand years, wondered whether even the promotion of natural philosophy was worth the risk:

Now I am extremely glad that these gentlemen in France are engaged in promoting natural philosophy with new experiments and speculations, but I also have some doubts and suspicions that, according to the ancient custom, the foreigners will make themselves the authors and discoverers of the inventions and speculations of our masters, and of those that we ourselves have found. This fear makes me go slowly in beginning this correspondence with those gentlemen of the Parisian academy, since in writing, one cannot do less than communicate something or other, and I fear that this may give those foreign minds an opportunity to rediscover the things; I am speaking of the causes, not the experiments. On the other hand, however,

---

[92] G. 275, 125r–26r. This letter published in *LIUI*, I, 115–18.

it seems to me that it would be well to be informed of what is being done and thought about in that academy; so that I find myself undecided, and I am therefore applying to Your Highness in order that you may instruct me how I should behave in this business . . . . Meanwhile I have replied equivocally to Signor Michelangelo Ricci, and when I have Your Highness' orders I will write to him decisively.[93]

Borelli was plainly worried; I do not know what he told Ricci, but only four days later he was again bothering Prince Leopold for instructions.[94] By December 1 he had had a reply from Ricci, and wrote again to the Prince:

Signor Michelangelo Ricci sends me a reply this week, and with many forceful and effective arguments succeeds in showing how much damage would be done to our Academy, and to all of Italy, by our keeping silent and not writing to those French academicians. In short, he would have it that the conclusions found and demonstrated by us should be revealed, but withholding and keeping secret the arguments and the demonstrations. In this manner, he says, we can be certain that the priority of the discovery made and disclosed by us cannot be taken away. Now I have replied that in a few days I shall have the opportunity to answer all the points contained in his letter. In the meantime Your Highness will be coming to Pisa, where you can determine and decide what to order me to do about this matter.[95]

To the twentieth-century reader it is not hard to imagine some of Ricci's unanswerable arguments; but Prince Leopold seems to have found it difficult to make up his mind what to do. His Christmas holiday at Pisa appears to have produced some formula that satisfied nobody. The next letter that we have from Borelli on this subject is dated April 19, 1659, and begins:

In the enclosed account sent to me by Signor Michelangelo Ricci, Your Highness will see described the procedure by which the new academy of philosophers at Paris is governed.[96] It will be seen after all that besides experimental observations these gentlemen also manage to philosophize about them . . . . I now await Your Highness' commands about the way, *and the precise words*, in which I am to answer, in agreement with what has been done before; because it seems to me that neither Signor Michelangelo nor those French gentlemen have understood the meaning of the first reply that I wrote on Your Highness' orders.[97]

---

[93] *Ibid.*, 126v–27r.

[94] *Ibid.*, 128v.

[95] *Ibid.*, 130v–31r.

[96] This must be the nine by-laws set up by Samuel Sorbière and sent to Thomas Hobbes in a letter dated February 1, 1658, and printed in Huygens, *Oeuvres*, IV, 513–16.

[97] G. 275, 146r–v. My italics.

The words that I have emphasized in this extract show that Borelli was quite determined that any future mistakes were to be made by the Prince; but I greatly regret that I have not found what was actually written on either occasion.

Meanwhile, it seems that not very much experimental work was being done in Paris at the time, as Leopold was told in so many words by Ismael Boulliau on December 19, 1659.[98] Leopold's correspondence in that direction during the year 1659 was largely with Boulliau and (through Boulliau, or more indirectly through Carlo Dati and Nicolas Heinsius of the Hague) with Christiaan Huygens. A great deal of this has really nothing to do with the Accademia del Cimento, as for instance the celebrated argument regarding the invention of the pendulum clock, which would have caused ill feeling except for the tact of Boulliau.[99]

The Accademia del Cimento became much more closely involved with the question of Saturn, which we have already discussed briefly in chapter 5, where it was noted that Huygens' *Systema Saturnium* was dedicated to Prince Leopold and sent to him by way of Nicolas Heinsius and Carlo Dati.[100] A copy was also sent to the Italophile French literary man Jean Chapelain, who wrote to Huygens on October 15, 1659 in terms that indicate how greatly Leopold was esteemed in Paris:

The dedication is grave and eloquent, and certainly you could not choose a subject more worthy of it than Prince Leopold of Tuscany, who is very knowledgeable in these matters and is regarded in Italy as the only support of learning. As to the injustice that he has done you,[101] it must be forgiven him because he has made amends for it so well. All Florence is so biassed in favor of the merit of Galileo that they think the only good astronomers are those whom he taught or who have learned from him. It is an effect of self-respect which is only to be condemned when it is stubbornly maintained.[102]

Huygens expected to be thanked for his gift and its dedication, but no thanks or acknowledgment came, and he complained to Chapelain, who thought at first that Dati must be responsible for the delay,[103] but by March 4, 1660 came to believe the worst of the Prince:

---

[98] G. 275, 169r–70r.
[99] The documents are in volume III of *Oeuvres de Christiaan Huygens*.
[100] It was posted on August 19, 1659 (Huygens, *Oeuvres*, II, 453, note 1).
[101] A reference to the dispute about the pendulum.
[102] *Lettres de Jean Chapelain*, ed. Ph. Tamizey de Larroque, 2 vols. (Paris, 1880, 1883), II, 58.
[103] *Ibid.*, pp. 68–69.

I am scandalized at Prince Leopold of Tuscany. It is absurd not to have thanked you yet for the honor that you have done to him and to his nation. It would be puerile if, because the late Galileo had thought of it, they were annoyed that you were the first to publish the pendulum clock, as it is certain that your idea was in no way borrowed from his.[104]

The imaginary difficulty was not disposed of until, on July 13, 1660, Dati told Heinsius the real reason that Huygens had received no letter from the Prince; it was because there had been no letter accompanying the book, "and I do not think it is usual to reply to letters [i.e., dedications] printed in books."[105] A week later Leopold wrote to Cosimo Brunetti, a Florentine living in Paris, in exactly the same vein,[106] and on October 4 he sent Huygens the reports by Borelli and Magalotti referred to on page 259,[107] and there was never afterwards any suspicion nor any reason for such a thing.

In the meantime, Leopold was proceeding cautiously in establishing a correspondence with the Montmor Academy. On July 5, 1660, Magalotti wrote to Ricci saying that the Prince was anxious lest the French should think that he had broken off relations and had ordered that a description of the experiment showing the descent of smoke in a vacuum should be sent to Ricci for forwarding to Thevenot.[108] Ricci sent it off, and on July 10 wrote to Leopold as follows:

The protection that your Serene House has at all times given to scholarship and its professors makes me hope that one day I shall see united, by way of a learned correspondence, those gentlemen of yours interested in natural things, and those of France. Having written to Signor Borelli several times about this in order to help Signor Thevenot, who asked me to do so, I now hear that at last Your Highness, being freed from the indisposition that hindered it, condescends to promote such a work . . .[109]

by sending the account of the behavior of smoke in a vacuum. We are again reminded by this letter that Leopold's health was never robust, and in the early months of 1660 he seems to have been ill for much of the time,[110] which may have been the principal reason for his apparent vacillation in this matter. In the twentieth century it is easy to forget the state of bad health or at least great discomfort in which

[104] *Ibid.*, p. 80.

[105] Huygens, *Oeuvres*, III, 502–3.

[106] *Ibid.*, pp. 176–77.

[107] *Ibid.*, p. 151, the covering letter. These documents all went by way of Dati and Heinsius.

[108] *LIUI*, II, 88–89.

[109] *Ibid.*, p. 92.

[110] See Gaetano Pieraccini, *La stirpe de' Medici di Cafaggiolo*, 3 vols. (Florence, 1924–25), II, 622.

all but those of strong constitution often found themselves three centuries ago.

It is evident that Ricci was thoroughly convinced of the desirability of international co-operation and scornful of secrecy, for he seems to have talked about the doings of the Academy to acquaintances in Rome. On August 1, Pierre Guisony, an Avignon physician who busied himself with experimental physics, wrote to Huygens from Rome that Ricci had told him about the experiment with the smoke in the vacuum,[111] and on August 27[112] he passed on a brief description of Borelli's experiments with a model of Saturn,[113] in spite of the fact that only on the 22nd he had told someone in Florence, probably Magalotti, that he would not dare to do so without permission from Prince Leopold.[114]

Thevenot wrote enthusiastically and, apparently for the first time, directly to Leopold on September 3, 1660, thanking him for the communication through Ricci of the activities of "the Academy of Pisa."[115] He did not conceal the news of the experiment, and someone passed it on to Oldenburg, who told John Beale about it on September 4, 1660, with the further information that Boyle claimed to have done it better.[116] Thevenot also wrote, at greater length, to Ricci, who reported to Leopold on October 14 as follows:

There has finally arrived the reply of Signor Thevenot concerning the experiment sent to him by me on the orders of Your Highness. The reason for the long delay, he says, was the solemn entry of the Queen,[117] which had drawn to itself the curiosity of the entire Kingdom, and even diverted those gentlemen[118] from their usual application to their studies. They had an extraordinary meeting afterwards in order to show the very elegant experiment, as Signor Thevenot calls it, to these people, who wished to try the experiment afresh, and to send Your Highness a Discourse about it as quickly as possible, and at the same time to represent to you the universal delight, the reverence, and the humility with which they have received both

---

[111] Huygens, *Oeuvres*, III, 104. Guisony ascribed the experiment to an academy at Pisa, apparently misunderstanding Ricci. The court was clearly at Florence during the summer of 1660.

[112] *Ibid.*, p. 116.

[113] See p. 260 above.

[114] G. 283, 83r.

[115] G. 315, 1030r–v (secretarial copy). This letter had been sent through Ricci, who again must have said something about Pisa.

[116] *Correspondence of Henry Oldenburg*, I, 386.

[117] Louis XIV had married Maria Theresa, the Infanta of Spain, and brought her to Paris in August 1660.

[118] The Academicians of the Montmor Academy. Ricci did not realize what a small matter scientific experiment was to nearly all of them.

the aid and the honor of the correspondence to be carried on between the two Academies.[119]

Although we have not the letters, we know that correspondence continued between Ricci and Thevenot during the following winter.[120] On December 3 Prince Leopold wrote to Thevenot the apologetic letter that I have translated on p. 66 above, which I interpret to mean that he really did not want much correspondence until his Academy could make a formal publication. But on April 7, 1661, Thevenot wrote from Paris enclosing a list, in French, of thirty-seven "observations" that had been made in Paris, and of fourteen more to be made "on the peak of the mountain of Teneriffe, in the Canary Islands,"[121] these last being a partial adaptation of the list of twenty-two prepared by Lord Brouncker and Robert Boyle, at the Royal Society. This document has the honor of being the very first entry in the *Register Book* no. 1, under the date January 2, 1660/1.[122] In his letter, Thevenot states that it was sent to him from England.

The thirty-seven observations, followed in fact by six more that are not numbered, are entirely concerned with the effects of what we should now call surface tension. They were made with very simple apparatus or none at all, and show excellent powers of observation. It is outside the scope of this book to examine them, and I shall confine myself to the remark that while many of them must have been made in the Accademia del Cimento, they were probably considered too trivial, or obvious, to record. I am inclined to suppose that Thevenot in his turn was unwilling to impart anything more interesting or at any rate more elaborate. These experiments on capillarity seem to have been a sort of scientific currency in France, for as late as April 25, 1664, Petit sent a description of twenty-one of them to the Marchese Malvagia,[123] and we have already seen that they were being written about in 1658.

The other list is quite different in detail from the one at the Royal Society and is not a translation. One can only suppose that someone in London wrote to a friend in Paris giving what he remembered of the list. Its only importance is as another indication of the development of international co-operation in science.

Unfortunately these documents arrived during the interminable

---

[119] G. 276, 81r.

[120] See, e.g., Huygens, *Oeuvres*, III, 248.

[121] The holograph letter is in G. 276, 124r. The list of 37 is in G. 270, 139r–41v, and the additional 14 in G. 270, 155r–56r. Printed in Italian translation in *TT* 2, 716–21.

[122] Printed in Birch, *History*, I, 8–10.

[123] A copy by Viviani is in G. 246, 17r–18v.

preparations for the wedding of Prince Cosimo and Margaret of Orleans, and as we saw on page 63 above, Leopold apologized for being unable to reply in kind.[124] At this time Viviani was also corresponding occasionally with Thevenot, and in a letter dated May 6, 1661, he says: "Regarding the experiments of this Academy of ours, an account is still being prepared of those already conducted before His Highness the Prince, who I believe will just recently have sent you a good many of them, with the intention of continuing."[125] There is nothing to support this last statement, and I am of the opinion that Leopold was optimistic that the *Saggi* would soon be published—as indeed it might have been—and was simply being evasive. On September 11 Leopold wrote again to Thevenot that he was sending him, by way of the newly appointed Resident in Paris, Giovanni Filippo Marucelli, several copies of Borelli's *Apollonius*—one for himself, the rest for members of the Paris Academy—but said nothing whatever about an exchange of information.[126]

The Duke's Resident, Marucelli, made friends with numerous literary and scientific people and, apart from his diplomatic duties, had a considerable direct correspondence with Leopold during the next half-dozen years,[127] until he was allowed to come home on account of his bad health. But this correspondence is mainly about books and works of art. Apparently Ricci kept up a correspondence with Thevenot, though we have none of the letters, but it does not seem to have been very productive, and on April 13, 1664, we find the former telling Leopold that he was "receiving no information whatever from the transalpine countries regarding mathematics and philosophy"[128]—that is to say, natural science.

To find the next exchange between Thevenot and Leopold that is important for our purpose, we have to pass to the year 1666, when the Académie Royale des Sciences was being formed by the minister Colbert. On August 2 of that year Thevenot wrote as follows:

Our Academy, which Your Highness condescended to mention,[129] had already been reduced to practical work and experiments, as we wished, but Mr. Colbert, to whom the design had been proposed as something worthy of the King's patronage, and of His Excellency's, has taken a strong fancy to the Academy, and in the hope of making much more of what we had thought of and proposed, is making several classes of academicians. He has

---

[124] Leopold to Thevenot, April 21, 1661; the draft is in G. 282, 50r–v.
[125] G. 252, 70r (Draft).
[126] G. 282, 57r (Draft).
[127] In G. 276, 277, and 282.
[128] G. 277, 8v.
[129] This probably refers to a letter of June 18 (G. 282, 134r–v).

already begun to form that of the geometers, with the intention of form-ing another of physicists, and still others as time goes on. They are already discussing instruments for astronomical observations, and a building in which to put them, and calling in very gifted people. They have begun with Signor Huygens . . .[130]

who indeed was given a pension of 6,000 livres a year and a fine suite of rooms in the Royal Library on the rue Vivienne.[131] Poor Thevenot was treated much less well, not being even made a member of the Académie until 1685.

Leopold replied on September 3:

I cannot adequately express to you how glad I am to hear of your Academy being honored by the high and powerful protection of such a King. Nor will anyone have room to doubt that both in this respect, and by it being really composed of the most scrupulous (*purgati*) and erudite talents, there can come from it anything but works deserving of eternal praise, and of singular benefit to the Republic of Letters.[132]

Thevenot came back on October 16 with more information about the new Academy in a letter already referred to on page 52.[133]

This correspondence, and especially Thevenot's letter of August 2, should dispose of the legend, occasionally heard, that the Académie Royale des Sciences was founded "in imitation" of the Accademia del Cimento. This sort of *post hoc, ergo propter hoc* argument reached its nadir in a letter dated February 3, 1662, from Marucelli to Leopold, saying that the Montmor Academy had a "colony" in London, "of which His Britannic Majesty does not disdain to call himself the head!"[134]

Apart from his rather tenuous correspondence with the Montmor Academy, Leopold seems to have been on the lookout for ways of helping scientific men on the other side of the Alps by various means, besides sending them books from Tuscany. There is, for example, an exchange of letters in 1661 and 1662 with the Elector Carl Ludwig of Bavaria, in which Leopold commends to the Elector the astronomer Ezekiel Spannheim.[135] He sent various instruments to Boulliau, in-

---

[130] G. 315, 1038v–39r (secretarial copy). On June 22 Huygens had already sent Leopold news about the beginnings of the new academy (Huygens, *Oeuvres*, VI, 53–55; G. 277, 316r–v).

[131] Gauja, "Les origines . . ." (1967), p. 46.

[132] *FAS*, filza M. del P. 5575a, document 102.

[133] G. 277, 347r–48v. Printed by R. M. McKeon, *Rev. Hist. Sci.* 18 (1965): 1–6.

[134] G. 276, 159r.

[135] G. 315, 9r–v, 11r–v, 13r–v.

cluding a valuable telescope as early as 1651[136] and Florentine thermometers in 1658.[137] Thermometers were apparently being sent as
gifts as late as 1667.

But it is only by a rather far-fetched association of ideas that all
this miscellaneous correspondence of Leopold's can be related to the
Accademia del Cimento. It shows instead that Leopold's interest in
natural philosophy, and astronomy in particular, was strong enough
for him to take a great deal of trouble to keep in touch with what was
being done elsewhere at a time when scientific publications were unknown. We must remember that the very first regular periodical of
this sort, the *Journal des Sçavans*, began to appear on January 5, 1665,
and the second, the *Philosophical Transactions*, on March 6 of the
same year.

[136] *BN*, fonds français 13039, fol. 38r.
[137] *G.* 282, 8r–v.

7.

THE DISSOLUTION OF THE ACADEMY

1. Introduction

The diary of the Accademia del Cimento ends suddenly on March 5, 1667. Targioni Tozzetti, who saw the original diary, tells us that many blank pages remained and felt that there was a mystery behind this sudden end.[1] Nevertheless, I believe that a careful examination of the documents makes possible a reasonable description of the process of decay that led to its dissolution.

In many textbooks one may find a facile explanation for this sudden demise, namely, that Prince Leopold was obliged to give up his scientific activities as a condition for his election as cardinal. This probably stems from a passage by Angelo Fabroni in his short biography of Lorenzo Magalotti:

While the Academicians strove more and more each day to increase [the universal praise of their work], an extremely tragic thing happened to the Academy. In those evil times there were many sworn enemies of the direct way of philosophizing, just as if it were opposed to religion, and Prince Leopold found himself obliged to accommodate himself to the taste of these people in order to proceed to a new honor. Thus in the year 1667 this so reputable Academy ceased to exist, and at the end of the same year he was created Cardinal.[2]

It is less well known that Fabroni himself changed his mind about this only a few years later. In this chapter I shall try to show, not only that there are documents to contradict such an assertion but also that the Academy was not the sort of organization that could be expected to survive.

---

[1] *TT* 1, 461.

[2] *Delle lettere familiari del Conte Lorenzo Magalotti e di altri insigni uomini a lui scritte*, 2 vols. (Florence, 1769), I, xviii.

2. EARLY SIGNS OF DISSOLUTION

The reader who has perused chapter 3 may have noticed the great difference between the organization of the Accademia del Cimento and that of a modern learned society or research laboratory. It might have been expected that the more energetic and original spirits would be the first to become restless in such surroundings, and indeed the most original and energetic of them all, Giovanni Alfonso Borelli, seems to have been disillusioned almost from the beginning, for on October 10, 1657, we find him writing from Florence to Paolo del Buono in these terms:

I have your most courteous letter,[3] which has brought me not a little consolation. And first, concerning our Academy, which you call a Lyceum, I wish that it might have the laws that you have invented for it; but the trouble is that only disorder is to be found there. This comes from the excessive ambition of one of the Academicians[4] who, although he is a rotten and mouldy peripatetic, wants to appear in a toga borrowed from a free and sincere philosopher; and because in the end of a bottle pours out the wine that it is filled with, the strangest monsters and chimeras seem to come out of him. . . .

Nor, therefore, have I any hope that in future we shall do better than in the past, and so here I am with the greatest desire that these few days of October may pass quickly, so that I may go away to Pisa and occupy my time there in getting on with the studies that I enjoy.[5]

Several pages later he refers to the instrument devised by Del Buono for trying to compress water.[6] Three have broken, and one was to have been constructed chiefly of copper, but: "This instrument has never yet been made and has had the same ill-luck as many other fine and curious things that have made no progress because of puerile things that are being done with great solemnity, expense, and loss of time."[7] Naturally Borelli would easily see through Marsili, but I think he was using him as a whipping boy on whom to vent his impatience with restraint. Borelli took care that the Prince should not suspect his disaffection, writing to him on October 29 that "distance has not taken my mind from the most sweet and virtuous conversation of the philosophers of our Academy."[8]

Much has been made of the quarrel between Borelli and Viviani,

---

[3] I have been unable to find the letter from Del Buono.
[4] This can be no other than Alessandro Marsili.
[5] LIUI, I, 94–95.
[6] See p. 219 above.
[7] LIUI, I, 98–99. The metal apparatus was actually tried on October 16 (G. 262, 40r–v), but also broke.
[8] G. 275, 80r.

which certainly can have done the Academy no good, although I am convinced that it cannot have been a major factor in its demise. Luigi Tenca has given us a perceptive account of this quarrel, illustrated by some of the most notable of the pertinent letters.[9]

By the end of October 1657 Borelli and Rinaldini were back in Pisa. The Academicians had been experimenting with artificial freezing in October, and Prince Leopold had noted that when a bulb with a long neck filled with water was first put in ice, there was a sudden jump in the level of the liquid in the neck. Similarly, when it was put in hot water, there was a sudden fall. Borelli at once saw the reason for this, but there were doubters, for the thermal expansion of solid substances was a novel idea at the time. The now familiar experiment with the heated ring and the mandrel was suggested; but it was not obvious that both the inner and the outer diameter of the ring would be increased by the heat, and Viviani did the necessary theory in two ways, sending one to Borelli on November 17 and the other to Rinaldini on the 26th.[10]

Borelli's reply of November 21[11] made Viviani furious, as well it might, and on the 26th he repeated the essence of it, with his own underlining, in a postscript to this letter to Rinaldini, of which we have Viviani's holograph draft.[12] This begins:

I really must unbosom myself to you. Signor Borelli, to whom, as I told you, I had sent the enclosed demonstration, replied in the following extremely deceitful words: "I have received your very kind letter with the demonstration enclosed, which I have found admirable and as perfect as can ever be desired. But there may have been a chance that you saw in Florence the many propositions on this subject that I outlined when I was there; nevertheless it is well that you too should have had a share of the pleasure in finding one of the causes of this phenomenon, which is entirely correct."

This is really a reply that has disgusted not only me, but everyone else to whom I have shown it, for they recognized in it very clearly his distress at never having come across such a demonstration and his extreme desire to appropriate this one, which I should otherwise have thought of as a bagatelle, but now esteem as something, seeing that those who are reputed richer than the king of Spain do their best with tricks to despoil others

---

[9] L. Tenca, *Rendiconti Ist. Lombardo Sci. e Lett.* 90 (1956): 107–21.

[10] G. 252, 36r. Both demonstrations are copied by Viviani at the end of a draft of the letter to Rinaldini dated November 26, 1657. Tenca (p. 112) is in error in referring this correspondence to G. 258.

[11] G. 283, 19r.

[12] G. 252, 41r–v. This is printed by Tenca, pp. 112–13, whose reading of Viviani's exceedingly difficult handwriting differs a little from another one, kindly made available to me in typescript by Signor Giuseppe di Pietro of Florence.

of what few goods have fallen to the lot of him who knew or believed himself to be very poor . . .[13]

But deceit often falls back on the deceiver; and indeed in my case not only I, but every one of the other gentlemen, assert that they never heard him say that he had found such a demonstration, which he would not have failed to mention to the Prince, if not to the others . . .

We may agree that Viviani's anger was more than justified, but how can we justify, or even explain the conduct of Borelli? At the risk of a digression I shall present a little more of this postscript, more, in fact, than we are given by Tenca, in order to throw some light on this question:

. . . since he said he had found a demonstration, or rather had made a calculation (given the inner diameter of the flask, the diameter of the graduated tube, and the distance that the water rises in it when the bulb is immersed in ice, or that it falls in hot water) of how much the contraction of the flask should amount to, and he found it less than a twentieth of the thickness of a piece of paper—a calculation that anyone who applies himself to it will be able to make. But there is nobody who has heard from him that he had demonstrated that the flask or the ring must necessarily become larger; only the talk about all those penetrating wedges of fire.[14] If he had reached such a conclusion, why not tell at least the Prince, to whom he had talked about these things earlier than to the others? That's enough! It takes a lot to get to know human nature, even at some cost to oneself. But say nothing about it, because I don't intend to come to an open rupture, although I don't know what answer I might have made to him in hot blood. I am at your command, and you will honor me by not spreading this abroad, for reasons that I do not want to put on paper. Therefore you will be able to tear off this half sheet and burn it as I ask, or else send it back to me.

One thing that is clear from this is that Viviani was not one to pick a quarrel. But from the passage about the flask I think that we can see how Borelli may have come to believe, in all honesty, that he had done the demonstration himself, especially as his extremely lively mind would have taken it in at a glance, making him feel that his calculation concerning the flask had really solved the problem. Furthermore, he knew that the other Academicians were more easily convinced by a good experiment than by geometrical reasoning, as he had just told Prince Leopold in a letter dated November 14 enclosing a note describing an experiment to show "the Peripatetics" that heat really increases the internal capacity of a glass vessel.[15]

[13] I have tried to reproduce the sense of Viviani's somewhat excited syntax.
[14] See note 240 on page 215 above.
[15] G. 275, 84r–85v; printed in *LIUI*, I, 92–94.

There has been much talk about the "enmity" or even "hatred" between the two men.[16] I think that such words are much too strong, and as evidence of this I would adduce Borelli's exemplary conduct in the matter of the lost books of the *Conics* of Apollonius of Perga. As was briefly noted in chapter 2, Viviani, who was in much closer touch with the mathematics of antiquity than with that of his own century, had spent a great deal of his spare time in an imaginative reconstruction of the lost books of Apollonius. Consider his feelings when he heard that Borelli had discovered an Arabic manuscript of these very books in the Grand Duke's library and that he was having these translated by an expert in Arabic, Abramo Ecchelense. As the latter knew no geometry, Borelli, who knew no Arabic, spent the summer and autumn of 1658 in Rome on this business, and Viviani, whose manuscript was about ready for the printer, obviously needed to prove that he had never even set eyes on the Arabic codex. His book[17] was published in 1659, and in the preface he went to much trouble to establish this fact. He had no need to convince Borelli of it, for in a letter from Rome, dated July 20, 1658, the latter writes as follows: "I also concur in and approve your decision and that of all your friends to send to the printer your discoveries about the conic sections; and I shall be able to testify along with the others that you have had no knowledge of these last books." He continues with a typical outburst: "So you should go ahead and enjoy this advantage, which I doubt whether I can enjoy, although I have by me a very compendious treatise on the conics, all written out in my own hand at a time when I was not rushed as I am now; that is to say, at a time when I could use my hands and my pen."[18] Having got rid of a little envy, he comes back with a salutation that in the context of the seventeenth century seems more than merely formal: "Meanwhile look after your health, and keep me in your good graces, while I embrace you with all my heart."[19]

In the preface of his book Viviani quotes the phrase "and I shall be able to testify" though not exactly, for he makes Borelli say "that you have not had *the least* knowledge" (*minima notizia*).[20]

I submit that this is not a letter written by a man poisoned by hate. I also believe that the second of the above passages shows, although

[16] E.g., *TT* 1, 211.

[17] *De maximis et minimis, geometrica divinatio in quintum conicorum Apollonii Pergaei adhuc desideratum*, etc. (Florence, 1659).

[18] This may have been the addition that was made in the 1679 edition of his *Euclides Restitutus*, first published at Pisa in 1658.

[19] G. 254, 107r. Printed in part by Tenca (1956), pp. 116–17.

[20] *De maximis et minimis*, preface, fol. b3v.

not more clearly than many others in his writings, the source of the self-assertion that his contemporaries found so intolerable. This source was envy. Consider his position at the Tuscan court; among these noblemen he was, socially, a nobody, the son of a plain soldier, and a Spanish soldier at that. Although he knew that in intellect he was greatly superior to all the Academicians but Viviani, and in energy far above Viviani as well, yet he had to jump to attention whenever Prince Leopold or the Grand Duke noticed his existence, even though there cannot often have been despots as benevolent as these. We have seen how he longed for the moment when he could escape from the Court, and from the Academy that was part of the Court, to his study at Pisa. When the Court came to Pisa, as it often did, did he wish he were in Florence? Men on Borelli's intellectual level are seldom oblivious to their own superiority, and this knowledge must have made the awareness of being an outsider doubly galling.

Thus I do not believe that Borelli hated Viviani, or that he even envied him more than he did the others. He quarreled with Viviani because Viviani was the one with whom he had occasion to quarrel.

At any rate the *Conics* of Apollonius was not the cause of any damage to the Academy, for both men seem to have behaved with entire correctness, as indeed it was to their advantage to do.[21] The translation of the Arabic manuscript found by Borelli, begun at the end of June 1658,[22] was published about the middle of 1661[23] on the orders of the Grand Duke,[24] who is said[25] to have delayed the publication until the learned world had had the opportunity to appreciate Viviani's very considerable achievement, but an examination of Borelli's letters to the Prince during the intervening period shows no sign of such a forced delay. Although the translation was substantially finished in the summer of 1659 and Borelli expected it to be published during the following winter,[26] there had to be a good deal of rewriting, the illustrations had to be made, and in the spring of 1660 there was even trouble from the Inquisition, because the book contained

[21] After coming to this conclusion I find it confirmed by G. Giovannozzi, who made a detailed study of the publication of Borelli's *Apollonius* (*Mem. Pontif. Accad. dei Nuovi Lincei*, ser. 2, vol. 2 [1916]: 1–32).

[22] Borelli to Viviani, from Rome, June 29, 1658, G. 254, 105r–6v.

[23] *Apollonii Pergaei conicorum lib. v.vi.vii. paraphraste Abalphato Asphahanensi nunc primum editi*, etc. (Florence, 1661).

[24] This is clear from a letter (G. 282, 57r) dated September 11, 1661, from Leopold to Melchisadec Thevenot.

[25] E.g., by Giamattista Corniani, *I secoli della letteratura italiana commentario*, etc., 9 vols. (Brescia, 1804–13), VIII, 33.

[26] Borelli to Viviani, from Rome, April 27, 1659 (G. 254, 110r–11r; printed by Tenca [1956], pp. 117–18.) This letter is entirely friendly and unaffected.

quotations from the Koran, and so forth,[27] which took months of diplomacy to surmount, not to mention some deletions.

We now come to the most serious of the quarrels between the two men, which may indeed have contributed to the decline of the Accademia del Cimento. In the spring of 1665 Viviani, after making some experiments with a four-pounder gun (*saltamartino*), had made for the Grand Duke a *tariffa* or charge table relating the range of the shot to the charge of powder. The Grand Duke had showed this to Borelli, who made a snap judgment that it was incorrect. There was a long discussion among the experts, at the end of which Borelli was shown that he was wrong, but instead of frankly admitting his error he denied that he had ever seen Viviani's tables, thereby flatly contradicting the word of the Grand Duke, and causing Viviani to write the following very uncharacteristic letter, printed by Tenca[28] from a manuscript in the Laurenziana.[29] It was written on March 31, 1665, probably to Bruto Molara, a courtier.

I read the crafty and lying letter of the man who claims to be sincere and sets himself up as a philosopher, or lover of truth. Even though his twistings are no news to me, I am greatly shocked, not only by being treated so badly (for I shall have to think about getting myself out of this in case His Highness is not sufficiently aware of the ignorance, as well as the malignity, of this fellow), but by his having taken such a liberty with a Grand Duke who is as acute as Campanella or Pico della Mirandola[30] and not inferior to them in memory, in feigning ignorance in His Highness' presence and pretending not to remember having heard of and seen my charge-tables (*tariffe*). And in having contradicted him, saying, "No, Sir, I was doing it this way too, but I soon became aware that I was going wrong, because the times of these trajectories and other things like that were out of place there." This is called trying to confuse the lynx-eyed, or to have him transformed into a booby.

Oh, dear God, what effrontery! what impertinence! . . . They say at the Palace that they expect His Highness to return before Easter. I shall speak with him then and tell him what I think. Really I ought to be satisfied with the outcome of this business up till now, because I am on top, my charge table is all right, he admits it even though he denied it at first . . . . But I still owe him this, and worse, for on this subject I must remember . . . that on one occasion His Highness, perhaps out of compassion and to point out my excessive stupidity, said these exact words about that man, "he is quick, he knows more about things of this kind than any of us; he

[27] G. 276, 19r–20r, Carlo Dati to Prince Leopold, March 22, 1660.
[28] Tenca (1956), pp. 119–20.
[29] *Cod. Ashburnham* 1811, fol. 14.
[30] Viviani's choice of examples is an interesting indication of his philosophical standpoint.

is a Sicilian[31] and I know that he likes to play the tyrant," and on that account, when I felt that this enmity was to be brought into this business, I had either to get out of it or—even I—to deal with it by means of trickery or superior force.

Ferdinand's judgment of Borelli showed his usual perspicacity, and it is greatly to the credit of that excellent prince that he did not take severe measures with him for this really disgraceful conduct. But Viviani never forgot it.

Nevertheless I think that the main effect of this quarrel was to discourage Prince Leopold, whose continued interest was essential to hold these centrifugal talents together. "I assure you," he wrote to Michelangelo Ricci at the end of May 1665, "that if I had a little more time, and if discord had not come in between these virtuosi, I should hope that under the protection of the Grand Duke, and with greater diligence, I could give them such help as would produce something good."[32]

It is interesting, none the less, that this deplorable incident seems to have set Borelli thinking about the problem of percussion,[33] and this resulted two years later in his *De vi percussionis* (Bologna, 1667).

3. THE DEPARTURE OF THREE ACADEMICIANS

In my opinion, the effective cause of the dissolution of the Accademia del Cimento was the departure of three of its members from Tuscany during the year 1667, and especially that of Giovanni Alfonso Borelli. The others were Rinaldini and Uliva. To substantiate this judgment, let us suppose that when Leopold returned from the long visit to Rome that he made after receiving the purple, he had tried to revive the Academy; who then would have been its effective members? Only Viviani and Leopold himself. There is evidence that Leopold did indeed try to get someone to replace Borelli in the Chair at Pisa; this is a copy of a reply from René Sluse of Liège to a letter from an unnamed correspondent, possibly Giovanni Filippo Marucelli, who had been the Grand Duke's Resident in Paris. It is dated July 1, 1667.[34] Sluse says in part:

I think that the resolve of Sig. Giovanni Alfonso Borelli to retire to the leisure of his own country will do great harm to the University of Pisa, a harm proportionate to the advantage that it received from the erudition of such a man. Prince Leopold honors me too much and contrary to any merit of mine in wishing to hear my opinion about a successor. I should

---

[31] As we have seen, Borelli was really born in Naples.
[32] Quoted in *TT* 1, 424–25.
[33] See *G.* 277, 152r–v, Borelli to Prince Leopold, April 6, 1665.
[34] *G.* 315, 977r–980v.

deem myself happy if I could serve His Highness, but to tell the truth I know of nobody—I will not say in these parts where few or none pay any attention to mathematics, as you know—but even in France, who would dare to undertake this task.[35]

A handsome tribute to Borelli, at least.

There was talk of taking the Danish geologist and anatomist Nicholas Steen (Steno) into the Academy. Magalotti wrote to Prince Leopold from Antwerp on January 6, 1668:

Really in the present scattering of our Academy by the departure of Borelli, Uliva, and Rinaldini it is my belief that nothing more desirable [than the acquisition of Steno] could happen. If the other two places were filled in this proportion, it would seem to me that we should have some reason to console ourselves in our loss . . . Borelli was a troublesome man—I had almost said intolerable—but in essence he was a virtuoso to give splendor to a court, for he had sound judgment.[36]

There is a surprising unanimity regarding Borelli among those who knew him best. No one could either forget his faults or deny his great talents.

It is evident that before the end of 1666 Borelli had thoughts of leaving, for on December 23 of that year he wrote to Uliva from Pisa, asking him to recover a number of books and other things that had been left in Florence, pack them in a certain box, well nailed up and tied, and send them down the river by boat.[37] On March 18, 1667, he wrote to Prince Leopold asking to be relieved of his post at Pisa and saying that he desired to return to Messina because of his advancing years and because the weather at Pisa did not suit him. He asked that he be allowed to retain his title of Lecturer at Pisa, and that a small part of his stipend be continued.[38] Leopold immediately wrote to his brother the Grand Duke, who was in Pisa at the time. There was a rapid and efficient postal service between Pisa and Florence, and on March 23 Ferdinand was able to reply that he had made inquiries and had found that the weather at Pisa was only an excuse, and that Borelli had been offered 200 scudi per annum at Messina, as well as a small villa outside the town. The Grand Duke drew the conclusion that Borelli's restless nature was what made him want to leave and decided that he should be allowed to keep his title but not any of his stipend. "But I cannot deny him leave to go, nor do I ever deny this to anyone who seeks it." [39]

[35] *Ibid.*, 979r–v.
[36] *LIUI*, I, 295. The original is in G. 278, 119r–20v.
[37] Quoted in *TT* 2, 242–43.
[38] *LIUI*, I, 133–34. Original in G. 281, 92r–93r.
[39] The letter is in *LIUI*, I, 135–36.

Riguccio Galluzzi suggests a plausible reason for the departure of the three Academicians which might indeed apply to Borelli—the generous pensions given by Louis XIV to Dati and to Viviani.[40] He says that this was as if the King and his minister, Colbert, had decided on the superior merit of these two, "so that it vexed the others, whose self-respect did not allow them to admit to inferiority."[41] Borelli might have felt a little better about it if he had known that Jean Chapelain, who was asked for recommendations by Colbert, had given him the names of both Borelli and Viviani in the same letter and with equal praise,[42] so that the decision seems to have been quite arbitrary.

Borelli was on his way to Messina that summer, for he wrote to Prince Leopold from Naples on July 12[43] and from Messina on August 23.[44] While he was in Naples he performed some of his experiments before the Accademia degli Investiganti, which had been reorganized in 1663 by Tommaso Cornelio and Sebastiano Bartoli.[45] He had been made a member of this Academy.

The excuse about the climate of Pisa was also used by Rinaldini, and he too was given leave to go away and accepted a position at Padua with a salary of 900 ducatoni[46] a year, arriving early in December 1667. On December 15 he wrote to Leopold, "After many days of travelling I arrived in Padua where, either from the quality of the air or that of the wines, I have obtained singular benefits in a short time."[47] We may be sure that Leopold took this with a grain of salt, and indeed he seems to have taken Rinaldini's departure rather badly, for on January 6, 1668, we find the latter writing to the bibliographer Antonio Magliabecchi, complaining that he had not been sent a copy of the "libri dell' esperienze," although Leopold had sent a copy to Cardinal Barbarigo.[48] Borelli, on the other hand, was sent twelve copies and distributed them among the Sicilian letterati.[49]

Uliva went to Rome and to his death.[50] I have not found any re-

[40] See p. 37 above.

[41] Galluzzi, Istoria del Granducata di Toscana sotto il governo della Casa Medici, etc., 5 vols. (Florence, 1781), IV, 170–71.

[42] Chapelain to Colbert, June 22, 1663, in Lettres de Jean Chapelain, ed. Ph. Tamizey de Larroque, 2 vols. (Paris, 1880 and 1883), II, 312–13.

[43] G. 278, 35r–36r.

[44] G. 278, 42r.

[45] Max H. Fisch, in Science, Medicine, and History. Essays . . . in honour of Charles Singer, 2 vols. (London, Oxford Univ. Press, 1953), I, 536. I am indebted to Professor Thomas B. Settle for this reference.

[46] "A silver coin . . . worth 5s. to 6s. sterling." (S.O.E.D.)

[47] LIUI, I, 187–88. Original in G. 278, 115r.

[48] Cited in TT 2, 280.

[49] Borelli to Prince Leopold, from Messina, May 14, 1668, G. 278, 174r–v.

[50] See p. 36.

liable account of the circumstances of his departure, which at any rate can have done no further damage to the Academy.

4. THE QUESTION OF THE CARDINALATE

On December 12, 1667, nine months after the last recorded meeting of the Accademia del Cimento, Prince Leopold was made a cardinal by Pope Clement IX, who had been elected on June 20 of the same year. Of the popes of that epoch, Clement IX was probably the least unfriendly to science and learning.[51] He was also the most friendly to the Medici family, and they hailed his elevation with joy, as well they might.[52] It will soon become clear to the reader that Leopold's cardinalate was a political, one might almost say a family, matter.

A priori one would not think of Leopold as a man likely to be made a cardinal. Too many of his acquaintances knew about his libertarian views for them to have been unknown to the hierarchy. They are familiar material in the letters; for example: "[Antonio Uliva] believes that he can convict the Book of Genesis of error in several places. Joking apart, he must be right this time, because Signor P. L. believes it."[53] He was interested in Jansenism, hardly in favor at Rome. Segni writes from Paris on April 2, 1666: "Your Highness commands me only to provide you some books from the Jansenist list already sent you.[54] As there are a considerable number of these, and they are heavy, I think it well to send them to Your Highness in a parcel, by sea, and not otherwise, by the courier. They will perhaps contain some doctrine not entirely approved by the Apostolic See."[55] Sergio Camerani believed that Leopold was always interested in religious questions, even if he did deal with heretics. "A friend of Calvinists, a reader of prohibited works, an attentive and not hostile observer of Jansenism, even after it was first condemned in 1653, he revealed tendencies that may have worried the Church."[56] Thus, thinks Camerani, it is permissible to suppose that Clement IX made

[51] See his biography in Angelo Fabroni, *Vitae Italorum doctrina excellentium*, 20 vols. (Pisa, 1778–1805), II, especially pp. 123–27.

[52] See *FAS*, filza Medici del Principato 5539, letter 89, Torquato Montauti to Leopold, June 25, 1667. Montauti was the Grand Duke's Resident in Rome.

[53] Magalotti to Segni, from Florence, Oct. 25, 1665. Printed by Ferdinando Massai, *Rivista delle Biblioteche e degli Archivi* 28 (1917): 126.

[54] On February 6 (*G*. 315, 841r–v, the covering letter).

[55] *G*. 277, 265v–66r.

[56] Camerani, *Arch. Stor. Ital.* (1939, vol. I), p. 28.

Leopold a cardinal to keep him away from the Jansenists and "from being able to establish in Tuscany a dangerous center of anti-Catholic propaganda."[57] Although it may be true that "the fundamental problem of Clement IX's reign was Jansenism,"[58] the facts, which are remarkably clear, are far different. I have already mentioned[59] that Fabroni, who in 1769 believed—possibly invented—the story about the exchange of a cardinal's hat for an Academy, later changed his mind. This occurred at some time before 1778, when during the composition of his biography of Clement IX, he found in the Medici Archives three letters between Torquato Montauti and Ferdinand II, whose Resident in Rome Montauti was for many years. His attention was probably drawn to them by the historian Riguccio Galluzzi, who was employed in reorganizing the Archives and referred briefly to these letters in his *Istoria del Granducato di Toscana*, pubished in 1781.[60] In the notes to his biography of Clement IX, Fabroni reprinted the first two of these letters and an extract from the third,[61] and in view of their great importance to the history of the Accademia del Cimento I reproduce them here in translation. After an extensive search in the Archivio di Stato I have been able to find the third of these letters, and also four later ones exchanged between the Grand Duke and Montauti on the same subject, which serve only to confirm the conclusions that can be deduced from the earlier ones.[62] The originals of the two earliest letters have eluded me.

On July 11, 1667, only three weeks after the elevation of Clement IX, Count Torquato Montauti wrote an urgent letter to Ferdinand, as follows:

MOST SERENE PRINCE:

I hope that my presumption in having my own letters delivered to Your Highness[63] will not displease you, as it is an act of obedience that I owe to the orders of His Holiness. After he had graciously listened to me, he asked me whether Your Highness had commanded me to submit anything else to

---

[57] *Ibid.*

[58] *Encycl. Ital.*, s.v. "Clemente IX."

[59] See p. 309.

[60] Galuzzi, *Istoria del Granducato*, IV, 169.

[61] *Vitae italorum* . . . , II, 182–86.

[62] *FAS*, Miscellanea Medicea, filza 334, inserto 1. The five letters are: Montauti to Ferdinand, July 19; Ferdinand to Montauti, July 26; Montauti to Ferdinand, August 2; Ferdinand to Montauti, August 9; Montauti to Ferdinand, August 13. Those from Montauti are autographs, the others drafts in Ferdinand's handwriting. The entire correspondence is being printed elsewhere.

[63] I.e., rather than to the secretariat.

him. When I told him that perhaps I had bothered the Holy See too much, he remarked that he had to deal with me in important business that needed secrecy and should therefore pass between His Holiness and Your Highness through me. Under these conditions he acquainted me with this, ordering me to write to Your Highness and to bring him the reply.

He told me that to the suitability of giving a cardinal's hat to the most Serene and so worthy House of Tuscany was added his desire that the Holy See should demonstrate the grateful memory that it keeps of Your Highness for the continual favors bestowed by you on him and his House. Therefore without waiting for any other urging he declared that he would very readily be satisfied with one of Your Highness' brothers, since he could not at present go farther afield, as he might perhaps have the opportunity to do at some other time. But as in carrying out this intention of his he was meeting with a considerable difficulty, that is to say the desire of both these princes for the cardinalate, he begged Your Highness to overcome it with your authority and prudence. He assumes that Pope Alexander[64] had reflected deeply on this circumstance, but said that it gives him much more to think about, because he esteems both of these princes equally, and knows that his family is under equal obligation to both, so that the gratification of one to the displeasure of the other would be but an imperfect satisfaction to His Holiness. Nor, for this reason, would he desire you to answer that Your Highness and his brothers are resigned to the wishes of His Holiness, but rather that you should seriously declare to him which would give you the greater satisfaction and be of greater service to your House; and that you should also assure him that the other would be resigned to it. In this manner he said that he deems it his very good fortune that his predecessor had left him this opportunity to show his feeling for your House by a just and appropriate act.

It seems to me that I have sufficiently expressed His Holiness' meaning, and now I take the liberty of pointing out to Your Highness that I do not think we have any time to lose, because the Pope needs assistance, so that it must be assumed that he will soon want to make a cardinal of his nephew, which is credible, and it almost seems that he may say that he will promote the Prince as well. I shall not omit to suggest humbly that all the servants and friends of your House are impatient to see those people rebuked who have neglected this duty, and also to see you bright with all that splendor that they know is your due. This much is desired for you by your true servants. Here I humbly bow to Your most Serene Highness, from whom I shall await orders as to how to behave in future in writing about this business.

Your Highness' most humble
servant and vassal

Rome, 11, July 1667   TORQUATO MONTAUTI

---

[64] Alexander VII had died on May 22, 1667 (*Encyl. Ital.*).

Ferdinand replied almost by return of post:

TO COUNT MONTAUTI   15 July, 1667

The very gracious expressions with which His Holiness has been pleased to declare his generous and cordial sentiments towards us and our family, just as they have exceeded our expectations, though these were great, have likewise filled our soul with holy consolation and increased our obligation to such an extent that we do not know how to find fitting words to attest the duty that we owe to the Holy See. And indeed, this finding ourselves forestalled in what we had been considering asking His Holiness when time should have ripened the opportunity, and being invited with feelings of paternal benevolence not to lose the present occasion for having one of the princes our brothers promoted to the cardinalate, is a reward that will remain forever impressed on our memory. We shall await impatiently the chance of showing our gratitude towards His Holiness, as the sublime position in which the divine Providence has placed him leaves us little hope of serving him, other than with a most reverent humility and the total submission of our will.

In the meantime, reverently accepting the favors that His Holiness spontaneously shows towards us, we shall be glad to reveal with complete sincerity and confidence what has occurred up to now in this matter, and what has formerly delayed and still is delaying our decision. You should therefore know that following the death of the senior cardinal our uncle, we devoted some thought to getting the hat for one of the princes, our brothers. But while we remained in some uncertainty, since one seemed preferable because of his age, and the other because of his nature and his calling, and both desired it, Prince Mattias came down with a very troublesome malady that made him more needful of rest than fit to embroil himself in new pursuits. But he has not ceased to try various remedies, and at present it is the showers at the Baths of San Filippo, to see if the illness can be successfully cured, so that we should not yet give up hope. Thus, as we are in a dilemma of this kind, we have been putting off our decision in order not to disturb the more than ordinary good will that has always existed between our brothers, by giving the older one a pretext for taking offense because the younger one had been given the preference. But since we are now on the point of seeing the effects of the baths, we do not intend to put off making up our mind, or receiving the favors of His Holiness, any longer, only begging him to be pleased to give us a few days time to declare which of the two princes we shall put forward. We desire this short time to undeceive the one who must stay behind, also following in this the concern of His Holiness, that one of them should not reasonably be offended by the promotion of the other. But should time press and not give us room to delay the business further, as we wish wholly to enjoy the favors of His Holiness, let couriers be sent specially and with all haste to warn us of the urgency, since we shall send you a reply in the same way, and orders as to how you must behave, being confident that in favoring us His Holiness will have chiefly in mind that which we deem to be of the greatest service to our House, and in consequence to his own, since our interests must have so much in common.

In dealing with such a serious business, do your best to be at His Holiness' feet as quickly as possible, and as long as you do not think you are bothering him, you might read him this letter so as to be more certain to report our exact ideas. And may God our Lord bless you and preserve you.

Fabroni finally prints an extract from the letter written by Torquato Montauti to the Grand Duke on July 19, 1665:

He [Clement IX] added that he might even have enlarged his plans, if this should have been necessary to Your Highness, as he particularly desires your satisfaction; and on this point he repeated to me at length the favors he had received from Your Highness, both personally and for his family, assuring me that they wished solely to depend on your House, and that their greatest merit would be to keep themselves subject to Your Highness,[65] thinking by this to provide you with a suitable house here, and a vine a little better than ordinary; so that at any time whoever might wish to apply to the Court of Rome should find these advantages; but he did not wish anything more grand. This idea of His Holiness may vary with unforeseen events, but the disposition he demonstrates to favor the interests of Your Highness and to maintain a real correspondence with you seems very likely to last.

A week later, on July 26, Ferdinand informed Montauti that the baths of San Filippo seemed to be of no help to Prince Mattias, and that the doctors were pessimistic, so that he had decided to ask for the promotion of Leopold. The Pope was told of this on the first of August and agreed, as Montauti reported the next day.

It seems to me that these letters show clearly that it did not matter at all to the Pope whether Mattias or Leopold was elected; he was determined only that there should again be a Medici cardinal. As a matter of fact it is clear that in 1655 Pope Alexander VII had wished to make a cardinal of one of the Medici princes, for there is a file in the Miscellanea Medicea with the title "Overtures made in Rome towards the choosing of a prince of the Most Serene House, either Mattias or Leopold. September 1665."[66] Some of the letters from Montauti to the Secretariat in Florence are in cipher, with translations between the lines, and in one of these dated September 5, 1665, we read, "... the Pope inclines more to Prince Leopold, saying that he wears the cassock, that he has studied and made himself a churchman, and that the other has not done so, has not studied, and is a soldier."

[65] Clement IX, born Giulio Rospigliosi, was from Pistoia, and his family were therefore subjects of the Grand Duke.

[66] *FAS*, Miscell. Med., filza 98, inserto 4.

Prince Mattias had frequently been in poor health, especially after a bad case of malaria in 1639. In 1667 he was suffering from gout— endemic among the Medici—and would scarcely have been the better choice. In any event Ferdinand's problem was solved by the death of Mattias on October 11, and on December 12 Leopold was elected and the faithful Montauti wrote to him enthusiastically: "God be thanked, Your Highness was created cardinal this morning with such universal jubilation and applause that Your Highness must really think that the whole Court rejoices in your new dignity, . . ." and goes on to list in his own writing the important people who had sent their congratulations. Finally, he had been to the *berrettaio*, who is sending some red hats to Florence in a hurry.[67] The next day Montauti wrote again, sending the congratulations of many more people; and the hat-maker is sending two more *berretti*, as those he sent yesterday pleased neither Montauti nor himself.[68]

I suspect that Leopold enjoyed this pomp, but he did not go to Rome until March 1668, staying there until June. His many letters to his brother the Grand Duke[69] show his enthusiasm for the city, its works of art, and its aristocratic society. Later[70] he became involved in ecclesiastical politics, and the letters change their tone, but it is worthy of note that he made only two visits to Rome, the other being at the end of 1669 to attend the Conclave that elected the successor of Clement IX.[71]

5. Leopold as Cardinal

There is abundant evidence that although the Academy had ceased to function, Leopold continued his correspondence with his scientific friends abroad, particularly the astronomers, and maintained an interest in science, though at a lower level. His extensive correspondence with Michelangelo Ricci[72] shows his abiding love of astronomy. He retained his protective attitude toward natural philosophers, even people like Fabri, the staunch defender of the geocentric system, who, abandoned by his Jesuit colleagues because he had busied himself with science, was tried by the Holy Office in 1671 and incarcerated until on February 20, 1672, Leopold wrote a letter to Cardinal Altieri asking compassion for him.[73] In spite of Borelli's defection, Leopold con-

---

[67] *FAS*, filza M.d.P. 5539, letter 115.
[68] *Ibid.*, letter 116.
[69] *FAS*, filza M.d.P. 5508.
[70] *Ibid.*, letters 40, etc.
[71] A. Favaro, *Atti R. Ist. Veneto* 71 (1912): 1173–74.
[72] There is a good sample in *LIUI*, II, *passim*.
[73] The draft is in *FAS*, filza M.d.P. 5508, fol. 101.

tinued their scientific correspondence.[74] He even made a few experiments, for instance some on bio-luminescence in July 1669.[75]

It is clear that he had no expectation whatever that the Academy would be revived. Even on February 10, 1668 he could write to Christiaan Huygens as follows:

Even if my occupations had allowed me to go with my experimental Academy vigorously and with the aid of a respectable number of brilliant men, I should not have omitted to desire sincerely that the Academy set up by His Most Christian Majesty[76] may operate with the greatest possible profit to everyone. You may consider how much more greatly I wish for this at this time when I can expect little, and when three of the best men that there were in the Academy have left our service; but it will be precious to me to feel that the work goes on where you are, with those two purposes, so important: to look over the great book of nature by means of experiments, and find new things never heard of before, and to purge books of those experimental errors that have been too easily believed, even by the most esteemed authors.[77]

Later in the same letter there is a passage that makes me wonder whether Leopold finally had doubts about the entirely empirical program of his late Academy: "One must finally conclude that he who wishes to be a good philosopher must also be a good geometer."

There is an interesting entry in the *Journal Book* of the Royal Society under the date December 17, 1668, as follows:

At this meeting were present two Italian Gentlemen; the one the Marquesso Ricardi, the other Sign.r Alexandro Segni, both introduced by Count Ubaldino. They acquainted the Society of the singular respect, the Cardinal Leopoldo de Medicis had for them, and he desired to have his Excuse made for not having himself returned his acknowledgments for the Book sent to him, (Viz.t The History of the R. Society)[78] which he had been hindred to do, by his lately received dignity of Cardinal, but that since that time he had desired, and already *obtained the Pope's permission to correspond with this Society*, which he now intended to make use of, to let them see the esteem he had of them and their Institution.[79]

Whereupon the President made a suitable reply, welcoming any correspondence that the Cardinal might send. It seems to me that the passage I have put in italics drives another very strong nail into the coffin of the "red hat for an academy" theory.

[74] Borelli's letters to the Cardinal are in G. 278.

[75] Leopold to Borelli, July 25, 1669 (*LIUI*, I, 143–45). Original in G. 282, 171r–72r.

[76] The formal title of the King of France. Leopold is referring to the new Académie Royale des Sciences.

[77] G. 282, 150r–v (Draft). Printed in Huygens, *Oeuvres*, VI, 106.

[78] By Thomas Sprat (London, 1667). See also p. 289.

[79] Royal Society, *Journal Book*, IV, 14–15. My italics.

6. THE DEAD ACADEMY

One cannot conscientiously report that the dissolution of the Accademia del Cimento made a great stir in the learned world, although Italian authors have made a tragedy out of it. Lorenzo Magalotti, who certainly had a vested interest in the Academy, seems to have been unwilling to admit that it had ceased to exist. Adelmann cites a letter from Magalotti in Paris to Leopold, dated June 1, 1668, rejoicing in the fact that the place of Borelli in the Academy had been so well filled by Steno;[80] it is therefore of interest that Steno himself had thanked Leopold for his copy of the *Saggi* in a letter[81] containing the following passage:

the greatest talents in all the experimental academies show a special eagerness to celebrate the noble achievement of the Accademia del Cimento, not without a very great sorrow in seeing stopped by unforeseen accidents a study whose very great advances in researches into nature, made in a short time, give everyone clear indications of what might have been hoped for from its continuation.

Magalotti even wrote a sonnet, which I shall not attempt to translate, with the title "To the Most Serene Prince Leopold of Tuscany. An invitation to take up once more the study of scientific experiments, and to continue his correspondence with the Academies of Paris and London."[82] But the Accademia del Cimento was entirely extinct.

The impact of its ending was undoubtedly greatly lessened by the circumstance that it was not terminated by any edict or announcement, but simply ceased to function, so that its passing made no noise. On May 9, 1668, Francesco Lana wrote from Brescia to the Accademia del Cimento the only letter I have discovered that was written to the Academy as a body.[83] I do not know whether he received an answer. On June 29 Oldenburg wrote to Adrien Auzout from London, "We hope that that great prince of the Roman Church, Cardinal de' Medici, will never leave off philosophizing, or making his academicians philosophize."[84]

---

[80] H. B. Adelmann, *Marcello Malpighi and the Evolution of Embryology,* 5 vols. (Ithaca, N.Y., 1966), I, 142. Adelmann seems to believe that the Academy kept on, perhaps because he relied too much on Caverni's *Storia del metodo sperimentale in Italia,* 5 vols. (Florence, 1891–98).

[81] Steno to Prince Leopold, from Pisa, January 14, 1668; G. 315, 1014r–15v (copy).

[82] *LIUI,* I, 312.

[83] G. 284, 11r–13r.

[84] *The Correspondence of Henry Oldenburg,* ed. & transl. by A. Rupert Hall and Marie Boas Hall (Madison & Milwaukee, 1965–[in progress]), IV, 482.

Nevertheless, the Academy was dead. An abortive attempt to revive it was made by the French government of Joachim Murat in 1801. In the official *Gazzetta Universale* for March 14 there appeared the following notice:

Florence, March 13

The government of Tuscany, ever intent on the promotion and reproduction of those establishments that can bring honor and benefit to Tuscany, has decreed the re-establishment of the Accademia del Cimento, so celebrated under the rule of the Medici, and which the well-remembered Leopold[85] was thinking of restoring to its ancient glory. As this Academy has therefore been re-established, a number of scholars have also been nominated who will be admitted to it for the time being, and among these our illustrious mathematician Dr. Pietro Feroni has been pleased to assume the burden of being Secretary, which was formerly carried by the immortal Magalotti. The said Academy will open its first session next Monday in this Museum of Natural History.[86]

The only sequel I have found to this abortive attempt is a letter written a fortnight later by Francesco Chiarenti stating that he was honored to be asked to join.[87] But the majority seem to have been unwilling to co-operate.[88]

7. Summary

I think that we may safely conclude that the Accademia del Cimento came to an end simply because there were not enough men in the Court with an urge to carry on its work. This may be a unpalatable conclusion to those who subscribe to the conspiracy theory of history, but on the other hand it has the advantage of removing from both Prince Leopold and the Pope any possible blame for the sad event. If this explanation is accepted, however, it merely moves the problem back one stage, because one may reasonably enquire about the motives of Borelli and Rinaldini in breaking away from what looks like an enviable position.

Anyone who has worked in a twentieth-century research laboratory, especially during the period of the second World War, will have a very plausible answer to this problem. He will remember the frustra-

---

[85] This was Leopold I, son of Francis II, who reigned from 1765 to 1790 and is considered to be one of the most enlightened of the rulers of Tuscany.

[86] *Gazzetta Universale* No. 21 (March 14, 1801): 166. This is the entire entry.

[87] *Ibid.*, No. 26 (March 31, 1801): 207.

[88] A constitution had been printed by the government press in a thirteen-page pamphlet dated February 27, 1801, a copy of which is in the library of *FMSS*.

tion of obtaining scientific results that were on no account to be published abroad; of being anonymous, in fact. It will be easy for him to imagine the impatience of a Borelli, whose immense talent had to remain hidden, not just for the duration of an emergency and as a patriotic duty, but for years and years and for no good reason whatever, his results not even to be published anonymously until a charming poet-out-of-Arcady had finished his preposterous literary labors. This was no way to foster the enthusiasm of talented researchers, and it would have been entirely out of character for Borelli to have remained.

8. THE FATE OF THE INSTRUMENTS AND APPARATUS

All authorities agree that it is impossible at this time to distinguish between the instruments and apparatus that belonged to the Accademia del Cimento—or perhaps we should say to Prince Leopold—and those associated with the experiments made by or for his elder brother the Grand Duke. It would not, in any event, be very interesting to do so, in view of the close relations between the brothers. At the death of Prince Leopold there seem to have been 1,282 pieces of glassware, but many of them were undoubtedly nothing more than bottles, jars, beakers, and so forth.[89]

Most of them still existed, according to Targioni Tozzetti, in 1740, when he saw them in magnificent cupboards in a large room next to the library in the Pitti Palace, the room, he says, in which the Academy used to hold its meetings.[90] Targioni Tozzetti recognized many of the pieces of apparatus from the illustrations in the *Saggi*. Some of them, he later discovered, had been taken home by one Vayringe, the royal instrumentmaker,[91] and it seems that Vayringe had never heard of the Accademia del Cimento and was astonished when he was shown a copy of the *Saggi*. After the death of Vayringe, his royal master, Francis, by this time Emperor, had part of the instruments packed up and sent to Vienna.[92] These have apparently all been lost or destroyed.

In 1765 Francis' younger son Leopold of Lorraine succeeded to the throne of Tuscany, and, unlike his father, came to live in Florence. In 1776 he had an inventory made, in which 547 pieces of the glassware were numbered.[93] Since then, breakage and loss has reduced the

---

[89] Museo di Storia delle Scienza, *Catalogo degli strumenti* (Florence, 1954), p. 27.

[90] *TT* 1, 464.

[91] Tuscany was ruled at this time by Francis II of Lorraine.

[92] *TT* 1, 465.

[93] *Saggi*, 1957 ed. (See Appendix A, no. 13), Appendix, p. 3.

total to about 233 pieces, some damaged.[94] The survival of many of these delicate instruments, and especially of a large number of thermometers, through the period of the Napoleonic wars was due to their having been carefully packed away in a case that was forgotten until 1829,[95] by which time Tuscany was under the rule of still another Leopold, an enlightened man interested in the history of the Province and sympathetic towards science.

Today the surviving apparatus is among the treasures of the Museo di Storia della Scienza. Many of the finest pieces are described and illustrated by Maria Luisa Bonelli in the appendix to the 1957 edition of the *Saggi*.[96] Fortunately the glassware of the Academy was out of the reach of the waters in the disastrous flood of November 1966.

[94] Museo di Storia della Scienza, *Catalogo* (1954), p. 28.
[95] G. Libri, *Ann. Chim. et Phys.* 45 (1830): 359.
[96] See Appendix A, no. 13; also her article in *Vetro e Silicati* 2 (1957): No. 6, pp. 21–26.

8.

CONCLUSIONS

There remains the task, difficult even though it is fascinating, of coming to viable conclusions about the motivation for, and the results of, the actions of the Accademia del Cimento. In doing this it is necessary to adopt an independent attitude, and this is probably less difficult for one who is not Italian. Except for Targioni Tozzetti, who wrote nearly two centuries ago, and the editors of *DIS*, who had in view a particular scholarly obligation which they discharged admirably, no one has written at length about the Academy; and few people have made an extensive study of the relevant manuscripts. I shall therefore try in this concluding chapter to present from a fresh point of view an estimate of what, with a little stretching of language, might be called the ecological relationship between the Accademia del Cimento and its century.

First it is necessary to discuss the so-called Baconianism of the Academy. It is fatally easy for English-speaking writers to assume that whatever influence they happen to think Francis Bacon had on seventeenth-century science in England—and their opinions on this vary widely[1]—this influence was exported to the whole world as if it were woollen cloth or iron bars. South of the Alps, at any rate, this was not so. I have already noted[2] that Bacon's works did not figure in the long list of books prepared by Rinaldini in 1656 at the Prince's orders. There is another piece of evidence even more demonstrative of this contention. In the truly excellent general index in volume 20 of the National Edition of Galileo's works, there are only two references to Francis Bacon. And what are these? Letters written to Bacon in 1616 and 1619 by Tobie Matthew, the first[3] sending him a copy of part of a letter from Galileo to Castelli, the second[4] concerning some books by Galileo. As Matthew translated Bacon's *Essays* into Italian in 1618, and as

[1] See p. 3 above.
[2] See p. 56 above.
[3] *G.OP.* XII, 255.
[4] *Ibid.,* 450.

there is no sign that Galileo read these or any of Bacon's Latin works either, we may assume that on Galileo, at least, Bacon had no influence at all, as we might indeed expect in view of the entirely different outlook of the two men.

But we must examine how "Baconian" the Academy was in the wider sense that assumes that any more-or-less systematic course of experimentation will lead to a body of scientific knowledge, and also redound to the good of mankind, the latter an important part of the Baconian philosophy. Even on such a definition the Accademia del Cimento was in no way Baconian, for a large proportion of their experiments were directed towards the demonstration of hypotheses, and those that were not were founded on intellectual curiosity and certainly not, with perhaps one exception,[5] on altruism. In saying this I am not forgetting the several places in Prince Leopold's correspondence in which he expressed the hope that the work of the Academy might be useful to the world of learning (*repubblica letteraria*). This utility was to natural philosophy, not to the general human condition.

It is more difficult to be sure of the influence of Robert Boyle, who is cited four times in that part of the *Saggi* that deals with experiments in the vacuum. Two of these, the one on the boiling of lukewarm water in the vacuum,[6] and the one on the solution of pearls and coral in vinegar,[7] were directly credited to Boyle; but they were performed after the Latin edition of Boyle's *New Experiments* had reached Florence, or at least several months after copies were sent to Prince Leopold and to Viviani.[8] It is quite clear that Boyle's book greatly impressed the Academicians, but by the time it arrived most of their own pneumatic experiments had been made—in July and August 1660—although several were repeated later. It is unlikely that they had seen the English edition by this time, or been informed of Boyle's experiments in some other way, for none of the references to Boyle occur in the document known as Draft A,[9] and only one of the relevant experiments, the one on sound, is there at all.[10] The Academy's experiments in the Torricellian vacuum began only in June 1660, in contrast to their experiments on atmospheric pressure, which are given a separate section in the *Saggi* and were virtually completed in 1657. The document translated on page 264 might seem to indicate an earlier knowledge of Boyle's works, but, in view of the fact that all

[5] See p. 234 above.
[6] *Saggi*, p. CVIII.
[7] *Ibid.*, p. CXI.
[8] See p. 263 above.
[9] See p. 68 above.
[10] *Saggi*, p. LXXXXVII. This is unlike Boyle's experiment on this subject.

but one of the experiments with animals in the vacuum date from 1662, the evidence is against it.

Indeed there exists a letter written by Borelli on July 21, 1662, probably to Malpighi, which shows that Boyle's *New Experiments* came as an unpleasant surprise. "As to Boyle," writes Borelli, "I have seen it with displeasure, because there are many of our Academy's things in it."[11] He could clearly see that a great deal of priority had been snatched from his beloved Italy.

The mainspring of those activities of the Academy that were not merely a result of curiosity was neither Baconian nor Boylean, but something much more to their credit, a desire to continue the work of Galileo in demolishing the authoritarian pseudo-science of the schools. This desire was particularly strong in Prince Leopold himself, who, as we have seen, held very advanced views and was no doubt supported by the Grand Duke, if we can judge by the sort of men that were appointed to the staff of the University of Pisa during his reign.

As I emphasized in the first chapter, no discussion of any aspect of the Italian natural science of this period can ignore the fact of the Counter-Reformation, especially after the condemnation of Galileo in 1633. The ecclesiastical authorities, under the leadership of the Jesuits, were doing their best to stamp out any innovation, and scientific men were under a restraint similar to that imposed on intellectuals in totalitarian countries in our own day, with the occasional but uncertain mitigation afforded by the protection of powerful rulers. Cosmological speculation was particularly dangerous, and the entire field of astronomy was suspect. Experimental physics and chemistry might have been expected to be harmless, but they could, of course, be used to attack some of the sacrosanct doctrines of Aristotle (as interpreted by countless commentators), and this was considered by the authorities to be subversive of public order. If you had to make experiments, it was better not to discuss the results to any extent. The peculiar empiricism of the Accademia del Cimento can be understood only in such a context.

In considering how adequately the Academy carried out its purpose, we must separate two fields of enquiry—the influence of the *Saggi*, at the time and later, and the success and pertinence of the experiments themselves, many of which, it must be remembered, remained unpublished until 1780 and had no more influence in their own time than the notebooks of Leonardo da Vinci in his.

---

[11] Modena, Biblioteca Estense e Universitaria, *Autografoteca Campoli*, s.v. Borelli. "Intorno al Boile io l'ho visto con mio dispiacere perche vi sono molte cose della nostra Accademia."

It would be pleasant to report[12] that the appearance of the *Saggi* caused a sensation in Italy and beyond the Alps and brought joy to the enemies of peripatetic dogma. It would also be the wildest sort of exaggeration. Unfortunately the long delay in the publication of the *Saggi* was almost fatal to its purpose; if it had appeared in 1660 or even in 1662, it might have created a much stronger impression. As a matter of fact, a very good book could have been made of the experiments that had been performed by the time the Academy took its long break in September 1658,[13] and by August 1660 it could have been written in almost the form we have it. To judge by the works that appeared during the controversy about Saturn,[14] publication could be very speedy in 1660, and the delay that actually occurred is an example of how "the best is the enemy of the good."

As we have seen, Leopold gave the book away with princely munificence, and none were available for sale. This may be the reason that actual reviews of the work are hard to find, for an editor might see little point in printing a review of a book that his readers could not buy. Oldenburg published only a list of the chapter headings in the *Philosophical Transactions*,[15] which journal, it should be remembered, was his own private venture at this time.[16] I have not been able to find any mention of the *Saggi* in the *Journal des Sçavans*. In fact, the only formal review that has come to my notice was printed by Francesco Nazari in the very first number of his *Giornale de Letterati*,[17] which began to appear in Rome early in 1668. This review begins as follows:

Not only as regards its style, which may serve as a model to students of good Tuscan idiom, but also by the nobility of its subject, which deals with many experiments by which philosophy may be enriched with at least a thousand truths by someone who knows how to manage them and use them in his treatises.[18] Nor must we fail to consider the merit of the virtuosi who have contributed to this work their sublime intellect and mature judgment, and much skill, industry, and diligence in making such experiments; and finally that which gives the most singular excellence to the whole, the contribution

---

[12] As was indeed done by J. C. Poggendorf in his *Geschichte der Physik* (Leipzig, 1878), p. 353.

[13] See above, p. 61.

[14] See above, p. 260.

[15] Vol. 3 (1668): 640.

[16] In 1670, an anonymous translation of a passage in the *Saggi*, pp. CXXXXVII–CLIII, appeared in the *Phil. Trans.* 5 (1670): 2020–23. (Note that there are two pages 2020 in vol. 5, which is paged very erratically.)

[17] Vol. 1 (1668): 4–6.

[18] I have followed the peculiar syntax of the original.

made by Prince Leopold, today a cardinal of the Holy Church, with his authority and protection, his sublime judgment and profound intelligence.

The review continues to its uninspiring end with a rather pedestrian summary of the instruments and experiments, especially those on the vacuum. No philosophical standpoint is detectable.

The transalpine academies do not seem to have been impressed. I can find no reference at all to the *Saggi* in the first two volumes of the *Registres* of the Académie Royale des Sciences in Paris, although, as we saw in chapter 6, there had been a great deal of communication between Paris and Florence during the lifetime of the Accademia del Cimento.

The Royal Society was polite but unenthusiastic. On March 19, 1667/8:

"Some account was given by Dr. Pope and Mr. Hooke of the book of the experiments of the academy *del Cimento*, which was, that the many subjects and experiments treated of in it had also been considered and tried in England, and even improved beyond the contents of that book; but that they were delivered in it with much accuracy and politeness, and some of them with an acknowledgment of the origin, whence they were derived."[19]

Two days earlier Oldenburg had written to Boyle:

At our last meeting, the Florentines . . . presented, in the name of prince Leopold, the pompous book of their experiments to the Society; for which they received solemn thanks, which the secretary [i.e., Oldenburg] was commanded by the president to deliver in Latin . . . . Meantime, I understand there is nothing new in it, as to us, except it be perhaps some experiments of amber, and a way of making a map of a country by sounds; I perceive by your last, you have not left the former untried.[20]

This was an exaggeration, but not a great one in 1668, so that even if we make every allowance for their belief that "we do things better at home," there was little reason for the English natural philosophers to cheer.

It is unlikely that Leopold ever became aware of these opinions, and the ones he gathered from individuals to whom he had sent copies

---

[19] Thomas Birch, *History of the Royal Society*, 4 vols. (London, 1756–57), II, 257.

[20] Boyle, *Works*, ed. Birch (1772 ed.), VI, 274. The word "pompous" must not be allowed to pass without comment, as it has been taken in its present-day pejorative sense by Professor A. R. Hall in his introduction to the 1964 reprint of Waller's translation of the *Saggi* (see appendix A, no. 14). The usual seventeenth-century meaning of the word was "magnificent" (O.E.D.), and I am convinced that this is what Oldenburg intended to convey.

of the *Saggi* are of course very different. I shall translate part of one of these letters of thanks; it is from the French savant Habert de Montmor, of whom we have already heard in chapter 6. It is undated.

I have also just received . . . the gift that it has pleased you to make me, of the first samples of the experiments of your illustrious Academy, where everything is excellent, magnificent, and worthy of your genius and that of those great personages who assemble under the protection of your Sublime Excellency. I can assure you that this work has received general approval, and all the scholars and interested people to whom I have shown it have praised it highly to me. Just now it is still in the hands of these new Academicians whom His Majesty has brought together in order to form his Academy. I wish that they themselves may be stimulated by such noble examples to produce some fine discoveries in physics.[21]

It is amusing to wonder whether experienced princes ever tired of such entirely predictable compliments.

Meanwhile the compiler of the work was on his diplomatic mission, and in spite of his well-documented disgust with the whole project, seems—like any author—to have been avid for news of its reception, so that we find him writing to the Prince, now cardinal, from Antwerp on January 6, 1668, a letter that is also of interest because it shows how ignorant of English even learned Italians were at the time. I quote a short extract:

I am in a state of inexpressible curiosity until I begin to hear the judgments that the world will give of our experiments, but now I shall not hear it discussed until Paris, as I imagine that Your Highness has no literary correspondents in Brussels. I begin to form a poor opinion of Florence and Rome, because by now it must be about two months since the book came out, and none of my friends writes me anything either good or bad about it. I am writing to one of them that I appreciate their discretion in telling me nothing, since it must be that they can tell only bad news, but that I should be incomparably more grateful for their honest opinion. But I console myself that the blame falls on me, for the work may well be good, but put together badly.

I hear that there has appeared in London a sort of experimental history[22] in which there are very fine things, but so far it has not even been translated into Latin. That man Hooke, if I am not mistaken, who makes those admirable microscopes, has also printed a large folio volume,[23] with the

---

[21] G. 314, 968r–69r (copy).

[22] This must be Thomas Sprat's *The History of the Royal Society of London for the Improving of Natural Knowledge* (London, 1667).

[23] *Micrographia, or some Physiological Descriptions of Minute Bodies, made by Magnifying Glasses, with Observations and Inquiries thereupon* (London, 1665). It is interesting that the fame of Hooke's microscopy had spread to Italy. Hooke did not himself make micrscopes.

observations of things seen with their aid, all excellently engraved on copper but this too as far as the explanations are concerned is in English, and again has not been translated.[24]

Nor could Prince Leopold read English. When Sir Thomas Baines presented him with Sprat's *History* on behalf of the Royal Society, he "received it wth Expressions of respect, but withal complaining that it was in English, to w.[ch] the D.[r] [Baines] replied that his Eminence by publishing his Experiments in Italian had made the same Example."[25] Nor could the English scholars generally read Italian, to say nothing of the Germans and the Dutch. Prince Leopold would have served his cause better if he had published the *Saggi* in Latin, but this would have gone against the example of Galileo.

The fact that the *Saggi* was translated several times during the century following 1667 (see appendix A) may seem to contradict this appraisal of its effect. It is especially important to account for the translation by Waller,[26] which was issued under the auspices of the Royal Society, as stated in its half-title, at a time when none of the working scientists in that Society can possibly have needed it.

I think there is a simple and yet adequate reason for the appearance of the Waller translation in 1684. While it was of no use to the professionals, there was an enormous interest in the new natural philosophy among educated laymen, and nowhere more so than in England. This "popular" audience was more greedy than critical, and cheerfully purchased such trivia as the translated reports of the fantastic academy organized in Paris by Théophraste Renaudot.[27] The interest of the Royal Society in the translation can be referred to the Society's urgent need to keep its name before the public in the seventeenth century, especially after the long interruption of the *Philosophical Transactions* that followed the death of Henry Oldenburg.[28]

Both the incomplete editions in French are clearly part of collections of memoirs "boiled down" for popular consumption. The Latin editions[29] of 1731 and 1756 present a different problem—not to account for their appearance, but to explain the long delay in providing

[24] G. 278, 119v–20r. Printed in *LIUI*, I, 296–97.

[25] Sir John Finch to Oldenburg, from Florence, July 24, 1668, in Royal Society MS. F1, no. 44; printed in *The Correspondence of Henry Oldenburg*, ed. & transl. by A. Rupert Hall and Marie Boas Hall (Madison & Milwaukee, Wis., 1965–[in progress]), IV, 541–42.

[26] Appendix A, no. 14.

[27] See p. 8 above. The second volume of this vapid stuff (1665) claimed to result from the great success of the first.

[28] The above discussion draws heavily upon a very useful conversation with Professor Stillman Drake.

[29] Appendix A, nos. 15 and 16.

an edition in a language useful to educated Teutons. No one seems to have thought this worth while in the seventeenth century, and in considering this we should remember that Boyle's *New Experiments* was at once put into Latin for continental perusal,[30] and that at least four further Latin editions appeared on the Continent by 1680. My guess about the 1731 edition of the *Saggi* is that Musschenbroek, who was a born experimenter, thought it would be interesting to repeat and improve upon the experiments of the Accademia del Cimento and presently found himself with a collection of laboratory notes that simply cried out for a book. Such a work might even have had a market in Dutch, if we can judge by his later *Beginselen der Natuurkunde*;[31] but he wisely chose the more internationally understood Latin. According to J. C. Poggendorf,[32] the translation was more common than the original in nineteenth-century German libraries, though to judge by the Union Catalog this is not the case in the United States and Canada today.

Musschenbroek's explanation of his decision to translate the *Saggi* would have us believe that it was pure altruism:

this treasury of Florentine experiments was written in the Italian tongue, so that it was unsuited to the use of the students to whom I daily commended it, unless it were translated into Latin, a labor that I was loath to undertake, knowing full well how much experience in both languages was required for such a work, which by far exceeded my powers. But to serve my readers, I was unwilling to refuse the task, nor to spare myself any labor, from which I seek neither praise nor glory of any kind, but aim only to serve those who cultivate philosophy and to stimulate them more and more.[33]

But he does not make it clear why he commended the *Saggi* daily to his students; indeed, he scarcely makes it credible, admitting as he does that "philosophy" has made no small progress in the years since the Academy published their work.[34] Up to the present I have not been able to find any letters that would help in disentangling his motives for making the translation.

Musschenbroek cannot be blamed for not being a historian, but in this same preface he contributed a good deal to the legend, dear to his Italian contemporaries, that the Royal Society, the Académie des Sciences, and even the St. Petersburg Academy were founded in imitation of the Accademia del Cimento.

[30] Cf. p. 289 above.
[31] (*Elements of natural science* [Leyden, 1736].)
[32] Poggendorf, *Geschichte der Physik* (Leipzig, 1879), p. 353.
[33] *Tentamina*, preface, sig.**, fol. 3r–v.
[34] *Ibid.*, fol. 3v.

And this brings me to the last part of my task—to evaluate the experiments and observations of the Academy in the context of scientific enquiry as it was during the years 1657 to 1662, when nearly all its work was performed.

As noted in chapter 1, the work of the Academy was by no means the first attempt at formal experimentation in physics, but it was the most sustained effort up to that time and certainly had the widest scope, bounded apparently only by the ability of the members to think of things to test by experiment. This eclectic approach naturally had its disadvantages, the most notable of which was their failure to carry lines of investigation far enough, if we except the experiments on artificial congelation that seem to have obsessed Prince Leopold and bored his coadjutors.

Considering the multitude of things on which experiments could be made, it is neither surprising nor important that some of their activities were trivial. What is important is the motivation for nearly all their more successful investigations, which, as I have hinted above, was to demolish Aristotelian dogma. Their best experiments, both published and unpublished, were tests of hypotheses, and it soon becomes clear to the student that they knew very well what sort of results ought to be obtained. Let us review some of these anti-peripatetic projects, beginning with a few of those in the *Saggi*.

The most important of these is the demonstration of the possibility of making a vacuum and the related proofs that the mercury in the Torricellian tube is sustained by the pressure of the atmosphere. The experiments directed to this end take up more than forty pages of the *Saggi*; many of them, such as the brilliantly simple one with the narrow-mouthed tube,[35] seem to have been original; others had been done elsewhere. The Academicians were by no means flogging a dead horse, for opposition to the idea of a vacuum continued in ecclesiastical circles for the rest of the century,[36] leaving out of account the followers of Descartes.

Another concern of the Academy was with the dogma of "positive levity," and the experiment on the motion of smoke in the vacuum showed that smoke, at least, did not possess any such property. The elaborate experiments described on pages CCVII to CCXVI of the *Saggi* were also designed to negate this dogma, and it is interesting that they are introduced by an idea from the *Timaeus* of Plato.

One of the supposed proofs of Nature's horror of a vacuum was the bursting of closed vessels by the freezing of water in them. Nearly all

[35] *Saggi*, p. LII and fig. XII.
[36] See Lynn Thorndike, *A History of Magic and Experimental Science*, 8 vols. (New York, 1923–58), vol. VII, *passim*.

substances contract on freezing, and water is one of the few exceptions. It was therefore necessary to put beyond doubt the fact that water expands when it freezes, and so the Academicians proceeded to burst vessels of increasing thickness until they finally succeeded in making one that resisted. Then they tried measuring the extent of the expansion and by two different methods found it to be about one-eighth.

One of the most strongly held of Aristotelian dogmas was that heat is a "quality" that has no physical existence, and some of the most interesting of the experiments were attempts to prove that it is a material substance. Associated with the idea that heat is a quality are the related idea that cold is another quality, its opposite, and the dogma of "antiperistasis," the intensification of one of the two by being surrounded by the other. This idea almost certainly arose from the circumstance that caves and cellars *seem* to be warmer in winter than in summer.

In experiments on artificial freezing it was observed by Prince Leopold himself that at the moment when a long-necked flask of liquid— or a thermometer—is plunged into ice, the level in the neck makes a sudden leap upwards before beginning to descend. The opposite behavior follows immersion in hot water. This of course looked extremely like a demonstration of antiperistasis, until Borelli suggested the correct explanation, namely, the expansion of the material of the flask itself, and the Academy made the excellent experiments described in the *Saggi*.[37] But this raised the whole question of the thermal expansion of solids, a phenomenon better known to the wheelwrights than to the virtuosi, and led to the well-known experiment with the ring and the mandrel, and several others, probably the most original and valuable experiments made by the Academicians. The expansion was ascribed by Borelli to the "corpuscles of fire" opening up the pores of the solid, and when it was found that a boxwood ring was enlarged by the absorption of water, the Academicians were confirmed in their belief that heat is indeed a substance. To demolish antiperistasis they performed two excellent experiments described on page CCLIX of the *Saggi*. On the other hand, the experiment suggested by Prince Leopold for the examination of the idea of "atoms of cold"[38] evidently failed to convince even the writer of the published account. The contraction of solids when cooled was a better demonstration.

It is thus clear from a perusal of the *Saggi* that the Academicians were highly successful in many of their attempts to refute the errors of the schools. There are equally good unpublished experiments, for

[37] Pp. CLXXIX–CLXXXI.
[38] *Saggi*, pp. CCLIX–CCLX.

example the one performed on September 14, 1657 to disprove the supposed conversion of air into water by cold.[39] It is surprising that this simple and well-designed experiment was not put in the *Saggi*.

The struggle against the Peripatetics was so important to the Academicians that it sometimes led them, I am sorry to say, to forget the ideals on which the Academy was founded. This appears only when we consider the unpublished papers, for these contain experiments, the results of which seem at first sight to support, rather than to contradict, scholastic dogma. None of these appear in the *Saggi*.[40] However, we must not forget that this was the period of the Counter-Reformation, when the technique of ignoring awkward facts was brought to a fine art by the writers of historical and religious works. The Academicians had an example among their friends in Cardinal Sforza Pallavicini, whose history of the Council of Trent was published during the lifetime of the Accademia del Cimento. It is unreasonable to expect twentieth-century standards of scientific objectivity in such an age. Indeed, the conduct of experiments and their interpretation is necessarily conditioned by hypothesis in any age; even the "Baconianism" of the Royal Society could not keep it from doubting the result of an experiment that seemed to introduce an inelegant complexity into a common natural phenomenon, the change in the volume of water with temperature.[41]

A very interesting example of the way in which the outlook of the Academicians was affected by the spirit of the times has recently been discussed by the writer.[42] Apart from the belief that heat is a quality and not a substance, the Peripatetics also maintained that "fire," one of the four elements, always seeks to go upwards. On September 7, 1657, the Academy made an experiment that could have a bearing on both these assertions, as follows:

September 7, 1657. In order to find out whether the expansion of cold and heat is spherically uniform, after trying many ways of making sure of this, we put small 10-degree instruments [i.e., spirit-in-glass thermometers] around two metal balls, that is to say beneath, above, and to the sides. One of the balls had been buried in ice for 46 hours, the other heated red-hot in the fire. It was found that of the said instruments around the hot ball, the one above was changed the most, those at the side less, and the one below still less than these. On the contrary, of those around the cold ball, the one underneath changed more than the others, those at the sides less, and the one above less than these.[43]

---

[39] See p. 268 above.
[40] The reader may refer to pages 271 and 274 above for examples.
[41] See p. 296 above.
[42] W. E. K. Middleton, *Physis* 10 (1968): 299–305.
[43] G. 262, 31v–32r.

Rinaldini, who proposed this experiment, was not at all happy with it, and on November 11 of the same year he wrote about it from Pisa to Prince Leopold:

Having reflected on the experiment with the iced ball and the instruments for heat and cold, and also on the one with the heated ball, I suspect that one matter was lost sight of, that is, to prevent the air that was cooled, and thus made heavier, falling on the lower instrument, or rather surrounding it, just as the heated air, made lighter, came to warm the upper one; so that we could not come to any conclusions about the diffusion of the hot and cold corpuscles. Therefore I beg Your Highness to have this experiment repeated in such a way as to remove this difficulty; but to make myself better understood I shall explain it a little with the attached figure.[44] AB being the ball, CD and EF the little instruments, when the ball is cooled the little instrument below is affected (i.e., contracts) more than the upper one, not because an effusion of corpuscles takes place, which, being heavy, descend, so that the cold works downwards more than upwards, as heat on the other hand ascends, because its lighter corpuscles go upwards, etc; but because the cold of the ball cools the air around it, makes it heavier, and makes this heavier part descend. So, as the colder air surrounds the little instrument EF, it is no wonder that it makes it cooler, making the liquid inside contract more. On the other hand if the ball is heated it warms the air and makes it lighter, so that it rises and surrounds the instrument CD, heating it more than the instrument EF can be heated.[45]

Rinaldini, it will be noted, had discovered and described the phenomenon of convection. He went on to suggest what he thought would be an improvement, namely, to surround the cooled or heated ball with a fitted jacket of wood, made in two parts so that it could be quickly placed around the ball, and containing chambers for the thermometers. The ball would not, of course, be heated red-hot, but "so that it could not be touched with the fingers without injury."

The next exhibit is entitled "An experiment, made in the way that will be described, for finding out whether heat diffuses spherically," and begins:

It is commonly said among the Peripatetics that a natural agent diffuses its action equally in every direction, this proceeding with a uniformly difform motion, as they assert. It therefore occurred to us to find out about this by experiment, as far as this might be possible.

The principal object, then, was to see in some manner whether heat is a quality, or whether it is nothing but corpuscles, of any shape whatever. [The experiment, in air, is then described, referring only to the heated ball.] From this it appears that we might gather that heat does not diffuse

---

[44] Not found; but the intention is perfectly clear.

[45] G. 275, 82r–83v. Printed in LIUI, I, 184–87.

equally in every direction, but more upward than downward. From this it similarly appears that we might conclude what was chiefly being sought [!], namely, that heat is not a quality, because if it were, it would appear that it should diffuse equally in every direction, exactly as is asserted by the Peripatetics. But if it consists of corpuscles, as is claimed by Democritus, it does not seem difficult to understand that they would move mainly upward.[46]

Nor does it seem (to us) axiomatic; but the document goes on:

It appeared to one of the Academicians that he could reply to this experiment by saying, in favor of the Peripatetics, that the quality of heat, warming the ambient air, makes the hotter parts ascend, and the cooler remain below; so that the warmer parts surrounding the upper thermometer make it move more than the lower one, which is surrounded by the parts of the air that are less warm.

In order to demonstrate this reply to be of no importance and render valid the experiment just described, its author repeated the experiment in this form:

There follows a lengthy description of the ball and thermometers encased in their wooden sheath, as Rinaldini had suggested in his letter of November 11.

It seems clear that the Academicians had painted themselves into a corner. Let us analyze what happened. First the experiment was made on September 11, 1657, and recorded in the diary without comment. Rinaldini, who had devised it, was able, thinking about it, to see how it really worked. At this time he saw no reason for postulating corpuscles, either of heat or of cold, but was satisfied with simple transfer of heat (whatever heat may be) from the ball to the air. But then, we may suppose, the Prince considered Rinaldini's letter and ordered the experiment to be repeated and the variation using the wooden sheath also performed. Rinaldini was presumably in Pisa, and Prince Leopold may have supervised it himself, though it is probable that Viviani had a hand in it. By this time they knew what they were trying to prove—a splendid recipe for being mistaken—and interpreted the experiment in favor of corpuscles of heat which, they assumed without proof, would move "mainly upward." But the member with a leaning toward Aristotelianism, who may have been shown Rinaldini's letter, took the convection theory and interpreted it in favor of the dogma of heat as a "quality," which warmed the air and caused it to move. The Academicians were so disconcerted with this apparent setback that not one of them recognized the explanatory value of Rinaldini's insight. Nor did they seem to consider that the postulated upward motion of their particles of heat was a variant of

[46] G. 263, 57r–58v. Published in *TT* 2, 703–4; also in *DIS*, 390–91.

the Peripatetic dogma of the positive lightness of fire. Unfortunately the experiment with the wooden sheath gave a positive result, perhaps because of an inexact fit, so that they were not constrained to think again, at least not at once. But remembering the experiment nearly three years later in July 1660, they tried to make another of this sort in an evacuated space, with equivocal results.[47] This latter experiment was made (to quote its title) "in order to find out what may be the motion of the invisible exhalations of fire in the vacuum." This time the exhalations were postulated in advance.

Another illuminating example of the effect of their hypotheses on their experimentation is the experiment in which they sought to investigate whether "cold" could be reflected from a concave mirror.[48] They were committed to the position that cold is nothing more than a deprivation of heat, and when they obtained a result that seemed to contradict this hypothesis, they tried to explain it away as a probable experimental error and then naively admitted that they had not repeated this very interesting experiment. This piece of self-deception did not escape the acute eyes of Benjamin Thompson, Count Rumford, who in 1804 was sure that he had proved the existence of radiations both frigorific and calorific. In a paper read to the Royal Society on February 2 of that year he wrote:

it is not a little curious, that the learned academicians who made that experiment . . . were so completely blinded by their prejudices respecting the nature of heat that they did not believe the report of their own eyes; but, regarding the reflection and concentration of cold (which they considered as a negative quality) as *impossible*, they concluded that the indication of such reflection and concentration which they observed must necessarily have arisen from some error committed in making the experiment.[49]

It seems to me that a consideration of these happenings throws a bright light upon the motivation of the Academicians, reinforcing the impression produced by the *Saggi* and many of the other papers and leading to the conclusion that at least the strongest spring of their action was a desire to put down the bigots who had, as they thought, so shamefully treated their revered Galileo. There can be little doubt that Viviani, the "last pupil" of the Master, upon whom must have devolved much of the actual conduct of the experiments, would have done his best to forward this plan.

[47] *Saggi*, pp. LXXXIX–LXXXXI.
[48] *Saggi*, p. CLXXVI.
[49] *Collected works of Count Rumford*, ed. Sanborn C. Brown (Cambridge, Mass., 1968–[in progress]), I, 373. The italics are in the original.

Apart from this question of motivation, we must remember the assertion in the preface to the *Saggi* that the "only task" of the Academy was "to make experiments and to tell about them." So carefully did they limit themselves in this regard that even in the diaries there is very little speculation and almost nothing that could be called theory. Three centuries later, therefore, we are obliged to consider them purely as experimenters, and in these closing paragraphs they must be compared with other experimenters of the period. The comparison that at once suggests itself is with Boyle, assisted, we may well suppose, by Hooke; and this, of course, must be the Boyle of the *New Experiments Physico-Mechanicall.*

First of all, it is extremely hard for the twentieth-century student to appreciate how very difficult any experimental work was three hundred years ago. In the first place, the experimenter, trying to measure some property of matter, would have no idea of the numerical magnitude of the results to be expected. In the second place, there was no clear understanding of the nature of air, or of vapors—the word "exhalations" covered an infinity of ignorance—or of liquids, or of solids, in short no guiding principles for the design of experiments. Nor were there any clear ideas of pressure or force or momentum; *forza* and *virtù* could stand for any of them. The distinction between temperature and quantity of heat was far in the future. Experimenters were asking questions of Nature with little knowledge of what language she would understand. Add to this the unreliability of most materials and the impurity of most reagents, and experimentation becomes a sort of leap into a dimly-lit and bottomless pool.

Therefore one has to admire the experimenters greatly. I find myself unable to award first prize to either Boyle (plus Hooke) or to the Accademia del Cimento, which for all practical purposes really means Borelli plus Viviani plus Rinaldini, as far as the actual conduct of the experiments is concerned, though in saying this I am forgetting about the endless semi-quantitative experiments on artificial freezing that fascinated Leopold, bored Viviani,[50] and took up so many pages of the *Saggi*. The greatly different styles of the two "laboratories" were largely determined by the availability of apparatus, the Italians having the advantage of being able to order glassware of almost any shape from a glass-blower of unequaled skill, whereas Boyle had to improvise nearly all the time and did it very well. I have already suggested more than once that the unlimited availability of apparatus was not always an unmixed blessing to the Academicians and resulted in some very badly thought out experiments; this substitution of gadgetry

[50] See appendix C, notes 31 and 33.

for hard thinking is not entirely foreign to the twentieth century either.

The exposition of the Academy's experiments in the *Saggi*—apart from the unconscionable delay in its publication—undoubtedly suffered from Lorenzo Magalotti's distaste for the task, an aversion that I have documented in chapter 4. Yet the Secretary's diffidence was probably less of a liability than the fact that the book was really written by a committee, as has been made clear by the editors of *DIS* in their exposition of the various drafts and the comments of the active experimenters on these. A better book could have been written about the experiments by Borelli, but a book by Borelli in Italian would not have been approved of by the Accademia della Crusca, and the esteem of the princes for the great literary genius of Galileo simply demanded that a book designed to carry on his battle should appear in the vernacular.

Tuscany had been in the front rank of this battle for sixty years, but the publication of the *Saggi* was the last shot from the valley of the Arno. The leadership in science was moving north, and the accession of Cosimo III in 1670 ensured that it should not return.

Appendix A.

A BIBLIOGRAPHY OF THE
SAGGI DI NATURALI ESPERIENZE

As far as I have been able to ascertain, there have been thirteen editions of the *Saggi* in its original language, one (up to now) in English,[1] two in Latin, and two in French.

I shall deal with all the Italian editions first.

1. *Editio princeps* (1667)
Saggi di naturali esperienze fatte nell'Accademia del Cimento sotto la protezione del Serenissimo Principe Leopoldo di Toscana e descritte dal segretario di essa Accademia. [Copper engraving 95 × 140 mm showing 3 crucibles on a fire, within an elaborate border, with the motto "Provando e riprovando"] *In Firenze per Giuseppe Cocchini all'Insegna della Stella. MDCLXVII. Con licenza de' Superiori.*

Collation: [ ]², engraved portrait, [ ]², ✠⁴, A – Z⁴, [Aa] – Ll⁴, *⁶, [ ]².

The portrait, an engraving after an unknown painter, is of Ferdinand II. It bears the legend:

> Se al gran genio real, scoperse il cielo
> Gli arcani intatti dell'eccelsa sfera
> Oggi la terra emulatrice altera
> Toglie ad ignote maraviglie il velo.

and is signed "Franciscus Spierre Lotaringus Sculpebat. 1659." Head- and tail-pieces throughout seem to have been selected from eighteen engravings on copper, each about 140 mm wide and 70 mm high. Several are used more than once. Most of these were made *ad hoc* (for they include representations of instruments); the rest are conventional

---

[1] Reproduced photographically, however, in the twentieth century. See no. 14.

baroque decorations. Some of each class are signed with the monogram of Valerio Spada.[2] The rather oversize woodcut initials are *ad hoc*.

The chief variant is in the title-page, for there exist a number of copies in which the title-page is dated 1666 instead of 1667. A comparison of these two title-pages shows that the 1667 version was reset, though in the same types. As the documents concerning the license to print are dated 18 Sept. 1667, 5 Oct. 1667, 7 Oct. 1667, and XI Octobris 1667, and the dedication 14 July 1667, and the text is, as far as I can determine, completely identical in all the copies, the only reasonable assumption is that some title-pages had been printed in advance in the hope that the book would be ready in 1666, and that the printer could not bear to throw them away. It is a mistake (frequently seen) to refer to an edition of 1666.

There are two varieties of the 1667 title-page, some copies lacking the apostrophe in "de' Superiori" in the last line. A comparison makes it appear that the omission was noticed during the run and corrected simply by moving the last word to the right to make room for the apostrophe.[3]

As the illustrations and the head- and tail-pieces are copperplate, they were of course struck after the letterpress had been done. Different copies vary in the order in which the head- and tail-pieces are placed, and this order does not depend on the year appearing on the title-page. Occasionally one may even be put upside-down. We may conclude that the workman struck all or most of these decorations copy by copy and paid no attention to having them in any particular order. The illustrations are on numbered pages, the (Roman) numbers being incorporated in the copper plates. This of course resulted in the absence of a good many signature letters and the apparent misplacing of catchwords, which would have worried the binders had not the book ended with a "notice to binders" explaining this and suggesting that they follow the Roman numbers.

2. *Second Edition* (1691)
Saggi di naturali esperienze fatte nell'Accademia del Cimento sotto la protezione del Serenissimo Principe Leopoldo di Toscana e descritte dal Segretario di essa Accademia. Seconda edizione. [Copper engraving as in *ed. Princ.*] *In Firenze nella nuova Stamperia de Gio. Filippo Cecchi. MDCXCI. Con licenza de' Superiori.*

[2] According to an account by Magalotti in the Ginori Venturi archives (now dispersed), quoted in *DIS*, 28–31.

[3] Bartolommeo Gamba, *Serie dei testi di lingua e di altre opere importanti nella italiana letteratura*, etc., 4th ed. (Venice, 1839), p. 258, notes a title page reading *Saggi . . . descritti dal Sollevato*, with the date MDCLXVI. I have not seen one of these.

Usual collation: [ ]², portrait engraving, [ ]², ✠⁴, A – Z⁴, [Aa] – Nn⁴.

The engraving, from Arnold van Westerhout's portrait of Cosimo III, is usually sewn with either the first or second section. The dedication is to Cosimo III and is signed by the printer Gio. Filippo Cecchi. Some copies (I have seen one in the Wellcome Library) have both portraits and both dedications. In the library of the University of British Columbia there is a copy with the portrait of Ferdinand II but without that of Cosimo III.

Apart from the dedication, it was clearly intended to make this edition an exact copy of the editio princeps. The plates were re-engraved with remarkable accuracy and are on the average better than in even a very good copy of the earlier edition (e.g., 31.k.1 in *BM*), being more strongly delineated and better struck. The errata noted in the first edition, including those referring to the plates, have been incorporated. The remarks about the head- and tail-pieces also apply to this edition.

3. *Saggi di naturali esperienze*, etc. . . . . Napoli, Giacomo Raillard, 1701. "Third edition in folio, 269 pages. Except in certain subsidiary parts, it coincides perfectly with the Florentine edition of 1666." [4]

The existence, date, and format of this edition are confirmed in a notice of a later one in the *Giornale de Letterati d'Italia* (Venice) 4 (1710): 446–47. I have not seen it.

4. *Saggi di naturali esperienze fatte nell'Accademia del Cimento sotto la protezione del Serenissimo Principe Leopoldo di Toscana. Dedicati al molto illustre Signor Anello di Napoli eccelente Dottore di Filosofia, e Medicina.* [an engraved intaglio copy of the original coat of arms in the first ed.] *In Napoli, nella stamperia di Giacomo Raillard, MDCCIV. Con licenza de' Superiori. A spese del medesimo.*

Collation: [ ]², *⁴, A – Z⁴, Aa – Mm⁴, Nn². 220 × 340 mm.
This edition has woodcut head- and tail-pieces, intaglio illustrations, and is otherwise intended to agree with the first edition as far as possible; but for the benefit of the binder the signatures are printed with type on the sheets carrying the figures.

5. *Saggi di naturali esperienze, fatte nell'Accademia del Cimento sotto la protezione del Serenissimo Principe Leopoldo di Toscana e descritte dal Segretario di essa Accademia. Quarto edizione Consacrati all' illustrissimo & eccellentissimo Signor Girolamo Flangini patrizio*

---

[4] S. Fermi, *Bibliografia Magalottiana*, etc. (Piacenza, 1904), p. 1.

veneto, &c. [Vignette] *In Venezia, MDCCXI.*[5] *per Domenico Lovisa. Con licenza de' Superiori.*

Collation: ✠[8], A – S[8].

This quarto edition corresponds page by page and almost line by line with the edition of 1667. The illustrations are copperplates, and, of course, on a smaller scale. 185 × 250 mm.

6. *Saggi di naturali esperienze fatte nell'Accademia del Cimento sotto la protezione del Serenissimo Principe Leopoldo di Toscana. Dedicati all'Altezza Serenissima di D. Cesare Michel-Angelo Davalos d'Aquino, d'Aragona,* [there follows a very long list of his titles, and then an intaglio copy, not very accurate, of the engraving on the title-page of the first edition] *In Napoli. Nella stamperia de Barnardo-Michele Raillard, M.DCC.XIV. Con licenza de' Superiori.*

Collation: [ ][2], *[4], A – Z[4], Aa –Mm[4], Nn[2]. 215 × 350 mm.

The letter of dedication is from the editors, Cellenio Zanclori and Lorenzo Cicarelli, to Davalos. Of two copies at the New York Academy of Medicine, one has a woodcut tailpiece on *4 recto, the other not.

7. *Saggi di naturali esperienze fatte nell'Accademia del Cimento descritto dal Conte Lorenzo Magalotti. In questa edizione si aggiunge la sua vita scritta del Signor Domenico Maria Manni Accademico Fiorentino.* [woodcut 52 × 35 mm with the motto "La felicità delle lettere"] *In Venezia MDCCLXI. Presso Giambatista Pasquali. Con licenza de' Superiori, e Privilegio.*

Collation: a[4], b[4], c[6], A – M[4], with 28 fold-out plates.

In this edition there was no attempt whatever to imitate the first edition.

8. The work was reprinted in its entirety in the *Notizie degli aggrandimenti delle scienze fisiche accaduti in Toscana nel corso di anni LX. del secolo XVII. raccolte dal Dottor Gio. Targioni Tozzetti,* 3 vols. in 4 (Florence, 1780)[6], II, 377–599.

[5] Fermi, *ibid.*, p. 2, gives the date as 1710 and says that some examples bear the date 1711—as does the one I have examined at the New York Academy of Medicine. The date 1710 is substantiated by a notice in *Giornale de letterati d'Italia* 4 (1710): 447. This was in the last number for the year 1710.

[6] Concerning this work and its author see chapter 1 above and appendix B below.

Large extracts from the diaries and other papers of the Accademia del Cimento are interpolated.

9. *Saggi di naturali esperienze . . . descritte dal Segretario di essa Accademia. Milano dalla Società Tipografica de' Classici Italiani, contrada di S. Margherita, N°. 1118, Anno 1806.* 282 pp. + 1 leaf. Folding plates. (Classici Italiani, vol. 217.) This forms volume 2 of the *Opere de Lorenzo Magalotti.*

10. *Saggi di naturali esperienze, fatte nell'Accademia del Cimento. Terza edizione fiorentina preceduta da notizie storiche dell'Accademia stessa e seguitata da alcune aggiunte* [vignette] *Firenze dai torchi della Tipografia Galileiana 1841.* Large quarto, iv + 134 + iv + 184 + XC pp. 22 fold-out plates. 220 × 310 mm.

The introduction, 134 pages of elegant rhetoric in praise of Tuscan science, is by Vincenzio Antinori. The *aggiunte* are extracted from the papers of the Academy and provided with notes by G. Gazzeri, then professor of chemistry at Florence and pharmacist to the Florentine Hospital. It was printed at the expense of the then Grand Duke Leopold II as a gift for the members of the third congress of Italian scientists, held in Florence in that year. A few deplorable liberties were taken with the text of the *Saggi.*

11. In honor of the tercentenary of the death of Galileo, the work was reprinted from the first edition (with corrections from the second) in the volume that has been referred to throughout the present book as *DIS*: Giorgio Abetti and Pietro Pagnini, editors, *Le Opere dei Discepoli di Galileo Galilei,* Edizione Nazionale. Volume I. L'Accademia del Cimento, parte prima. Firenze, S. A. G. Barbèra Editore. 1942. xvi + 500 pp., Portraits.

This work, which is uniform with the National edition of Galileo's works, was intended to be one of twenty-one volumes completing that edition by a study of his "disciples"—a word interpreted more widely in this connection than many non-Italian scholars might prefer. The advent of World War II interrupted this project, and this invaluable volume remains as an earnest—it is to be hoped—of more to come. It contains: pp. 3–75, An introduction; pp. 77–270, the *Saggi di naturali esperienze* (not in facsimile); pp. 271–494, various documents from the Galileo manuscripts that deal directly with the compilation and printing of the *Saggi;* pp. 495–500, an index of names.

12. *Saggi di naturali esperienze fatte nell'Accademia del Serenissimo Principe Leopoldo di Toscana e descritte dal segretario di essa Ac-*

cademia: Lorenzo Magalotti. Ristampa integrale illustrata a cura di Enrico Falqui Edizione numerate per sè e per pochi di Colombo in Roma 1947. xxiv + 260 pp. + 5 folding plates. 140 × 190 mm. Three hundred numbered copies.

13. *I Saggi di naturali esperienze fatte nell' Accademia del Cimento e Strumenti e suppellettili della medesima Accademia conservati presso il Museo di Storia della Scienza di Firenze. Pubblicata a cura della Domus Galilaeana di Pisa e del suddetto Museo sotto gli auspici del Consiglio Nazionale delle Ricerche* [printer's device] *Domus Galilaeana Pisa* [1957]. Pages [a] – [h] + 18 not numbered + CCLXIX + 19 not numbered + 66.

This is a full-size photographic reproduction of the 1667 edition, made in celebration of the 300th anniversary of the founding of the Academy. There were 600 numbered copies. The volume has a short introduction by Giovanni Polvani, and an appendix by Maria Luisa Bonelli on the surviving instruments and equipment of the Academy, illustrated by excellent photographs.

Let us pass now to the translations.

14. *Essayes of Natural Experiments Made in the Academie del Cimento Under the Protection of the Most Serene Prince Leopold of Tuscany. Written in Italian by the secretary of that Academy, Englished by Richard Waller, Fellow of the Royal Society. London, Printed for Benjamin Alsop at the Angel and Bible in the Poultrey, over-against the Church. 1684.*

Collation: [ ]⁸, a – b⁴, A – Z⁴, with 19 intaglio plates added. 172 × 231 mm. This has been reproduced photographically with an introduction by A. Rupert Hall. (New York, Johnson Reprint Corp., 1964). [*Sources of Science,* no. 1.]

15. *Tentamina experimentorum naturalium captorum in Academia del Cimento sub auspiciis Serenissimi Principis Leopoldi Magni Etruriae Ducis[7] et ab ejus Academiae secretario conscriptorum: Ex italico in latinum sermonem conversa. Quibus commentarios, nova experimenta, et orationem de methodo instituendi experimenta physica addidit Petrus van Musschenbroek, L.A.M. Med. & Phil. D. Phil & Mathes. Profess. In Acad. Ultraj.* [printer's device] *Lugduni Batavorum, Apud Joan. et Herm. Verbeek, Bibliop. MDCCXXXI*

---

[7] This is, of course, an error.

8 leaves + XLVIII pp. + 6 leaves + 194 + 192 pp. + 7 leaves, 32 folding plates. There is a half-title introducing "pars altera," and we are told in the preface that this was done so that the work could be divided between two printers in order to save time.[8]

The translation is straightforward. Except in a few places the additions are of interest only to the *later* history of physics, and indeed serve to show how far experimental technique had developed in seventy years. Pages I–XLVII are a formal address delivered by Musschenbroek to the University of Utrecht on March 27, 1730, when he left his post of *Magister academicus*.

16. *Tentamina experimentorum* ... [as in no. 15] ... *Petrus van Musschenbroek, L.A.M. Med et Phil D. Phil et Mathes. Prof. in Acad. Ultraj.* [Printer's device, with the motto "Labore et favore"] *Viennae, Pragae, et Tergesti,*[9] *typis et sumptibus Johannis Thomae Trattner, Caes. Regiae Majest. Aule Typogr. et Bibliop. MDCCLVI.*

This edition was clearly intended to be a close copy of no. 15. The pages usually have the same catchwords, except in the front matter.

17. A partial translation into French is to be found in: *Mémoires de physique pure, sans mathématiques, de toutes les académies de sciences,* etc., Tome I (Lausanne, Antoine Chapuis, 1754), pp. 131–215 and plates VII–XXV. with the title: "Essais d'expériences naturelles, faites dans l'Académie del Cimento, sous les auspices du Serenissime Prince Léopold de Toscane, décrites par le sécretaire de cette Académie, Florence 1667."

This is an incomplete translation, taking a narrow view of what can be called physics, and larded with extracts from papers by Wilhelm Homberg.[10] On page 132 we find an explanatory footnote: "Much difference will perhaps be noticed between our French translation ... and the Latin one of Mr. Musschenbroek; but that illustrious savant has himself agreed that ours is preferable. He has felt this so strongly that he proposes to issue a second edition of his, corrected from this." The 1756 Latin edition (our no. 16) shows no signs of any such action.

The demand for "collections" of this kind was brisk in the age of the *philosophes,* for the very next year saw:

---

[8] Preface, fol. **4r.

[9] Trieste.

[10] (1652–1715). He had been a member of the Académie Royale des Sciences and was a versatile experimenter.

18. Another incomplete translation into French, in: *Collection académique composée des mémoires, actes, ou journaux des plus célèbres académies & societés littéraires étrangères*, etc., Tome I (Dijon, Francois Desventes; Auxerre, François Fournier, 1755), pp. 1–251, with introduction & notes, pp. lxj–lxxj.

This has no separate title-page for the *Saggi*, but at the head of the introduction we find: "Essais d'expériences physiques faités dans l'Académie del Cimento de Florence, avec des notes ou additions tirées de la traduction latine de Van-Musschenbroek." On p. 1 we have a woodcut headpiece 75 × 147 mm with the motto "Sparsa collegit" and a repetition of the above. The translation begins with "Expériences touchant la pression naturelle de l'air," i.e., on p. XXIII of the 1667 edition.

Appendix B.

NOTICE REGARDING GIOVANNI TARGIONI TOZZETTI

Giovanni Targioni Tozzetti was born in Florence on September 11, 1712, the son of Benedetto Targioni and Cecilia Tozzetti.[1] When he was a mere boy he was encouraged by his father, who was a physician, in the study of botany and, later, of mineralogy and paleontology. He became an avid collector of plants and fossils and recorded that his private museum at Pisa had grown to 50,000 specimens when in 1740 it was destroyed by a flood. Meanwhile he had obtained the degree of Doctor of Medicine in 1734 and had returned to Florence to practice with his father and to study botany with the eminent botanist Pierantonio Micheli, with whom he explored the flora of Tuscany. He excuses his failure to learn more from Micheli by the circumstance that during the same period he was studying practical and theoretical medicine, mathematics, and Greek—at home![2]

Micheli died at the end of 1735, and Targioni Tozzetti was named by the Grand Duke Giovan Gastone to succeed him as lecturer in botany and also to be assistant librarian of the Magliabecchi Library. He bought Micheli's specimens, library, and manuscripts, and agreed to publish these last, but lack of funds delayed this publication, and it was not until 1748 that he was able to publish Micheli's catalogue of the botanical garden, the direction of which he had given up in 1746.

In eight years he catalogued the Magliabecchi Library of 40,000 books and 1,110 manuscripts; and Lastri points out that he did not merely make a catalogue, but stored his mind with what he read. He later published numerous letters of famous people to Magliabecchi

[1] The chief source for his life is the obituary by Lastri, *Novelle Letterarie* 14 (1783): columns 97–105 and 113–17. See also *Biographie Universelle*, vol. 14, pp. 16–17. A charming short autobiography was left in manuscript and printed only in 1852 in the introduction to *Notizie sulla storia delle scienze fisiche in Toscana, cavate da un manoscritto inedito di G. Targioni Tozzetti* (Florence, 1852), pp. ix–xvii.

[2] *Notizie sulla storia . . .*, p. xi.

and others. From 1750 to his death on January 7, 1783, he worked indefatigably at a surprisingly large number of subjects, as may be deduced from the bibliography below, the completeness of which I cannot guarantee. He is best known for two works, the *Relazioni di viaggi* of 1751 to 1754 and the *Notizie degli aggrandimenti* of 1780, at the end of which he announced a grandiose project for no less than five such compilations, dealing with Tuscan science from Etruscan times until those of Giovan Gastone. But death interrupted this great plan, of which only a small part was completed, and published seventy years later. The distinguishing characteristics of this tireless scholar were his love of Tuscany and his wide-ranging curiosity. As will appear from the list of his works, these two incentives led him into questions of drainage, famine relief, public health, agriculture, the floods of the Arno, and above all the history and geography of his native province.

BIBLIOGRAPHY
OF GIOVANNI TARGIONI TOZZETTI

1. *Theses de praestantia & usu plantarum in medicina.* Pisa, 1734.
2. *Lettera sopra una numerosissima specie di farfalli, vedutasi in Firenze sulla metà di luglio 1741.* Florence, 1741.
3. *Clarorum Belgarum ad A. Magliabecchium nonnullosque alios epistolae*, etc. Florence, 1745.
4. *Clarorum Venetorum ad A. Magliabecchium nonnullosque alios epistolae*, etc., 2 vols. Florence, 1745, 1746.
5. *Clarorum Germanorum ad A. Magliabecchium nonnullosque alios epistolae.* Tom. 1 [all published]. Florence, 1746.
6. *Cl. Petri Antonii Michelii catalogus plantarum horti Caesarei florentini opus postumum iussu Societatis Botanicae editum*, etc. Florence, 1748.
7. *Lista di notizie d'istoria naturale della Toscana, che si desiderano.* Florence, 1751.
8. "Discorso del fiorino di sigillo della Repubblica Fiorentina," *Mem. Società Columbaria Fiorentina*, vol. 2. Leghorn, 1752.
9. *Prima raccolta di osservazioni mediche.* Florence, 1752.
10. *Prodromo della corografia, e della topografia fisica della Toscana.* Florence, 1754.
11. *Relazioni di alcuni viaggi fatti in diverse parti della Toscana per osservare le produzioni naturali, e gli antichi monumenti di essa*, etc., 6 vols. Florence, 1751–54. Second, enlarged edition, 12 vols., Florence, 1768–79.
12. *Relazione d'alcuni innesti di vaiuolo.* Florence, 1756.
13. *Ragionamenti sull'agricoltura toscana.* Lucca, 1759.

14. *Succinta relazione dell'ultima malattia, morte ed apertura del cadavere di Girolamo Samminiati.* Florence, 1760.

15. *Parere sopra l'utilità delle colmate di Bellavista, per rapporto alla salubrità della Valdinievole.* Florence, 1760.

16. *Considerazioni sopra il parere di Nenci intorno le acque stagnanti delle colmate,* etc. Florence, 1760.

17. *Ragionamento sopra le cause, e sopra i rimedj dell'insalubrità d'aria della Valdinievole,* 2 vols. Florence, 1761.

18. *Avvertimento circa alla scelta del grano da seminarsi nell' anno 1766.* Florence, 1766.

19. *Breve istruzione circa i modi di accrescere il pane col mescuglio d'alcuni sostanze vegetabili,* etc. Florence, 1766.

20. *Alimurgia, ossia modo di render meno gravi le carestie, proposto per sollievo de' poveri.* Florence, 1767.

21. *Disamini d'alcuni progetti fatti nel secolo XVI. per salvar Firenze dalle inondazioni dell'Arno.* Florence, 1767.

22. *Relazioni delle febbri chi si sono provate epidemiche in diverse parti della Toscana l'anno 1767.* Florence, 1767.

23. *Analisi e difesa dell'alimurgia.* Venice, 1769.

24. *Relazione della ricognizione del cadavere della fanciulla Anna Maria Cioni.* Florence, 1770.

25. *Istruzione al popolo circa ai tentativi da farsi per ravivare gli annegati, ed altri apparentemente morti.* Florence, 1772.

26. *Raccolta di teorie, osservazioni, e regole per ben distinguere, e prontamente dissipare le asfissie o morti apparenti,* etc. Florence, 1773.

27. *Notizie degli aggrandimenti delle scienze fisiche, accaduti in Toscana nel corso di anni LX. del secolo XVII,* 3 vols. in 4. Florence, 1780 [also with a different title; see p. 14 above].

28. *Notizie sulla storia delle scienze fisiche in Toscana, cavata da un manoscritto inedito di G. Targioni Tozzetti.* Florence, 1852.

29. *Notizie della vita e delle opere di P. A. Micheli botanico fiorentino ... pubblicate per cura di A. Targioni Tozzetti.* Florence, 1858.

30. *Dissertatione sopra una lucerna antica torvata col lume acceso.* Florence, 1878. [Written in 1736. Edited by E. Bechi.]

31. *Le collezioni di Giorgio Everardo Rumpff ... Estratto da un catalogo manoscritto dal Prof. G. Targioni Tozzetti per cura di U. Martelli,* etc. Florence, 1903.

32. *Vera natura, cause, e tristi effetti della ruggine, della volpe, del carbonchio, e di altre malattie del grano ... con presentazione, annotazioni e biografia dell'autore di Gabriele Goidànich.* Rome, 1943. [Reale Accademia d'Italia, Studi e documenti, No. 12.]

Appendix C.

CONCORDANCE OF THE DIARIES SURVIVING
FROM THE YEAR 1657

The five diaries in question are as follows (see chapter 3):
1. The copy of the "great diary," i.e., that in Codex G. 262.
2. Viviani's copy (G. 260, 226r–81r).
3. The diary denoted by the letter "K" (G. 260, 2r–27r).
4. The fragmentary account of experiments made by Rinaldini (G. 260, 34r–49r).
5. The diary that forms Codex G. 261.

In this appendix the heads of the first and last will be set out, and their relation to the others explained by occasional notes. In general, the first and second agree very closely in content and fairly closely in language, so that only material differences will be noted.

Date	G. 262	G. 261
June 19, 1657	Ashes in water increased its specific gravity. Salt in water increased its specific gravity. Attraction by a magnet through air and through water.	1. Ashes in water increased its specific gravity. 2. Salt in water increased its specific gravity. 3. Attraction by a magnet through air and through water. 4. Weight raised by an "armed" lodestone. 5. Experiments with amber and straw.
June 20	Specific gravity of a saturated salt solution. Niter dissolved in a saturated salt solution. Solution of coral in vinegar. Ashes in water: the specific gravity was increased	6. Specific gravity of a saturated salt solution. 7. Ashes in water: increased specific gravity. 8. Ashes put in water raised its temperature. 9. Specific gravity of a saturated solution of niter.

359

Date	G. 262	G. 261
	only when they were in suspension. Changes in the specific gravity of water after boiling for 8 hours. A thermometer [probably a *termometro infingardo*] opened after 16 years, to see whether the spirit burned.	10. Change of specific gravity of water by boiling it for 8 hours.
June 21	Change of specific gravity of water on boiling. Discussion of experiments to be made.[1] Solution of coral and marble in vinegar, using stronger vinegar.	[Nothing reported.]
June 22	Powdered coral takes the acidity from vinegar more than other powders. Oil of tartar as a test for water. Oil of sulphur as a test for wines. Solution of salts one after another in the same water. Water shaken in a vessel became cooler.[2] Water passed through ashes was cooled.[3] Relative rates of cooling of water mixed with vinegar and with wine. Experiments on capillarity.	11. Specific gravity of water, in which common salt, niter, and rocksalt have been dissolved. 12. Powdered coral and powdered marble take the acidity from vinegar. 13. Shaking water did not heat it.[2] 14. Capillary rise in small tubes. 15. Water passed through ashes was cooled. 16. Oil of tartar as a test for water. 17. Changes in turbidity of water by the action of various reagents. 18. Relative rates of cooling of water mixed with vinegar and with wine.
June 25	Gassendi's experiment of sprinkling salt on a slab of ice placed on a table.	19. Gassendi's experiment of sprinkling salt on a slab of ice placed on a table.

[1] See page 53.
[2] These experiments on heat were almost always poorly controlled.
[3] The temperature of the ashes was not measured.

Date	G. 262	G. 261
	Disproved the assertion that wax put into boiling water would not soften.	20. No vinegar has been found that floats on water.
		21. Sweet white wine does not go to the bottom of water.
	Heat produced by slaking quicklime.	22. Heat produced by slaking quicklime.
	Shape of a stream of water coming out of the bottom of a vessel.	23. No more common salt would dissolve in the water of experiment 11.
	No vinegar had been found that would float on water.	24. Disproved the assertion that wax put into boiling water would not soften.
	Crystallization from a solution of niter.	
	Experiments to find out whether moisture can penetrate glass.	
June 26	A discussion (no experiments made).	25. Attempt to de-salt water by filtration through clay.
	Report on the keeping qualities of various waters on shipboard.	26. Unsuccessful attempt to filter water through wax.
		27. Making tincture of roses, by two methods.
June 27[4]	Alcohol floats on oil. If a finger is plunged through the alcohol into the oil it will not get oily.	28. Color changes induced in tincture of roses by oil of tartar and spirit of sulphur.
	Oil of aniseed in alcohol does not change its color. Added water makes it milky, and further spirit clears it again.	29. Miscellaneous tests with tincture of roses.
		30. Alcohol on oil, etc.
	Distillation and heating of vitriol.	31. Water or vinegar in "olio detto infernale" heats it; oil does not.
	Making tinctures of roses, etc.	32. Shape of water droplets on arum leaves.
	Filtration of salt water through clay.	33. Wine enclosed in a freezing mixture. The spirit of wine is not pushed to the center.
	Tried to filter water through wax.	
	Repeated experiments to find out whether mois-	

[4] Four of the experiments on this date are noted by Rinaldini (G. 260, 35r).

Date	G. 262	G. 261
	ture can penetrate glass. Wine enclosed in a freezing mixture; the spirit of wine is not pushed to the center. Shape of water droplets on arum leaves.	
June 28	Distillation of water, vinegar, and oil. Spirit of sulphur heats water. Experiment with the armed lodestone of Cardinal Giovanni Carlo de' Medici. Change of volume when salt is dissolved in water. Saturated solutions by the use of heat.	34. Distillation of water, vinegar, and oil (experiment continued for 6 days). 35. Evaporation of water. 36. Volume change when salt is dissolved in water. 37. Another experiment to the same end. 38. Acceleration of bodies sinking in water. 39. Pieces of metal colliding under water produced no waves on the surface. 40. Water boiled down acquired a taste. 41. Change of specific gravity of water on cooling.
June 30	Water made turbid with saffron increases in specific gravity. Chemical experiments on turbidity. Various experiments with tincture of roses.	42. Specific gravity of salt solutions of various concentrations. 43. Specific gravity of water tinted with saffron; no change.[5] 44. Specific gravity of solutions of various gums and salts.
July 2	Experiments with asbestos cloth. A variety of experiments on color changes by the use of chemicals.	45. Some ashes found to be at the same temperature as the air. 46. Various experiments on floating bodies. 47. Experiments with asbestos cloth and cord. 48. Experiments with mother of pearl. 49. Change of specific gravity of water on heating.

---

[5] In G. 262 the water was made turbid.

Date	G. 262	G. 261
		50. Opening the "so-called philosophers' glasses" under water.
		51. Does the superincumbent water alter the specific gravity of that at the bottom?
July 3	Test of Gilbert's statement about the effect of spirit of wine on amber. Electrical properties of diamond and of steel.	52. Specific gravity of various solutions of common salt, with a very large apparatus (continued on July 4).
July 4	Relative rate of evaporation of solutions of various salts over a period of 6 days. Change in weight of saturated solutions of various salts over 6 days. Chemical experiments with vegetable dyes. Chemical experiments with "litharge of gold."	53. Boiling water found not to penetrate glass. 54. Experiment with 2 unequal floating bodies. 55. Water that had evaporated to $\frac{1}{4}$ its original volume had acquired a taste. 56. Moisture formed inside a hermetically sealed vase when it was cooled with ice.
July 5	Observation of vortical motion of air and water shaken in a spherical flask. An attempt to detect waves on the surface of water due to a sound coming from within it.	57. A vase of water weighed the same when hot as when cold. 58. Two plane surfaces of lead made to cohere. 59. Water felt warmer when stirred with the finger than when the finger was held still in the water. 60. Creeping of salt solutions over the edges of beakers in 12 days. 61. Shape of the stream of water coming out of a hole in the bottom of a vessel. 62. An attempt to detect waves on the surface of water due to a sound coming from within it.
July 7	A lighted coal extinguished by plunging it into alcohol, or spraying it with alcohol. Falling water sets itself into rotation, right- or	63. Elaborate experiment, begun on June 28, on the solubilities of various salts. 64. Further experiments on solution.

Date	G. 262	G. 261
	left-handed according to the impulse given to it.	
July 9	Distillation experiments.	(July 9 and 10) 65. Lighted coal extinguished by alcohol, but if alcohol is poured on to a flame it is ignited. 66. Brine can be frozen.
July 10	Magnetic experiments on the effect of interposing various substances. An alcohol flame melts lead and copper.	67. Bringing a hot iron near one pan of a balance disturbs the equilibrium. 68. Color changes, using senna.
July 12[6]	Weighing red-hot metal, and a balance disturbed by the proximity of a piece of hot iron.	(July 12 and 13) 69. Specific gravity of water from the top, middle, and bottom of an aquarium. 70. Specific gravities of natural water and myrtle water. 71. Fresh and boiled water compared as to specific gravity. 72. The water at the bottom of a cylindrical vase all came out through an opening at the base before wine above it began to come out.
July 15	The specific gravity of water from the top, middle, and bottom of the Boboli aquarium was exactly the same. Expansion of fire [i.e., of the gases from gunpowder].	[Nothing reported.]
July 16	The water at the bottom of a cylindrical vase all came out through an opening in the base before the wine above it began to come out. When water is greatly re-	73. Rate of flow of water from an opening at the base of a cylindrical vessel. 74. Motion of a floating needle pulled along a trough by a magnet at one end. 75. An observation contradicting the

[6] Rinaldini (G. 260, 40r) notes two experiments on this date. See also *TT* 2, 174.

Date	G. 262	G. 261
	duced in volume by boiling there is a slight increase in its specific gravity.	spontaneous generation of animals from raindrops. 76. Expansion of the gases from gunpowder. 77. An experiment with gunpowder (description incomplete).
July 18[7]	Observations of drops of oil ascending through a column of water.	(July 17[8] and 18) 78. A pneumatic experiment, not clearly described. 79. Temperatures measured in sand, ashes, etc. 80. Experiments with floating bodies. 81. Pieces of stone put into water do not alter its temperature.
July 20	The reaction between iron pyrites and nitric acid. Hot water falling from the same height as cold water makes a duller sound.	(July 20 and 21[9]) 82. Dissection of some fishes, and examination of their swim bladders. 83. Neither wine nor vinegar, each at the temperature of some water, alter its temperature when mixed with it. 84. Attempt to change the temperature of water by long stirring. 85. Hot and cold water made a different sound when poured on the ground. 86. A closed glass ball put over a fire exploded violently. 87. Water coming out horizontally from an opening in a vessel follows a parabolic path. 88. Temperatures at various distances above a surface of snow.
July 22[10]	Specific gravities of salt solutions.	[Nothing reported.]

[7] Viviani reports an additional experiment, like no. 74 in G. 261, noting that it belongs to July 17.

[8] Rinaldini reported some experiments on July 17 (G. 260, 41r).

[9] The diary "K" records seven experiments, not the same as those in G. 261 for that date, on July 21.

[10] Diary "K" reports as in G. 262, but adds an experiment on the specific gravity of water at various depths in the Boboli aquarium.

Date	G. 262	G. 261
July 23[11]	Timing water flowing out of a hole in the bottom of a tall vase.	89. Reaction of mercury with nitric acid and other reactions producing gases. 90. Specific gravity of various wines.
July 24	A pneumatic experiment (a figure needed).	(July 24, 26, and 27) 91. Chemical experiments with nitric acid. 92. Water left for 2 days in a closed room became a little colder than the air around it. 94. Capillary rise of very hot and very cold water and other liquids.
July 27[12]	Water left for 3 days in a closed room became a little colder than the air around it. Temperature at various heights above a surface of snow.	
July 28[13]	An instrument for collecting the "air" evaporated from a liquid, and introducing a fly or other small animal.	94. Various experiments on capillarity. 95. Capillary depression of mercury surfaces.
July 29	Specific gravity of various wines and infusions.[14]	[Nothing reported.]
July 30[15]	[Nothing reported.]	96. Capillary elevation and specific gravity of various liquids.
Aug. 2	Roberval's [or Auzout's] experiment.	[Nothing reported.]
Aug. 3	A variation of Roberval's experiment.	97. Roberval's experiment.

---

[11] On this date Rinaldini (G. 260, 42r) reports an experiment on the specific gravity of various wines, giving data (which is not done in G. 261). Diary "K" reports exactly what is in G. 262; also on July 27.

[12] Nothing in Viviani's copy on this date.

[13] Nothing in "K."

[14] Viviani adds the information that the sweeter wines are the heavier.

[15] Six experiments and observations in diary "K" for this date (G. 260, 5v–6v; also TT2, 176–77).

Date	G. 262	G. 261
Aug. 4[16]	The Torricellian experiment made, and the whole instrument covered with a bell jar.	98. The Torricellian experiment with a tall cistern, so that water could be added over the lower mercury surface. 99. The same, adding spirit instead of water. 100. Light observed through 10 ells [5.84 m] of water. 101. A thermometer read the same in air and in spirit, at any depth in the latter. 102. Insects observed under the microscope.
Aug. 6	The Torricellian experiment using a very small cistern, which was then sealed.	103. The experiment with the small deflated bladder that swelled when a vacuum was made around it. 104. Soapsuds in a vacuum; also a fly, which died.
Aug. 7	Capillary rise in small tubes. Capillary rise around the water-line of a floating body; also the "bank" that the water seems to raise round a very light, thin floating body.	105. Flow of water from a rectangular wooden tank through 3 tubes of various bores, timed when they were used singly and together. 106. Acceleration of a floating needle, pulled along a trough by a lodestone at one end.
Aug. 8	[Nothing reported.]	Continuation of no. 106, under different conditions.
Aug. 9	The experiment with the little bladder *in vacuo*.	107. Further experiments similar to no. 106. 108. Time to half empty a vessel, through various nozzles. 109. Horizontal projection of liquid from nozzles. [No number] Attraction of a floating needle along a channel by a magnet.
Aug. 11	Beaten white of egg *in vacuo*.	110. More experiments like no. 109, with discordant results.

[16] Rinaldini (G. 260, 49r) reported experiments similar to nos. 102 and 103 in G. 261.

Date	G. 262	G. 261
	Exceptional capillary behavior of mercury. Effects of surface contamination of mercury.	111. Repeat of the Torricellian experiment in the form of no. 98.
Aug. 12	Fourteen parts of water left above the mercury in the Torricellian tube depressed the mercury by one part. A similar experiment with spirit of wine.[17]	[Nothing reported.]
Aug. 13	Water, and later spirit, added to the cistern, and the rise of mercury in the tube noted.	112. Torricellian experiment in a tube 4 ells [2.3 m] long. 113. The same with 4 fingers of water in the tube. 114. Outflow of a jet of mercury from an opening near the bottom of a tall cylinder.
Aug. 16	Found glass impenetrable to a strong odor.	115. Repeated no. 113; also using spirit, finding results different from no. 113.
Aug. 17	Attempt to determine the maximum dilation of air, using a Torricellian tube with a large bulb at the top.	116. End of a 4-day experiment to see whether water would come to temperature equilibrium with the air in a locked room. 117. Experiments with floating wooden blocks. 118. A pneumatic experiment (description obscure). 119. Expansion of water with temperature demonstrated. 120. Oil solidified by the bubbles from the reaction between nitric acid and copper.
Aug. 18	Repeated the experiment of Aug. 17.	No. 120 continued. 121. Expansion of water again demonstrated. 122. Specific gravities of some solids, by weighing.

[17] Somehow they managed to find that mercury is 42 times as heavy as spirit of wine.

Date	G. 262	G. 261
Aug. 20	Again repeated the experiment of Aug. 17. Observed peculiar properties of silver from a mine in Saxony.	123. A little air left in a long Torricellian tube. 124. Wine and vinegar have equal cooling power. 125. The experiment of inclining the Torricellian tube.
Aug. 21	Rate of emptying of a cylindrical vessel through a hole near its bottom.	126. Attempt to swing a pendulum in a vacuum. 127. Observation of overnight cooling of equal volumes of water, wine vinegar, and oil.
Aug. 23	[Nothing reported.]	128. Attempt to find maximum dilatation of air. 129. Refraction of sunlight by a water surface.
Aug. 24	Candido del Buono derived from the experiments of Aug. 12 and 13[18] an apparatus for measuring the relative specific gravities of liquids.	[Nothing reported.]
Aug. 25[19]	A candle flame observed through 10 ells [5.84 m] of water in a wooden tube.	130. Experiment no. 128 was repeated. 131. An experiment to find out whether odors penetrate glass. 132. Weight of air relative to water, by a hydrometric method. 133. A covered silver vessel broken open by the expansion of water in freezing.
Aug. 26	[Nothing reported.]	134. Freezing water in a cast silver ball. 135. Transmission of light through water in a tube 10 ells long. 136. Waves produced by throwing heavy objects into water.

[18] In G. 262 these experiments are cited using the numbers given them by Viviani. Experiments are not serially numbered in G. 262.

[19] Rinaldini reports experiments no. 130, 131, and 133 of G. 261, and one other.

Date	G. 262	G. 261
Aug. 27	[Nothing reported.]	137. Repeat of no. 128. The highest temperature of the year was also noted.
Aug. 28	Oil solidified by the bubbles from the reaction between nitric acid and copper.[20]	138. An experiment with two tubes and some mercury, not clear.
Aug. 29	A Torricellian tube gradually inclined away from the vertical.	139. Thermometers held above, below, and to the sides of a large block of compressed snow.
		140. Weight of air by two weighings of a glass vessel.
		141. Repeat of no. 134. Results inconclusive.
		142. A magnetic experiment.
Sept. 1	Freezing of water in a covered silver beaker. Money, or any flat disk, pushed under mercury to the bottom of a vessel stayed there.	143. Thermometers above and below a metal ball that had been cooled in ice and then dried.
		144. Qualitative demonstration of the expansion of air by heat.
		145. They began another method of weighing air.
		146. Repeat of nos. 139 and 143.
		147. A variant of no. 144.
		148. Repeat of no. 135.
Sept. 2[21]	Freezing of water in a silver beaker with screwed top. Freezing of water in a cast silver ball.	[Nothing reported.]
Sept. 3	[Nothing reported.]	149. An attempt to compress water.
		150. A variant of no. 143.
Sept. 4[22]	Weight of air by a hydrometric method. Search for a liquid lighter than spirit of wine.	151. A repetition of no. 150 with means of protecting the thermometers from air currents.

[20] Viviani dates this experiment the 26th.

[21] On this date and the 3rd, diary "K" reports an experiment in which hot and cold water were allowed to come into equilibrium with the air in a small room. This is copied in *TT* 2, 177–78, as part of the "experiments made by the Grand Duke and some of his courtiers."

[22] The experiment mentioned in note 21 was repeated twice, ending on September 6.

Date	G. 262	G. 261
Sept. 6	Investigation of elm-leaf galls. Toads are not generated from rain.	152. A repetition of no. 151, interchanging the thermometers.
Sept. 7	Thermometers around a cooled metal ball.	153. Weighing of air. 154. The same, by a method somewhat similar to no. 132.
Sept. 10	Paolo del Buono's instrument for trying to compress water.	155. Like no. 152, with a ball heated only by hammering. 156. Differences in the specific gravity of liquids, using the Torricellian tube. 157. Change in the height of the mercury with temperature.
Sept. 11	Thermal expansion of air by a hydrometric method.[23]	158. Thermal expansion of air by a hydrometric method. 159. Another experiment for the same purpose. 160. The experiment with the syringe attached to the closed cistern of the Torricellian tube. 161. Observations on leaf galls. 162. An experiment to prove the existence of a vacuum.[24] 163. Another, for the same purpose.[25]
Sept. 13	The Torricellian experiment with a syringe attached to a small closed cistern.	[Nothing reported.]
Sept. 14	Proof that the moisture within a closed vessel that has been cooled is not condensed air.	164. The Torricellian experiment with various quantities of air left in the tube.
Sept. 15	Demonstration of the thermal dilatation of mercury. Warming the closed cistern	165. Inclining the Torricellian tube.

[23] Rinaldini reports a similar experiment on this date (G. 260, 48r).
[24] See *Saggi*, p. XXXXVI.
[25] See *Saggi*, p. L.

Date	G. 262	G. 261
	raised the mercury in the Torricellian tube.	
Sept. 17	The humidity of the air has an effect on the height of the mercury in the Torricellian tube. Report of measurements of atmospheric pressure made at various heights at Artimino, at Poggio di S. Giusto, and on the Palazzo Vecchio in Florence. Results of heating and and cooling the space above the mercury in the Torricellian tube. Investigation of water waves produced by throwing projectiles in. (Continued until Sept. 23.)	166. Acting on a suggestion sent by the Grand Duke from Artimino, they investigated the effect of humidity on the height of the mercury in the Torricellian tube. 167. Water waves were studied in the aquarium.
Sept. 19	[Nothing further reported.]	168. An experiment to find out whether light is bent by passing from air into a vacuum.
Sept. 20	[Nothing reported.]	169. The inclined Torricellian tube. 170. A Torricellian tube in a cistern having layers of mercury, water, wine, and spirit. After the vacuum was made the tube was gradually raised. 171. An attempt to make odors go through glass. [Unnumbered] Determination of the weight of air (not clear).
Sept. 21	[Nothing reported.]	172. Barometric measurements at the Palazzo Vecchio. 173. Humidity observations made at the top at the tower and at the ground.
Sept. 24	"On the 24th, with the departure of Prince Leopold for Artimino, the	[No entry. There are no experiments numbered 174 and 175.]

Date	G. 262	G. 261
	Academy was given leave until October 3rd, which was the first day that it met after the return of His Highness."	
Sept. 28	Report of observations of a water barometer at Artimino, made on that date.[26]	176. Report of observations of a water barometer made at Artimino on that date.
Oct. 3	Breaking a cast silver ball by freezing water in it. The inclined Torricellian tube.	177. An inclined Torricellian tube with a quadrant to measure the angles.
		178. Refraction of sunlight by a surface of water.
		179. Reaction between mercury and nitric acid.
		180. A balance brought to equilibrium with one pan in air, the other under water. When the level of the water was gradually lowered, the pan in water went down, surprising those who were "not well grounded in mathematics."
		181. The Torricellian tube, the vacuum made, was stopped with a finger and held up. When it was inclined the sensation produced in the finger varied.
Oct. 4	A bowl of water suspended by four strings united at the point of support could be swung without spilling; but if the strings were separately supported, even a small swing would spill the water. In freezing water, it first contracts and then expands. Bubbles form during the freezing.	182. Thick glass flasks full of water exploded on freezing.
		183. Oil contracted on congealing.
		184. The suspended bowls of water, as in G. 262 for this date (very different wording).
		185. Pendulum observations.
		186. Further observations on stopping the Torricellian tube with the finger.

[26] Nothing in Viviani's diary on this date.

Date	G. 262	G. 261
Oct. 5	Two thick glass flasks full of water exploded on freezing.	187. Water being frozen with a thermometer in the vessel (not clear). 188. In filing iron the file, being in motion, is heated less than the iron. If the file is held still and the iron moved, the iron is the less heated.
Oct. 6	A thick bronze ball[27] was broken open by the force of freezing water.	189. Rates of cooling of water while being frozen.[28] 190. Observations on spinning tops. 191. A brass (*ottone*) ball, made in two parts screwed together, torn apart at the screw by the force of freezing water.
Oct. 8	The experiments on the force of freezing water repeated.	192. The Torricellian tube being plunged into water, the height of the mercury rises.
Oct. 9	Observations on newts. A blow on the rim of a full conical glass broke the foot cleanly off, with no damage to the place struck. Observations with the microscope of bubbles in ice.	193. Extract of red roses looked at through the microscope. 194. Repetition of no. 189, with a vessel of different shape. 195. Freezing experiment in an elaborate vessel (not clear). 196. Timing a pendulum as its swings decreased. 197. No. 192 repeated, with variations.
Oct. 10	Examination of an aperture sight made by Eustachio Divini.[29] Examination of a crossbow. How to split a peach stone. Oil in freezing decreases in volume, but very little.	198. Further freezing experiments with still another vessel. 199. A conical glass of water was held in two fingers and the rim struck, causing the foot of the glass to break off. 200. How to split a peach stone. [Not numbered] Observation of lice with the microscope. 201. Description of an aperture sight [made by Eustachio Divini].

[27] This was not similar to the ball of no. 191 in G. 261.

[28] These observations were made by Rinaldini and by Borelli.

[29] A celebrated instrumentmaker, noted for his telescopes.

Date	G. 262	G. 261
Oct. 12	Water frozen in a cast bronze ball, the ball broke at a place where it had been weakened in turning it. Another ball with no such weak spot burst open just the same. Frozen oil was seen to sink in water. The sudden increase in the volume of water when it freezes has not been reduced to order.	202. Quantitative experiments on the freezing of water in a vessel with a long graduated neck. 203. Water frozen in two metal balls, one of bronze, the other of brass. Both broke open. 204. Frozen oil was seen to sink in water. 205. Comparison of a pendulum and a sand glass.
Oct. 13	More experiments on freezing. A pneumatic experiment [*Saggi*, p. XXXXVI and fig. VIII]. Another pneumatic experiment [*Saggi*, p. L and fig. X].	206. A pneumatic experiment [as in G. 262 for this date]. 207. Another pneumatic experiment [ditto]. 208. Freezing water in a flask with a long neck divided into 600 "degrees."
Oct. 15	Up to that time, oil was the only substance they had found that did not dilate on freezing. More observations about the freezing of water in a glass flask.	209. The freezing of white wine, vinegar and verjuice observed. 210. A small hollow ball of brass with very thick walls was filled with water and put in a freezing mixture. It did not burst. 211. Freezing of water in a flask with a graduated neck. 212. Comparison of thermometers of various lengths. 213. Mercury could not be frozen, nor could spirit of wine. 214. Spirit of wine poured on the hand feels colder than does water at the same temperature.
Oct. 16	Freezing of white wine, vinegar, and verjuice. The flask was found to break when the freezing of water began in the neck. Paolo del Buono's instrument for investigating	215. Experiment no. 211 was repeated. 216. Relative contraction of wine, vinegar, and verjuice. 217. Paolo del Buono's instrument for investigating the compressibility of water, made partly of metal, broke.

Date	G. 262	G. 261
	the compressibility of water, made partly of metal, broke.	218. [a] Peculiar properties of water near the freezing point. [b] Sudden change in the opposite direction when the bulb was put in hot or cold water. [Behavior of spirit of wine in these respects.]
		219. As experiment no. 209, but quantitative.
		220. A brass ball filled with water and having a screw stopper, placed on the fire, blew up. They had previously tried to burst it by freezing, and suspected a small crack in the metal.
Oct. 17	A brass ball, etc., as no. 220 in G. 261. The observation of the sudden change in level in an unexpected direction when a bulb with a long neck partly filled with liquid is put in hot or in cold water. Observation of the small bubbles of air that always seem to collect between the glass and the mercury in the Torricellian tube.	[Nothing reported.]
Oct. 18	[Nothing reported.]	221. Experiment on freezing water in a bulb with a long neck.
Oct. 19	Observations on artificial freezing, in progress [data recorded under Oct. 24]. Explosion of a brass ball filled with water and tightly closed, when put on a fire.	222. Weight of air compressed into a large copper ball.
Oct. 20	[Nothing reported.]	223. Thermal dilatation of water in a graduated flask.
		224. Another unsuccessful experiment for the same purpose.

Date	G. 262	G. 261
		225. A closed brass ball full of water put on the fire. It "exploded more terribly than the other."
		226. A Torricellian tube immersed in the Boboli aquarium.
Oct. 22	Reference to Borelli's quantitative experiments on the sudden jump of the liquid when its container placed in hot water. The freezing of water seems to take place instantaneously.	[Nothing reported.]
Oct. 23	They have devised the proper shape of flask for the experiments on artificial freezing. Freezing disengages air bubbles.	227. Observations on behavior of liquids in flasks of two sizes, put in ice. 228. When they are heated and cooled again, glass tubes go out of round.
Oct. 24	Confirmation of yesterday's experiments on freezing.	229. Quantitative experiments on artificial freezing. 230. The same, with a smaller vessel.
Oct. 26	Further experiments on freezing, starting from water at various temperatures.	231. Further freezing experiments. 232. Another experiment on freezing. 233. Cones of wood and lead of equal size, let fall into clay.
Oct. 29	[Nothing reported.]	234. Two brass balls were filled with water. One, put in a freezing mixture, burst open; the other, put in the fire, did not. [Not numbered.] Some more freezing experiments. 235. An attempt to compress water by the weight of mercury.
Oct. 30	Further experiments on freezing.[30, 31]	236. Freezing of water in a glass bulb. They found out how to prevent the bursting of the bulb.

---

[30] The original "laboratory notes" are in G. 260, 11r and 12r.

[31] Viviani merely notes (G. 260, 279r) that freezing experiments were continued, but "nothing different appeared in them."

Date	G. 262	G. 261
Oct. 31	Freezing of water, starting at various temperatures.	237. Freezing of water, starting at various temperatures. 238. An attempt to compress water by air pressure.
Nov. 3	Attempt to compress water by the weight of a column of mercury.[32]	239. As in no. 237, but with a larger bulb.
Nov. 4	[Nothing reported.][33]	[Nothing reported.]
Nov. 5	Further freezing experiments.	240. Further freezing experiments.
Nov. 7	[Nothing reported.]	241. Conical weights of wood and of lead dropped from various heights on to clay, and the depth of penetration observed. 242. Rate of cooling and warming of water in a bulb.[34]
Nov. 8	[Nothing reported.]	243. More observations as in experiment 242. 244. Weights dropped into a mixture of wax and lard. 245. Rate of ascent through water of small balls of cork and wax, also of hazel nuts, was found to be uniform.
Nov. 9	[Nothing reported.]	246. Water was frozen in flasks of thick and of thin glass. 247. The fit of a ring on a mandrel was changed by change of its temperature.
Nov. 10	Rate of cooling of equal beakers of water and of spirit placed in ice.	248. Rate of cooling of equal beakers of water and of spirit placed in ice. 249. Mercury in a flask with a long neck appears to increase sud-

---

[32] Viviani notes that the instrument "was invented by Captain Guerrini" (G. 260, 279r).

[33] But Viviani says (*ibid.*) that "On November 4, 5, 6, 7, 8, 10, 1657, the observations of the freezing of water were repeated several times, without finding anything different from what was found on previous days."

[34] Copied in *TT* 2, 182 from some other source.

Date	G. 262	G. 261
		denly in volume when first put in ice.
		250. A different ring-and-mandrel experiment.[35]
Nov. 11	[Nothing reported.]	[Nothing reported.]
Dec. 3	Rates of heating and of cooling of water and of spirit in beakers placed in hot water and taken out again.	251. Rates of heating and of cooling of water and of mercury.[36]
	Similar experiment with water and mercury.	252. Experiments with [badly annealed?] flasks.
	Similar experiments, comparing spirit and mercury.	253. The experiment with the ring and the mandrel.
	The experiment with the ring and the mandrel.	254. Rates of heating and cooling of spirit and of mercury.
	Experiments with [badly annealed?] flasks.	
Dec. 4	It was observed that spectacles fall out of their horn frames when heated.	255. Continuation of no. 254.[37]
		256. Repeat of no. 252.
	A chemical experiment with "tincture of coral."[38]	257. A pneumatic experiment, not clear.
		258. A chemical experiment with "tincture of coral."
	Rate of heating and cooling of spirit and of mercury.	259. Sealing the end of a thermometer tube does not alter the readings of the instrument.
Dec. 6	[Nothing, but there is a fig. XXXV on fol. 47v which corresponds to the figure for no. 260 of G. 261, and to fig. VIII and p. CLXXXVI of the Saggi.]	260. A hollow glass ring filled with hot water expands.
		261. Effect of humidity on rings of wood.

[35] This entry is followed by "Because of the arrival of Cardinal Antonio Barberini we remained until December 3 without doing any more."

[36] The numbers differ from those in G. 262 in one experiment, but agree in another.

[37] The data are the same as in G. 262.

[38] Viviani's diary ends at this point.

Date	G. 262	G. 261
Dec. 10	[Nothing reported.]	262. Abortive attempt to set up a water barometer.
Dec. 12	More quantitative freezing experiments.	263. Freezing experiments.[39]
Dec. 14	[Nothing reported.]	264. More freezing experiments, starting with both cold and hot water.
Dec. 15	[Nothing reported.]	265. Still more freezing experiments.
Dec. 17	[Nothing reported.]	266. An attempt to measure the maximum dilatation of air.
		267. Summary of experiments made in the past weeks, with discussions.[40]
Dec. 19	Breaking of very thin glass bubbles by submerging them deeper and deeper in water.	268. Breaking of thin glass bubbles.[41]
	A musket ball tied to one end of a balance beam by a cord one ell long, and allowed to drop, lifts a much larger weight hung from the other end of the balance.	269. A musket ball tied to one end of a balance, etc.[42]
	Freezing and thawing of water in open and in closed flasks.	
	It was observed that natural and artificial ice are different.	
	Freezing of myrtle water.[43]	
Dec. 20	[Nothing reported.]	270. Making a vacuum in a tube less than an ell and a quarter long [see *Saggi*, p. LIV and fig. XIII].
		271. A similar experiment with water in a tube less than 18 ells long.

[39] Experiments similar, but data differ.

[40] G. 261, 62v–63r.

[41] This entry is exactly as in G. 262.

[42] This experiment seems to have been started on December 17.

[43] In the margin, in another hand: "Experiment of Prince Leopold against frigorific atoms."

Date	G. 262	G. 261
Dec. 22	Water in a U-tube with arms of different diameters comes to equilibrium with the two free surfaces at different levels. [This is the last entry in G. 262 until July 22, 1658 (fol. 50v).]	272. Maximum dilatation of air in an apparatus ascribed to Paolo del Buono.[44] 273. No. 272 repeated.
Dec. 24		274. Another repetition of no 272.
Dec. 27		275. Still another repetition of no. 272.
1658 Jan. 8		276. [At Pisa] A test of Viviani's instrument, an air baroscope, on the Campanile.
Jan. 12		277. Experiments 270 and 271 repeated "in the presence of the Doctors" with success. 278. The Torricellian tube and its cistern immersed in water in a crystal glass tube 4 ells high.
Jan. 14		279. Expansion of boxwood by absorption of water.
Jan. 15		280. The second experiment against "positive levity" [see *Saggi*, p. CCXII and figs. III, IV, V].
Jan. 19		281. Experiments on the speed of sound. 282. Ditto, over the sea.
Jan. 23,	and shortly thereafter, at Leghorn and possibly Pisa.	283. Relative specific gravities of wine and sea water. 284. Eggs observed not to float in sea water. 285. Natural freezing of water, etc. 286. Water frozen in a vacuum.[45] 287. An experiment in which an inverted flask of a special shape was filled with water and im-

[44] But not that in the letter of which we only have a copy of a copy. See Middleton, *Arch. Hist. Exact Sciences* 6 (1969): 1–28.

[45] We are told that this "closed the mouths of those who were saying that water did not freeze without air."

Date	G. 262	G. 261

mersed in a larger vessel of water and watched for the time that the Court remained at Leghorn, to see whether bubbles would form.

288. A shot fired horizontally from a gun in the Fortress of Leghorn reached the water in about the same time as one dropped from its muzzle.

[End of this diary]

Appendix D.

ORIGINALS OF UNPUBLISHED MANUSCRIPT PASSAGES
TRANSLATED IN THE TEXT

For easy reference these passages are given in order and identified by
the number of the chapter and the footnote that refers to the translated passage; or in case the passage does not give rise to a footnote,
the page number has been given.

CHAPTER 2

Note 58
Ricevò la benignissima lettera di V.A.R.^{ma} nella quale si degna rin-
graziarmi di quello che V.A. mi ha si puo dire generosamente donato:
imperocche non poca parte delle cognizioni, e specolazioni contenute
in quel mio libro ebbero motivo dalle sperienze fatte nell accademia
di V.A., senza delle quali non avrei potuto perfezzionare tal opera.

Note 106
Et quand vous vous efforcerez de donner au public quelque chose à la
Gloire d'un si grand Prince, quoy que vous y soyez plutost engagé par
tout autre motif que par celuy de l'interest; Ce sera un moyen infallible de l'obliger a vous le continuer à l'avenir.

CHAPTER 3

Page 45
Si esperimentò, se fusse vero ciò che scrive il Gassendo, cioè che po-
nendo sopra una lastra di ghiaccio posta sopra una tavola una quan-
tità di sal comune, questo faceva sì in poco spazio di tempo, che detta
lastra rimaneva sì fortem^e attaccata alla tavola, che senza gran forza
non si poteva staccare. Si trovò vero, e trovasi inoltre, che l'acqua
colata in sulla tavola era salata. Si vedde poi, che il ghiaccio dalla
parte volta verso la tavola, cioè sottoposta alla faccia coperta di sale
era fatta opaca per una nuvola bianca composta di varie particelle di
sale, e per ultimo sperandosi al lume il ghiaccio si vedeva doppo
esservi stato il sale minutam.^e scabroso, et a puntoline di diamante
vagam.^e intagliato, in somma simile a quei bicchieri, che per la

similitudine, che hanno col diaccio, soglionsi chiamar diacciati. Si osservò finalm.e che spingendo detta lastra parallela all' orizonte era impossible distaccarla, come il contrario più facile riusci coll'alzarla con forza perpendicolarm.e al piano orizontale.

Page 45
Fu fatta la prova con una lastra di diaccio posta sopra un tavola e sopra quello del sal comune se si attachasse [*sic*] alla detta tavola, che si attaccò mirabilmente, e fatto con il salnitro nella medesima maniera non si volse mai attaccare.

Note 15
Mi avvisi intanto che strumenti ò altro che lei stimassi bene che io portassi meco a Pisa, e mi dica se brama che innanzi facciamo quà qualche osservazione di quelle però dove non si habbia a fare strumenti per mano del gonfia, il quale domanj parte per cotesta volta. Si prepari ancora per le osservazioni che si doveranno fare a Liv.o, e mi avvisi se anco per queste è necessario [*sic*] alcuna cosa.

Note 28
Il Ser.mo Arciduca ha voluto che fra gl'accademici eletti da S.A. sie annumerato ancor io, che non ho saputo disdire a'i di lui cenni tanto più che mi è nota la gran premura che mi tengono L'Imp.re e L'Imper.se Il n.o dell'Accademici è di dieci, e sono questi. Il Co: Raimondo Montecuccoli Gnle della Cavall.ria di S.M.Ces. 2.o Il Marchese Gnle Mattei Cavallerizzo Maggiore del Ser.mo Archid.a 3.o Il Conte Francesco Piccolomini Aragona 4.o Il Maris.e D. Giberto Pio di Savoia 5. Il Barone Horatio Buccellieni Cons.lo nel p.mo ordine del Regg.to di Vienna 6. Il Barone e Colonello Mattias Vertemuti 7. L'Abb.e Spinola 8. Il Conte Francesco d'Elci 9. Il Nobile Francesco Zorzi Veneto 10. Il Marchetti Resid.te der Ser.mo di Toscana.

Vi erano molti altri suggetti, che potevano esser ascritti a' questa Accad. ma S.A. disse, che S.S.M.M.ta et essa havevano pensato che fusse maggior honorevolezza dell'istesso Congreso l'ammettervisi solamente quelli di Nobilta conspicuo, e qualificata; onde non ho stimato a mio svantaggio l'esser anche per questo capo riconosciuto da tutti di Corte non indegno d'entrar in questa riga.

Ho parlato con V.S.Ill.ma di questa materia per supplimento, non havendo notitia più degna da participarle.

Note 29
Il Ser.mo Arciduca applicato al solito a gl'essercitii [*sic*] virtuosi ha procurato di far scielta [*sic*] d'alcuni Cav.ri Italiani per formare nel palazzo Imperiale un'Accademia di belle Lettere, e già ne sono eletti dieci a tal fine, a cui con molto applaso [*sic*] danno mano anche

LL.MM.^{tà}, et in effetto si è risoluto per la p.^{ma} Domenica dell anno la prima funzione da tenersi nella Camera med.^{ma} dell'Imp.^{re}

Note 36
A dì 20 Maggio
Si riaprì l'accademia, et in quella mattina+ furono consegnato *a me Lorenzo Magalotti,* le scritture tutte appartenenti a quella, sì de' diarij, come d'esperienze proposte, e da farsi, e++ insieme si disposero varie cose pel proseguimento dell'accademia.

+ *mi* is cancelled.
** These words are in the margin, indicated by a caret.
++ *con esse* struck out and replaced by *insieme*.

Note 51
Si determinò, che li Sig:^{ri} D. D. Rinaldini, Borelli ed Uliva dovessero ragunarsi ogni giorno a Palazzo alle 21. ore discorrere, e dare gli ordini necessarii per le esperienze da farsi di mano in mano il giorno appresso.

Note 60
Infino al dì 23. 7bre si consumo il tempo in osservare varii effetti dell'intersecazione, et espansione, che fanno i cerchi prodotti nell'acqua della caduta de' proietti in essa senza esservi per allora raccolta, o stabilita cosa alcuna di certo.

Venne in tanto avviso dalla Corte, la quale si ritrovava ad Artimino, che in conferma di quanto viene scritto di Francia per autenticare per vera cagione del sollevamento del mercurio la pressione dell'aria si era osservato, che calandosi dalla cima del Palazzo di Artimino sino alle radici del monte,... l'instrumento detto υγροσταθμιον* andava il mercurio continovam.^e [*sic*] alzando nel descendere... tornandosi in sù andava calando tanto sin che ridotto sulla cima alla sua solita altezza.

L'istessa esperienza si fece sul Poggio di S. Giusto, e sulla Torre di Palazzo vecchio di Firenze...

Il giorno poi de' 24. con la partenza del Sig.^r Principe Leopoldo verso Artimino, fu licenziata l'Accademia per infino al dì 3. Ottobre, che fu il primo che ella si ragunasse doppo il primo+ di S.A.S.

* This reading is doubtful. There is a space in G. 262 which someone has filled in tentatively in pencil. In G. 260 Viviani has simply left it out, writing "l'istrumento d.^{tto} andava."

+ So in G. 262. In G. 260 Viviani has (more plausibly) "doppo il ritorno di S.A."

Note 61
Il dopo desinare... si fece avanti al Seren.^{mo} Gran Duca l'esperienza

sud.ᵃ contro la leggerezza positiva, come anco quella del bollimento a diaccio . . .

Note 66

A dì 4.5.6.7.8.9.10. 9:ᵉ, 1657 Si replicò più volte l'osser.ᵉ dell'addiaciam.ᵗᵉ dell'acqua senza trovarsi diversità dall' operate i giorni innanzi.

Note 71

Sig. Lorenzo. Si ricordi di tener forte quello li ho detto di non introdurre alle esperienze, altri che quelli opereranno continuamente, perche si perde troppo tempo et io premo nella spedizione et à quei Sigʳⁱ Accademici che sogliono venire talvolta più di radi, dica che io li chiamero quando si proveranno le cose avanti a me, e saranno aggiustate e li do la buona sera. Pal.º li 30 Lugl. 1662

Note 73

Ho sperimentato nel Campanile lo strumento di VS. per pesar l'aria, ma ho auto difficulta nel caldo, e nel freddo percio convien avervi qualche considerazione. Ho lodato l'invenzione et ho auto occasione di parlar di VS. al Serᵐᵒ Leopoldo con tanta estimazione ch'ella non saprebbe desiderar di piu; l'ho fatto volontieri, e farò di nuovo conoscendo di dir il vero.

Note 76

Supplico poi umilissimamente V.A., che rappresenta al Ser.ᵐᵒ G. Duca la massima cautela, con la quale io ho parlato intorno à questo fatto, dalla quale non mi parto ne anco adesso: si che io non intendo aver' detto niuna cosa asseverantemente, se non per mera coniettura, esposta à poter'esser fallace: e però con ogn'umiltà la supplico, che riceva le cose dette da mè, non per pubblicarle, o impegnarmi à sostenerle; ma come dubbij indigesti, et imperfetti, communicati privatamente à V.A. Ser.ᵐᵃ, allettato dall'infinita benignità, con la quale suol' ricever' le mie fantasie, e compatire i miei difetti. E per fine la reverisco umilissimamente, come devo.

Note 80

Si è quì sparte voce, che il Sig.ʳᵉ Auditor Vettori sia per venire à Pisa. V.S. costi potrà meglio di noi sapere, se questo è vero, perche quando il d.º Sig.ʳᵉ non venisse à Pisa son necessitato à pregarla, che da parte mia faccia à d.º Sig.ʳᵉ istanza, che dia commissione à qualche dottore di Pisa, acciò che possa rivedere il mio libro, e concedere la licenza di stamparsi. E se ancora à V.S. paresse esser necessario adoperare l'Autorità del Ser.ᵐᵒ Principe per questo affare, la priego ancora, che gliene dica una parola, acciò ch'egli colla sua autorità renda facile questo negozio.

Note 86
La forza di varij accidenti è stata cagione che molti della mia Accademia sieno stati, e sieno separati in diversi luoghi. Onde per qualche tempo non si è applicato alle esperienze e alli studi incominciati.

Note 92
Il rimanente poi di questo mese,* e del seguente d'8bre insino ai 25* benche più volte si ragunasse l'Accademia, ciò fu solo in ordine alle osservazioni di Saturno, alla disputa sostenuta col P^re Fabri, e finalmente di eventilare le proposizioni dei discorsi che si mandarono in Olanda, come apparisce nel libro.

Nel mese d'Ottobre corsero le vacanze dell'Accademia insino al dì 25. che si riapri di nuovo.

---

* The passage between the asterisks is in G. 260 but is crossed out. It does not appear in G. 262.

Page 62
Nei giorni seguenti fino al dì 15 di Novembre si adoperò intorno alle due seguente esperienze, nelle quali, perchè s'incontrarono diverse difficultà, che resero più volte la fatica usatasi senza frutto, basterà il racontarle solamente in quella forma, che furono felicem.^e operate.

Note 96
Mi dispiace ben grandem:^te di non poter tendere così adesso il dovuto contracambio alla Sua gentilezza, come sarebbe il mio desiderio, perche stando [sic] noi tutti applicati intorno ai preparati per il ricevimento della Ser:^ma Sposa, et alle festi da farsi per le Nozze, si sono tralasciati e da me e da questi SS:^ri Accademici per questo tempo simili studi; Ma dato che sarà fine a questa funzione ripiglieremo il filo delle Esperienze, et spero poter corrispondere allora a V.S. . . .

CHAPTER 4

Note 2
Havevo destinato, come gl'accenai, d'inviarle altre mie lett:^e nelle quale dovevo significarle un mio concetto, et un opera, che io ho fra mano, che dovrà essere (se mi sortirà tirarla a fine, e se non m'inganno) in qualche parte di non poco utile alla Rupub.^ca litteraria [sic], nella q^le tenendo V.S. uno de' principali luoghi dovevo (come io farò) richiederla, et del suo consiglio, et del suo aiuto . . . .

Note 9
Si ragunò l'Accademia in casa del Sig.^r Lorenzo Magalotti a fine di replicare alcune esperienze che parevano più necessarie per dar compimento all'opera, che debba stamparsi, le quali tutte, quando ne

venga agevolata la pratica hanno a rifarsi alla presenza dell'A.S. Seren.ᵐᵃ

Note 10

...stimasi opportuno, che fino al mio arrivo in Firenze, non si dia ordine alcuno, di far metter mano à i disegni, al Sig. Falconieri solo si potrebbe accettare di verderne Una Mostra, ad effetto di vedere se si adattono gl'[uno?] o gli altri al Modo come Vorrebbero esser lavorati i Rami per il Bisogno nostro...

Note 22

Si volle vedere il progresso del'agghiacciam.ᵗᵒ altre volte, veduto nella solita palla, e di più osservare per via delle vibrazioni del pendolo i tempi, nei quali accadono le seguenti alterazioni di condensarsi e rarefarsi, che fa l'acqua naturale nel suo gelare.

Note 27

Sig. Lorenzo. Ho caro sia consegnato il Manoscritto per il Cd Rannucci, dal quale avviso che non si stamperà quello che non si vuole.

Page 72

Indovinar sentir [?]* originale è pericoloso—e io forse sono uscito dell'ufficio di Corretore di Stampa Rivegga adunque prima di darlo allo Stampatore e mi dia pur da fare se bene sono occupato.

---

* I owe this conjecture to Professor Felix Gilbert.

Note 30

Il Sʳ Magalotti mi dice d'aver lesto il primo trattato delle esperienze per darlo a' Revisore.

CHAPTER 5

Note 21

Sigʳ Lorenzo Aviamo come V.S. vedra dall' inclusa scrittura, e calculazione del Borelli una occasione di veder in Cielo una Bella Curiosita, che di rado accade circa Venere, ma bisognerebbe stare ad osservare in luoghi da scoprire [?] l'orizonte, ove nasce et ove tramonta il Sole, e dove non è il Mare, bisogna salire alla Montagna qua il Borelli vuole andare alla Verrucola, io ho ordinato sia fatta altra osservazione alla Gorgona. et ho prestato un occhiale à quel Castellano, ma à me non toccera poter fare questa osservazione Se ò V.S. la potesse fare, ò Antonio Maria del Buono ò Altri costa non saria che bene. Intanto le Auguro perfetta salute et ogni Contento.

Note 65

In esecuzione de comandamenti dell'A.V.; parmi di ricordare che l'oppinione dell'evaporazione dell'argento vivo non fusse da me pro-

posta come particolare di persona alcuna, essendo universaliss.º sut-
terfugio di tutti coloro, che negano il vacuo il ricorrere a queste
esalazioni del mercurio, estratte in un certo modo violentemente dalla
massa di esso, per riempire lo spazio lasciato nella sua caduta.

Note 66
... essendo questo fattosi più per prova, e per fermare il modo, più
che per praticar l'esperienza.

CHAPTER 6

Note 7
Trovandomi quà con incertezza di poter tornamene a servirla di
presenza avanti il di lei partire per Roma, non voglio mancare a me
stesso in tralasciar d'indurle a memoria l'intenzion data da VS: Ill:ma
di far godere alla nostra Toscana gl'inesausti tesori della sua Inghil-
terra, introducendo tra questa, e l'Accademia filosofica del Ser:mo
Principe Leopoldo una scambievole corrispondenza in materie fisiche,
e filosofiche fondata sopra esperienze fatte o da farsi per la cognizione
d'effetti naturali, e per lo scoprimento del vero, e con tal mezzo
stabilire una virtuosa e perpetua Amicizia tra quella nostra Adunanza
et alcuni dei p:mi Letterati di quel Regno. Cosa da noi Accademici
grandemente ambita, e desiderata, e non meno dall'istesso Ser:mo
Principe, il quale come è ben noto a VS. Ill:ma con sue cortesiss:e
instanze a lei med:a hà fatto conoscer quanto grato gli saria questo
nobil commercio, e principalm:te con quel Sig:r Ruberto Boyle da
VS: con tanti titoli d'eccel:za piu e piu volte esaltatoci, e con quanta
ansietà S. A. Ser:ma attenda opportuna congiontura di palesare la sua
generosa benevolenza verso Ingegno cosi peregrino e sublime; assicu-
randola che subito che per mezzo di lei si sia attaccato questo Commercio verrà ancora perpetuato forsi [*sic*] con sodisfazione di quei
SS:ri Inglesi non minore del frutto che dalla novità delle esperienze
loro, e delle lor salde dottrine ci promettiamo. Confidasi dunque
nell'impareggiabil prudenza et industria di VS: Ill:ma sperando che
questo per cosi dire Embrione in breve si abbia a veder nato, e per-
fezionato, anzi che venendo del suo favorevol Genio raccolto, assistito,
et educato, abbia ben presto a crescere, prendendo vigore e forza
sempre magg^{re}. A questa fine sarà da noi destinato in Lion di Francia
qualche Mercante, che ricevendo le lettere di Firenze, e di Londra, le
invij per il lor recapito quando occorra.

Io poi augurandole (in caso di mia assenza) fortunatissimo Viaggio,
ricordole il favorirmi in ogni altro particolare ricercatole, e la supplico a far mie scuse con d:º Sig^r Boyle dell'essermi preso ardire
d'inviarle l'Opera mia de Massimi e Minimi, et anco del non averla
accompagnata con mie lettere, pregandola a far passare un simile

offizio con gli altri SS: Barlow, Wallis, Wardo, et Oldenburg, a i quali tutti desidero inoltre che sia fatto nota la mia singolar osservanza verso l'egregia fama di cosi illustri soggetti; e la brama che tengo di venire onorato col titolo di lor Ser:re come io sono ossequioso Ammiratore del'Eminente dottrina loro nelle filosofiche, Geom che, et Astronomiche discipline.

Crederò che a quest'ora ella averà spedito a Livorno per Londra non solo i detti miei libri, ma quel ritratto ancora del mio tanto riverito Maestro Galileo, che VS: Ill:ma si compiacque di ricevere; e godero poi di sentire che per L'intercessione e liberalità di VS: il Sig: Barlow ne abbia voluto ornare la sua famosa Libreria d'Oxfort, onorando il ritratto non meno di quel che cosi grata Nazione onori, e stimi l'opere del Vero.

Soggiungo, pregandola di significare in specie al Sig: Boyle quanto grato sia pascer al Ser:mo Principe l'intendere qualche particolar del progresso di lui intorno a quella comparazione ch'ella disse a S. Alt:a aver egli alle mani circa alle Nozioni dell Atomi e gl'Esperimenti della Chimia, la quale detto Sig: và tuttavia meditando con averne già scritto non so che fogli. Ohime Beato Sig:r Ruberto se per mezzo di VS: io giungerò in si gran distanza a possedere la grazia di tanti Eroi! ma più se una volta mi toccasse in sorte di potermi rassegnare a loro di pretesa quali io sono di VS: Ill:ma

fedel:mo Amico e dev:mo Sre Obblig:mo
VINCENZIO VIVIANI

Pistoia li 6 Ott:e 1660

Note 23
Habbiamo in Londra per l'ordine Reale una Academia in Londra cominciata alla somiglianza di quella di VAS. per far delle esperienze; et Io ho già sapendo il genio di SAS: proposto ad alcuni di loro li piu virtuosi la corrispondenza con VAS: la quale quanto primo che Io intenderò di esser di suo aggradimento, istabilirò essendo questi piu che ambitiosi di tanto honore.

Note 24
Ho molto caro che cotesti virtuosi habbino cominciato a studiare sopra il libro della natura, per riconoscere la verità delle cose, e circa la corrispondenza che VS propone, mi sarà al maggior segno grata, e stimabile per tutti i conti, offerandomi sempre pronto in contracambiare il loro affetto con ogni applicazione possibile.

Note 93
Ora io godo sommamente che da quei Sig.ri in Francia si vada con nuove esperienze e speculazioni promovendo la natural filosofia; Ma

ho anco qualche sospetto e gelosia, che dell'inventioni e speculazioni de i nostri maestri, e di quelle, ch' habbiam trovate noi se n' habbino secondo l'usanza vecchia far autori, e ritrovatori gli stranieri. Questo rispetto mi far andar ritenuto ad attaccar questo commercio con quei Sig.^{ri} dell'Accademia Parigina; perche non si puol far di meno nello scrivere di non communicarle qualche cosa, e l'istesso dubito che dà campo a quegl'ingegni pellegrini di ritrovar le cose; tratto delle ragioni, non dell' esperienze. Dall'altra parte parmi che sarebbe pur bene esser informati di quello, che si va operando, e speculando in quell'accademia; si che io mi trovo irresoluto, e però ricorro a V.A. Ser.^{ma} perche mi commandi come mi debbo portare in questo affare . . . Ho risposto ambiguamente per ora al Sig.^{r} Michel Angelo Ricci, et hauto ch' havrò il commandamento di V.A. li scriverò con resolutezza.

Note 95
Il Sig.^{r} Michelagnelo Ricci mi replica questa settimana, e con molte raggioni vive, et efficace procura mostrare quanto prejudizio si faccia alla nostra accademia, et all'Italia tutta con il nostro tacere, e non scrivere a quei Sig.^{ri} dell'accademia di Francia: vorrebbe egli in somma che si palesassero le conchiusioni da noi ritrovate, e dimostrate tacendo pero, et occultando le raggioni, e le dimostrationi: in questa maniera dice egli potremo esser sicuri che non ci possa esser tolto il primo luogo dell' inventione, preoccupata e palesata da noi. Ora io in risposta ho detto che fra pochi giorni havero agio di sodisfare a tutti i particolari contenuti in detta lettera intanto verrà V.A. a Pisa, dove potià determinare, e risolvere quello che commanda si facci sopra questo particolare.

Note 97
Nell'inclusa relatione inviatami dal Sig.^{r} Michelagnelo Ricci vedra V.A. Ser.^{ma} descritta la forma con la quale si governa la nuova accademia de Filosofi di Parigi. Egli si vede in somma che oltr'all'osservationi esperimentali quei Sig.^{ri} procurano ancora di filosofarci [*sic*] attorno . . . . Aspetto ora l'ordine di V.A. Ser^{ma} circa il modo, e le parole precise che devo rispondere conforme lo fatto per il passato: perche non mi pare che il Sig.^{r} Michelagnelo ne quei Sig.^{ri} di Francia habbino capito il senso della prima risposta che io scrissi per ordine di V.A.

Note 119
Finalmente è venuta la risposta del Sig^r Tevenot in proposito dell'-esperienza da mè inviatagli per commandamento di V. Altezza Ser.^{ma}, e d'aver tanto prolungato ei dice esserne stato cagione la solenne entrata della Regina c'hà e tirato à sè la curiosità di tutto il Regno,

e divertita ancora que' Sig.^{ri} dalla solita applicaz.^e de' loro studj. Hanno poi straordinariam^{te} unita l'Accademia a fine di partecipare l'esperienza graziosiss.^{ma}, come la chiama il Sig^r Tevenot, à quei Sig.^{ri}, li quali vogliono provar di nuovo l'esperimento, e quanto prima mandare all'Altezza V'ra Ser.^{ma} sopra di quella un Discorso, et insieme rappresentarle il gusto un[ivers]ale, la riverenza, e l'ossequio col quale hanno ricevuto e l'aiuto e l'honore della communicaz.^e da tenersi fra le due Accademie.

Note 130

La nostra Academia, di cui V.A. Serma si degna di far menzione, s'era già ridotto secondo i nostri voti alla pratica, et alle esperienze, ma Monsieur Colbert a cui n'era stato proposto il disegno come cosa degna d'esser favorita del Re, e da S. Ecc.^{za} se n'è fortemente invaghito et da speranza di fare assai già [*sic*] di quello, che da noi era stato pensato, e proposto degl'accademici ne fanno piu Classi, già han cominciato a formare quella di Geometri, con intenzione di formar le altre di physici et altri di mano in mano. gia si tratta d'istrumenti per le osservazioni astronomiche, d'Edifici per collocarsi, di chiamarvi soggetti de' maggior valore, et hanno cominciato dal Sig^r Hugenio...

Note 132

Non possò a bastanza esprimere a V.S. il mio godemento in sentire cotesta loro Accademia onorata dell alta e potente protezione di un tanto Rè, nè resterà luogo ad alcun di dubitare, che et in riguardo di ciò, e per esser veramente composta de più purgati Ingegni, e più eruditi, non potranno da esso derivare che Opere degne d'eterna lode, e di singolar benefizio a la Republica Letteraria.

Note 134

"... si distende fin'a Londra in una colonia protettore, e capo di cui non isdegna chiamarsi S.M. B:^{ca} ..."

CHAPTER 7

Note 8

...la lontananza non hà punto rimosso l'animo mio della suaviss.^a, e virtuosiss.^a conversazione di ñri Sig.^{ri} Accademici Filosofi.

Note 19

Io similm.^{te} concorro et approvo la risoluzione di V.S. e di tutti i suoi amici di mandare ella alle stampe le sue invenzioni intorno a i conici, et io potrò testificare fra gli altri, ch'ella non ha havuto notizia di questi ultimi libri. Si che tiri pure avanti, e goda di questo benefizio, del q.^{le} io dubito non poter godere, ancorchè habbia appresso di me un compendiosiss.^{mo} trattato de' Conici disteso tutto di mano mia in tempo che io non era stroppiato come son hora, cioè in tempo, che potevo adoprar le mani e la penna. Attenda ella intanto alla sua salute, et a conservarmi nella sua grazia, mentre io l'abb.^o di tutto quore.

** The passage between the asterisks has been underlined, presumably by Viviani.

Note 35
Stimo, che la risoluzione del Sig. Gio. Alfonso Borelli di ritirarsi all'Ozio della Patria farà gran danno allo Studio di Pisa e proporzionato ai vantaggi che riceveva dalla dottrina d'un tal soggetto. Il Seren.^{mo} Principe Leopoldo m'onora troppo e contro ogni mio merito volendo sentire il mio parere circa la persona del successore: Mi stimerei felice se io potessi servire S.A.Ser.^{ma}, ma a dire il vero io non conosco nessuno, non diro in queste parti, ove poco, o niente s'attende alle mattematiche, come VS. Illma sà ma ne anco in Francia il quale avesse ardire di mettersi a questa impresa.

Note 55
Altro non mi commanda l'A.V: che il provvederle alcuni libri della lista Giansenistica già mandatale. Questi per essere e di numero considerabile, e di peso stimo bene, inviare in un' involto per V:A: per mare, e non altrimenti per il corriere. Conterranno forse qualche dottrina non interamente approvata dalla Sede Apostolica.

Note 66
... il Papa inclina più al Principe Leopoldo, dicendo, che già è in Abito,* che ha studiato, et ha rendito Eccles.^e; e che l'altro non ha tale rendito, non ha studiato, et è soldato."

* Leopold apparently wore the *abito talare*, or cassock, much of the time, as did many people who were thinking of taking orders but were putting it off for one reason or another.

Note 67
Sia ringraziato Iddio V.A. è Cardinale creato in questa mattina con giubilo, et applauso universale in maniera, che con la nuova Dignita deve V.A. stimar molto, che tutta la Corte gioivea [?].

Note 81
... li piu grandi ingegnij di tutte le Academie esperimentali mostrano una particolar ambizione nel celebrare la nobile impresa dell'Accademia del Cimento non senza grandissima dolore riguardando da non sperati accidenti impedito uno studio, del quale i grandissimi avanzi fatti in poco tempo nelli ricercamenti naturali danno ad ognuno

chiari indizi di quello che dalla sua continuazione sarebbe stato da sperare.

CHAPTER 8

Note 21

J'ay encore reçeu depuis peù ... le present qu'il Luy a pleù de me fair des premiers exemplairs des Experiences de Son Illustre Academie ou tout est excellent magnifique, et digne de Vostre Esprit, et du genie de ces grands personnages qui s'assemblent sous la protection de vostre E.S. Je la puis asseurer que cet ouvrage a receù une approbation generalle, et tous les savants et les curieux ausquels je luy communiqué qui m'en ont fait de grands eloges. Il est encore a present entre les mains de ces nouveaux Academiciens que S. Ma.^{te} à assembls pour en composer son academie. Je souhait quils se excitès par de si beaux Exemples a produire de leur part quelques belles connoissance Phisiques.*

---

* A badly garbled secretarial copy.

Note 43

A dì 7. 7bre 1657. Per riconoscere, se l'espansione del caldo, e del freddo fosse sfericam.^{e} uniforme doppo molte maniere di accetarsene con porre instrumentini di 10. gradi attorno due palle di metallo, cioè sotto, sopra, e dai lati, l'una stata 46. ore sepolata nel diaccio, l'altra arroventata nel fuoco si trovò dei detti strumentini del caldo più alterato quello di sopra, meno quei dei lati, e meno di questi quello di sotto, e per lo contrario, di quei del freddo si mutò più degli altri quel di sotto, meno quei dei lati, e meno di questi quello di sopra.

Appendix E.

NOTE ON THE ISOCHRONISM OF THE PENDULUM

The parenthesis on page 103 has caused some discussion, and Rafaello Caverni in his *Storia del metodo sperimentale in Italia*, 5 vols. (Florence, 1891–98), IV, 400, went so far as to declare that it is a deliberate lie, inserted by Viviani, "fearful for the honor of his adored master." Caverni frequently let his imagination take him beyond his sources, but it is of some interest to inquire into this matter more closely. Of three drafts of the passage still existing in G. 264, two of them (fol. 92r and 141r–v) have the parenthesis reduced to "as was first noted by Galileo," but there is a separate slip (fol. 141 bis) attached to fol. 141. This, in Magalotti's writing, has a longer parenthesis, as in the published version except for the phrase "before anyone else" (*prima di ogni altro*). It would there seem that this statement was gradually made more definite and inclusive.

Several passages in Galileo's works refer to the absolute isochronism of the pendulum, the most emphatic that I have found being the following from his last major work:

Let two equal leaden balls be suspended by two cords of equal length, about 4 or 5 ells long. Fastening the said cords at their upper ends, let the balls be removed from their perpendicular position; but let one be taken away 80 degrees or more, and the other not more than 4 or 5, so that when they are set free, one will descend and, going beyond the vertical, describe very large arcs of 160, 150, 140 degrees, etc., diminishing gradually, but the other, swinging freely, will pass over small arcs of 10, 8, 6, etc., also diminishing little by little. Now I say firstly, that the first will pass through its 180, 160, etc. degrees in just the same time as the other does its 10, 8, etc. . . . . If two companions set out to count the vibrations, one the very large ones and the other the very small ones, they will see that they will count not only tens, but hundreds too, without the difference of but one, or even of one point. (*Discorsi*, 4th day, *G.OP.*, VIII, 277).

This was published in 1638, and there is a similar passage, but without any numbers, in his *Operazioni astronomiche*, written in that year but published only in *G.OP.*, VIII, 452–66. The only passage that I know of in which Galileo hinted at any doubt about the absolute isochronism of the pendulum is in the second day of the *Dialogo sopra i due massimi sistemi del mondo* (1632): "... the same pendulum always makes its reciprocations in equal times, whether they be very long or very short, that is to say whether the pendulum is taken a very long way or a very little way from the perpendicular; and even if the times are not absolutely equal, they differ insensibly, as experiment will show you" (*G.OP.*, VII, 256).

The theory that Viviani was responsible for the parenthesis that so offended Caverni is hard to square with the circumstance that in a letter to Prince Leopold dated August 20, 1659, Viviani quoted verbatim the passage reproduced above from the *Discorsi* (see *G.OP.*, XIX, 649). But it is clear that the parenthesis in question, even if it is not mendacious, is ill-informed. The reader may also consult P. Ariotti, *Isis* 59 (1968): 414–26, on this subject.

The isochronism of the pendulum provided the subject of one of Lorenzo Magalotti's few forays into experimental science. In a letter to Viviani, undated but on internal evidence belonging to December 1661 or January 1662, Magalotti writes:

Signor Vincenzio, I must tell you that I tried today whether the large vibrations of the pendulum take longer than the small ones. I found that the latter are quicker than the former, the same pendulum made 50 large vibrations in 55 of Signor Bruto [Molaro]'s pulse beats, and 50 of the small ones were made in only 52 beats. I repeated this a number of times, and always found exactly the same effect (Florence, Bibl. Riccardiana, ms. 2487, fol. 5r).

BIBLIOGRAPHY

This short bibliography is intended merely to direct the reader's attention to works that have been useful in writing this book. Only entire volumes have been mentioned; references to articles in serial publications will be found in the footnotes. Where good collected editions are available (e.g., Boyle, Galileo) individual works have been omitted. For the editions of the *Saggi* see appendix A.

Acton, Harold, *The Last Medici*, 2nd ed. London, 1958.

Adelmann, H. B., *Marcello Malpighi and the Evolution of Embryology*, 5 vols. Ithaca, N.Y., 1966.

Andreae, Johann Valentin, *Reipublicae Christianopolitanae descriptio.* Strassburg, 1619; trans. by Felix Emil Held; New York, 1916 (with a historical introduction).

Antinori, V., editor, *Archivio meteorologico centrale italiano nell'I. e R. Museo di Fisica e Storia Naturale. Prima pubblicazione.* Florence, 1858.

Archi, Antonio, *Il tramonto dei principati in Italia.* Rocca San Casciano, 1962.

Bacon, Francis, *The Works of Francis Bacon*, ed. James Spedding, Robert Leslie Ellis, and Douglas Heath; 14 vols. London, 1857–74.

Barbensi, Gustavo, *Borelli.* Trieste, 1947.

Berigardus [Claude Guillermet de Berigard], *Circulus Pisanus Claudii Berigardi Molinensis*, etc. Udine, 1643. 2nd ed. (augmented), Padua, 1661.

Bianchini, Giuseppe Maria, *Dei Gran Duchi di Toscana della reale Casa de' Medici Protettori delle lettere, e delle belle arti, ragionamenti istorici.* Venice, 1741.

Bigazzi, Pasquale Augusto, *Firenze e contorni; manuale bibliografico e biografico*, etc. Florence, 1893.

Birch, Thomas, *History of the Royal Society*; 4 vols. London, 1756.

Borelli, Giovanni Alfonso, *Euclides restitutus, sive prisca geometriae elementa*, etc. Pisa, 1658. Trans. by Domenico Magni as *Euclide rinnovato, overo gl'antichi elementi della geometria*, etc. Bologna, 1663.

———, *Apollonii Pergaei conicorum lib. V. VI. VII. paraphraste Abalphato Aspahanensi nunc primum editi ... Abrahamus Echel-*

lensis maronita . . . reddidit. Io: Alfonsus Borellus . . . curam in geo-metricis versioni contulit, etc. Florence, 1661.

————, *Theoricae mediceorum planetarum, ex causis physicis deductae,* etc. Florence, 1666.

————, *De vi percussionis liber.* Bologna, 1667.

Boyle, Robert, *The Works*, ed. Thomas Birch; 6 vols. London, 1772.

Bronowski, J., and Bruce Mazlish, *The Western Intellectual Tradition from Leonardo to Hegel.* London, 1960.

Brown, Harcourt, *Scientific Organizations in Seventeenth Century France.* Baltimore, 1934.

Büchner, A. E., *Academiae Sacri Romani Imperii Leopoldino-Carolinae naturae curiosorum historia.* Halle, 1755.

Camerani, Sergio, *Bibliografia Medicea.* Florence, 1964.

Cassini, Jean Dominique, *Mémoires pour servir à l'histoire des sciences.* Paris, 1810.

Caverni, Raffaello, *Storia del metodo sperimentale in Italia*; 5 vols. Florence, 1891–98.

Celebrazione della Accademia del Cimento nel tricentenario della fondazione. Pisa, 1958.

Chapelain, Jean, *Lettres*, ed. Ph. Tamizey de Larroque; 2 vols. Paris, 1880, 1883.

Corniani, Giambattista, *I secoli della letteratura italiana commentario,* etc.; 9 vols. Brescia, 1804–13.

Crinò, Anna Maria, *Fatti e figure del seicento anglo-toscano.* Florence, 1957.

Crosland, Maurice P., *Historical Studies in the Language of Chemistry.* London, 1962.

Dante Alighieri, *The Divine Comedy . . . a new Translation into English Blank Verse by Lawrence Grant White.* New York, 1948.

Dati, Carlo, *Lettere a Filaleti di Timauro Antiate, Della vera storia della cicloide, e della famosissima esperienza dell' argento vivo, 24 gennaio 1662.* Florence, 1663.

Della Porta, G. B., *Magia naturalis libri XX.* Naples, 1589. Trans. as *Natural Magic in Twenty Books,* etc. London, 1658.

Fabroni, Angelo, *Lettere inedite d'uomini illustri,* etc.; 2 vols. Florence, 1773, 1775.

————, *Vitae italorum doctrinae excellentium, qui saeculis XVII & XVIII floruerunt*; 20 vols. Pisa, 1778–1805.

Favaro, Angelo, *Galileo Galilei e lo studio di Padova*; 2 vols. Florence, 1883.

F.C.S.D.O. [Colangelo, F.] *Racconto istorico della vita di Gio: Battista della Porta,* etc. Naples, 1813.

Fermi, Stefano, *Lorenzo Magalotti, scienziato e letterato, 1637–1712,* etc. Piacenza, 1903.

————, *Bibliografia magalottiana,* etc. Piacenza, 1904.

Firenze, Museo di Storia della Scienza, *Catalogo degli strumenti,* etc. Florence, 1954.

Frisi, Paolo, *Elogio del Galileo*. Leghorn, 1775.

Galilei, Galileo, *Le Opere, edizione nazionale*, ed. A. Favaro; 20 vols. Florence, 1890–1909 (reprinted, 1929–40).

———, *Le Opere dei discepoli di Galileo Galilei, edizione nazionale. Vol. I, L'Accademia del Cimento, Parte prima*. Florence, 1942 (no more yet published).

Galluzzi, Riguccio, *Istoria del Granducato di Toscana sotto il governo della Casa Medici*, etc.; 5 vols., 4°. Florence, 1781. (Also 8 vols., 8°. Leghorn, 1781.)

Gamba, Bartolommeo, *Serie dei testi di lingua e di altre opere importanti nella italiana letteratura*, etc., 4th ed. Venice, 1839.

Gassendi, Pierre, *Opera omnia*, etc., ed. H. L. Habert de Montmor and F. Henri; 6 vols. Lyons, 1658–75.

Gerland, E., and F. Traumüller, *Geschichte der physikalischen Experimentierkunst*. Leipzig, 1899 (reprinted, Hildesheim, 1965).

Geymonat, Ludovico, *Galileo Galilei: A Biography and Inquiry into his Philosophy of Science*. New York, 1965.

Gilbert, William, *De magnete, magneticisque corporibus, et de magno magnete tellure*, etc. London, 1600. Trans. P. Fleury Mottelay, London, 1893.

Guhrauer, G. E., *Joachim Jungius und sein Zeitalter*, etc. Stuttgart and Tübingen, 1850.

Hall, Marie Boas, *Robert Boyle on Natural Philosophy*. Bloomington, Ind., 1965.

Hooke, Robert, *Philosophical experiments and observations of . . . Dr Robert Hooke*, etc. (ed. W. Derham). London, 1726.

Huygens, Christiaan, *Oeuvres complètes de Christiaan Huygens, publiées par la Société Hollandaise des Sciences;* 22 vols. The Hague, 1888–1950.

Imbert, G., *La vita fiorentina nel seicento secondo memorie sincrone (1644–1670)*. Florence, 1906.

Italy, Ministero dell'Interno. Archivio di Stato di Firenze, *Archivio mediceo del principato. Inventario sommario*. Rome, 1951.

Kepler, Johann, *Gesammelte Werke;* 18 vols. Munich, 1937–59.

Koyré, Alexandre, *Etudes galiléennes;* 3 vols. Paris, 1939.

———, *La révolution astronomique. Copernic, Kepler, Borelli*. Paris, 1961.

Lassels, Richard, *The Voyage of Italy, or a Compleat Journey through Italy*, etc. (op. posth., ed. S. Wilson); 2 vols. Paris, 1670.

Lenoble, Robert, *Mersenne ou la naissance du mécanisme*. Paris, 1943.

Libri, Guglielmo, *Histoire des sciences mathématiques en Italie, depuis la renaissance des lettres jusqu'à la fin du XVII siècle;* 4 vols. Paris, 1838–41.

McMullin, Ernan, ed., *Galileo, Man of Science*. New York and London, 1967.

Magalotti, Lorenzo, *Lettere scientifiche ed erudite*, etc. Florence, 1721. (Further editions, Venice, 1734, 1740, 1756, and 1772.)

————, *Delle lettere familiare del Conte Lorenzo Magalotti e di altri insigni uomini a lui scritte*; 2 vols. Florence, 1769 (with a *Life* by A. Fabroni).

Maignan, Emmanuel, *Cursus philosophicus concinnatus ex notissimis cuique principiis*, etc.; 4 vols. paged as one. Toulouse, 1653.

Malpighi, Marcello, *Opera postuma . . . quibus praefixa est ejusdem vita à seipso scripta*. London, 1697.

Maugain, G., *Etude sur l'évolution intellectuelle de l'Italie de 1657 à 1750 environ*. Paris, 1909.

Mawry, Louis Ferdinand Alfred, *L'ancienne Académie des ɔciences*, 2nd ed. Paris, 1864.

Maylender, Michele, *Storie delle accademie d'Italia*; 5 vols. Bologna, 1927–30.

Mersenne, Marin, *Cogitata physico-mathematica*; 3 vols. Paris, 1644–47.

Monconys, Balthasar de, *Journal des voyages de M. de Monconys*, etc.; 3 vols. Lyons, 1655–66. (2nd ed.; 2 vols. Lyons, 1677.)

Moreni, Domenico, *Bibliografia storico-ragionata della Toscana*, etc., 2 vols. Florence, 1805.

Moscovici, Serge, *L'expérience du mouvement. Jean-Baptiste Baliani disciple et critique de Galilée*. Paris, 1967.

Nelli, Giovanni Batista Clemente, *Saggio di storia letteraria fiorentina del secolo XVII*. Lucca, 1759.

Neri, Antonio, *L'arte vetraria distinta in libri sette*, etc. Florence, 1612.

[Noël, Etienne], *Stephani Natalis . . . Gravitas comparata seu comparatio gravitatis aëris cum hydrargyri gravitate*. Paris, 1648.

Odescalchi, Baldassare, *Memorie istorico-critiche dell'Accademia de' Lincei e del principe Federico Cesi*, etc. Rome, 1806.

[Oldenburg, Henry]. *The Correspondence of Henry Oldenburg*, ed. and trans. A. Rupert Hall and Marie Boas Hall; 5 + vols. Madison and Milwaukee, Wis., 1965– (in progress).

Olschki, Leonardo, *Geschichte der neusprachlichen wissenschaftlichen Literatur*; 3 vols. Halle, 1927 (reprinted, Vaduz, 1965).

Ornstein, Martha, *The Rôle of Scientific Societies in the Seventeenth Century*. Chicago, 1928. (2nd ed., 1938.)

Pascal, Blaise, *Oeuvres de Blaise Pascal, publiées suivant l'ordre chronologique . . . par Léon Brunschvicg, Pierre Boutroux, et Félix Gazier*; 14 vols. Paris, 1904–14.

Pecquet, Jean, *Experimenta nova anatomica*. Paris, 1651.

Peregrinus of Maricourt, Peter, *Epistle to Sygerus of Foucaucourt, Soldier, concerning the Magnet* [1269], transl. Silvanus P. Thompson. London, 1902.

Pieraccini, Gaetano, *La stirpe de' Medici di Cafaggiolo*; 3 vols. Florence, 1924, 1925.

Poggendorf, J. C., *Geschichte der Physik*. Leipzig, 1878.

Procissi, A., *La Collezione Galileiana della Biblioteca Nazionale di Firenze. I. "Anteriori. Galileo."* Rome, 1959. (All published so far.)

Purver, Margery, *The Royal Society: Concept and Creation*. London, 1967.

Redi, Francesco, *Opere di Francesco Redi, gentiluomo aretino*, etc.; 7 vols. Venice, 1712–30. (2nd [?] ed., 7 vols. Naples, 1740–41.)

Renaudot, Théophraste, *Première* [. . . *quatriesme*] *centurie des questions traitées ez conférences du Bureau d'Adresse*, etc.; 4 vols. Paris, 1634–41.

Reresby, John, *Travels of Sir John Reresby*. London, 1904. (Ed. princ. 1813.)

Riccioli, Giambattista, *Almagestum novum astronomiam veterem novamque complectens*, etc. Bologna, 1651.

Schott, Gaspar, *Mechanica hydraulico-pneumatica*, etc. Würzburg, 1657.

Spink, J. S., *French Free-Thought from Gassendi to Voltaire*. London, 1960.

Sprat, Thomas, *The History of the Royal Society of London, for the Improving of Natural Knowledge*. London, 1667. (Reprinted with introduction and notes by J. I. Cope and H. W. Jones. St. Louis, Mo., 1958.)

[Steen, Nicolaus] *Nicolai Stenonis opera philosophica*; 2 vols. Copenhagen, 1910.

Sturm, Johann Christoph, *Collegium experimentale sive curiosum*, etc.; 2 vols. Nuremberg, 1676, 1685.

Targioni Tozzetti, Giovanni, *Notizie degli aggrandimenti delle science fisiche, accaduti in Toscana nel corso di anni LX. del secolo XVII*; 3 vols. in 4. Florence, 1780. (This also appeared with a different title; see p. 14 above.)

Taylor, A. E., *A Commentary on Plato's Timaeus*. Oxford, 1928 (reprinted, 1962).

Thorndike, Lynn, *A History of Magic and Experimental Science*; 8 vols. New York, 1923–58.

Tiraboschi, Girolamo, *Storia della letteratura italiana*, etc.; 9 vols. in 20. Florence, 1805–13. (There are several other editions.)

Vite degli Arcadi illustri, Parte prima. Rome, 1708.

Viviani, Vincenzio, *De maximis et minimis, geometrica divinatio in quintum Conicorum Apollonii Pergaei adhuc desideratum*, etc. Florence, 1659.

Young, George Frederick, *The Medici*. London, 1909. (2nd ed., 1911.)

INDEX

Page references to the translation of the *Saggi di naturali esperienze* are given in italics.

A

Abetti, Giorgio, and Pietro Pagnini: *Le Opere dei discepoli di Galileo Galilei*, 15, 351

Academia Secretorum Naturae, 7

Academicians, motivation of, 331–46 *passim*

Académie Française, 298

Académie Montmor, 1, 297–307 *passim*

Académie Royale des Sciences, 32, 298, 307, 325n, 338; formation of, 306–7; *Registres*, 335

Académie Thevenot, 298

Academy of Belles Lettres, Vienna, 48–49

Academy of Christina of Sweden, 30

Accademia Aldina, 6

Accademia degli Affidati, 6

Accademia, degli Investiganti, 47; influence on Accademia del Cimento, 9; Borelli a member, 318

Accademia dei Lincei, 7–8, 26, 47; influence on Accademia del Cimento, 9

Accademia della Crusca, 11, 53, 346; protected by Leopold, 50; *Vocabolario*, 11, 23, *161n, 236n*

Accademia Platonica, 50

Acton, Harold, 19n, 63

Adelmann, Howard B., 51, 326

Aggiunti, Niccolò (1600–35), 21n; on capillarity, *154n*; expansion of water in freezing, *168n*; freezing experiments, *198n, 199n*

Aguillon, François d' (1567–1617), 56

Air: weight of, 62, *243–44*, 269–70; maximum dilatation of, *116–20*, 368, 380, 381; supposed conversion to water, 268, 341, 371; specific gravity of, 269, 369, 370, 371; compression of, 270; thermal expansion of, 271, 371

Air baroscope, *131–37*, 381

Air pressure, 55, *105–31*, 332

Air pump, *154n*, 263, *264–65*

Alamanni, Antonio, *93n, 97n*

Alexander VII, Pope (1599–1667), 19, 72, 321, 323

Allestry, James, 282, 299

Altieri, Cardinal Alberto (1617–98), 324

Amalgamation, 64, *278–79*; in the Torricellian vacuum, 279

Amber: in a vacuum, *143–45*; experiments with, *230–33*; Plutarch on, *231*

Amontons, Guillaume (1663–1705), 273

Andreae, Johann Valentin (1586–1654): *Reipublicae Christiano politanae descriptio*, 8

Animals: in the vacuum, *159–65*, 333; digestion by, *250–51*

Antinori, Vincenzio, 27, *93n*, 256, 351

"Antiperistasis," 340; defined, *207*; experiment about, *246*

Apollonius of Perga, 37, 299; *Conics*, 313, 314

Archi, Antonio, 25

Archivio di Stato. *See* Florence

Aristaeus, 37

403

THE JOHNS HOPKINS PRESS

Designed by Gerard A. Valerio

Composed in Baskerville Text
by Monotype Composition Company

Printed on 60-lb. Sebago MF Regular
by Universal Lithographers, Inc.

Bound in Joanna Arrestox 38620
by L. H. Jenkins, Inc.